D0897828

Being Reasonable about the Economics of Health

Being Reasonable about the Economics of Health

Selected Essays by Alan Williams

Compiled and edited by

A.J. Culyer and Alan Maynard

University of York

Edward Elgar
Cheltenham, UK • Lyme, US

Published by
Edward Elgar Publishing Limited
8 Lansdown Place
Cheltenham
Glos GL50 2HU
UK

Edward Elgar Publishing, Inc.
1 Pinnacle Hill Road
Lyme
NH 03768
US

A catalogue record for this book
is available from the British Library

Library of Congress Cataloguing in Publication Data
Williams, Alan, 1927–
    Being reasonable about the economics of health: selected essays by Alan
Williams / edited by A.J. Culyer, Alan Maynard
    Includes index.
    1. Medical care—Cost effectiveness. 2. Health care rationing.
3. Health status indicators. 4. Medical policy. I. Culyer, A.J.
(Anthony J.) II. Maynard, Alan. III. Title.
RA410.5.W554   1997
338.4'761—dc21   97–12051
                                            CIP

ISBN 1 85898 648 6

Printed and bound in Great Britain by
Biddles Ltd, Guildford and King's Lynn

Contents

Figures

Tables

Acknowledgements

The following permissions to reprint previously published articles are gratefully acknowledged:

'Cost benefit analysis: bastard science and/or insidious poison in the body politick', *Journal of Public Economics*, **1**, 2, 199–225, 1972, with kind permission from Elsevier Science S> A>, P.O. Box 564, 1001 Lausanne, Switzerland;

'Measuring the effectiveness of health care systems', in Perlman, M. (ed.), *The Economics of Health and Medical Care*, Macmillan, 361–376, 1974, Basingstoke, with kind permission of the Macmillan Press Ltd;

'"Need" as a demand concept (with special reference to health' in Culyer, A.J. and Wright, K.G. (eds.), *Economic Aspects of Health Services*, Martin Robertson, 32–45, 1978, with kind permission from Blackwell Publishers, 108 Cowley Road, Oxford, OX4 1JF;

'The budget as a (mis-) information system', in Culyer, A.J. and Wright, K.G. (eds.), *Economic Aspects of Health Services*, Martin Robertson, 84–91, 1978, with kind permission from Blackwell Publishers, 108 Cowley Road, Oxford, OX4 1JF;

'One economist's view of social medicine', *Journal of Epidemiology and Community Health*, **33**, 3–7, 1979, with kind permission from the BMJ Publishing Group;

'The role of economics in the evaluation of health care technologies', in Culyer, A.J. and Horisberger, B. (eds.), *Evaluation of Health Care Technologies*, Springer-Verlag, 38–68, 1983, Berlin, with kind permission of Springer-Verlag GmbH & Co, KG;

'Economics of coronary artery bypass grafting', *British Medical Journal*, **291**, 1183–1186, 1988, from the BMJ Publishing Group;

'Health Economics: the end of clinical freedom?', *British Medical Journal*, **297**, 1183–1186, 1988, from the BMJ Publishing Group;

'Priority setting in public and private health care: a guide through the ideological jungle', reprinted from *Journal of Health Economics*, **7**, 2, 173–183, 1988, with kind permission from Elsevier Science – NL, Sara Burgerhartstraat 25, 1055 KV Amerstdam, The Netherlands;

'Cost-effectiveness analysis: is it ethical?', *Journal of Medical Ethics*, **18**, 1, 7–11, 1992, from the BMJ Publishing Group;

'How should information on cost-effectiveness influence clinical practice?', in Delamothe, T. (ed.), *Outcomes Into Clinical Practice*, BMJ Publishing Group, 99–107, 1994, from the BMJ Publishing Group;

'QALYs and ethics: a health economist's perspective', *Social Science and Medicine*, **43**, 12, 1795–1804, 1996, with kind permission from Elsevier Science Ltd, The Boulevard, Langford Lane, Kidlington OX5 1GB, UK.

Introduction

A. J. Culyer and Alan Maynard

The profession of economics, the professions involved in the finance and provision of health services and the clients of both have reason to celebrate the career of Alan Williams. For the editors there are additional reasons to celebrate his work and his seventieth birthday in June 1997. For Culyer, he was the first professional economist whom he had met. Williams admitted him as an undergraduate at Exeter University and first fired him with an enthusiasm for economics, sowing the seeds of a lasting love of both the geometric approach in economics and utilitarianism. For Maynard, exposure to the geometric acrobatics of his public finance text as an undergraduate preceded a brief exposure to him as a graduate student in York in the late 1960s (Alan, at the time, was seconded to the Treasury and beginning his work in the health area). For over a quarter of a century we have been colleagues in York and, for us both, Alan Williams has been a constant source of intellectual inspiration and encouragement, a frequent source of unattributed *bons mots* and, in short, an intellectual role model whose attempted emulation by us is but a shadow of the original.

Alan Williams read economics at the University of Birmingham and graduated B.Com. in 1951. He has taught at the universities of Uppsala and Stockholm in Sweden, the Massachusetts Institute of Technology, Princeton and Berkeley in the USA and at the University of Exeter before being appointed to a Senior Lectureship in York in 1964. After several years teaching his specialism, public finance, in York he was seconded to HM Treasury where he was Director of Economic Studies at its Centre for Administrative Studies. It was while he was at the Treasury that he became interested in health care.

In the late 1960s the Treasury, attempting to curb public expenditure, sent him to investigate the Ministry of Health's hospital building programme. He found that the principle determining hospital investment was 'Buggins turn' 'weighted' for the age of the hospital, the latter being ambiguous, since hospitals were the product of incremental building over centuries!

When he confronted the Ministry's investment criteria, he was his reasonable self and suggested that expenditure might be linked to benefit in terms of population health. This was clearly a novel idea in the Ministry and he tells how, when he presented this notion to senior officials, he was received unsympathetically by the largely medical audience who assured him that hospital investments could and should be determined by those (that is, they) who had the 'experience'!

Lurking, however, in the Ministry's audience was another intelligent and determined radical, Professor Archie Cochrane. Cochrane contacted Williams and encouraged him to develop his idea of measuring health care outcomes in terms of improvements in the length and quality of life. It was this encouragement which led Alan to develop his work on social indicators, which was the precursor of QALYs and the EuroQol.

Alan returned to York as a specialist in public finance and his new-found interest in health economics. Around him grew up a group of colleagues who devoted increasing parts of their academic life to health economics both in York and elsewhere. He admits to making the decision to be a health economist more fully during the time he served on the Royal Commission on the NHS in the late 1970s. With financial support from the Nuffield Provincial Hospitals Trust, who 'bought out' his time from York so that it could be devoted to the Commission, he sought to develop a thorough analysis of the NHS by carefully commissioned and targeted research.

He threw himself into this work but failed to receive the support necessary for it to be of the quality he required. The frustration this generated led to physical exhaustion and his resignation from the Commission. The Royal Commission's impact was limited, reporting as it did in the first year of the Thatcher Government. The immediate impact of his new enthusiasm was thus limited. The impact of his decision to be a health economist has of course been far greater.

This is demonstrated in the papers included in this volume. The task of compiling and editing the book was straightforward. Alan was asked to select his favourite papers and to suggest his preferred way of organising them. We added some items which we thought he should not have omitted, edited the text lightly to avoid repetition and made the style consistent. The book is divided into four parts and the papers in each section are ordered consecutively to reflect the growth of his thinking on each theme.

The result is what we – Alan and the editors – consider to be his key papers. Some of these have not appeared before in accessible places and they are offered to readers, both those who have already been inspired by his work and those yet to be, in the form of an autobiographical *Festschrift*.

The resultant record of the man's achievement is a testimony to an unusually creative imagination and evidence of a dogged determination to see a set of related projects through from their theoretical inception to, if not completion, at least to their having become generally seen as essential parts of the core of health economics, of our way of thinking about health and health care.

Although some major contributions, such as his work on public finance and local government (themselves pioneering), have been omitted so that the book can be a comprehensive account of his health related work for the past 30 years, the full range of application of the ideas within these covers is actually far wider than might appear. His approach to the general issues of normative economics represents a liberation from the confines of the Pareto principle which has yet to be fully realised in fields of public policy beyond health but which underlies what many now call 'extra–welfarism' – the ideas that the assessment of the goodness or badness of social policies ought to transcend 'utility' (especially utility from goods and services) and that the sources that yield benefits and opportunity costs might properly be decision makers and the clients of the systems in question; his clarifications of the most useful way to do cost–benefit analysis (useful, that is, to society, rather than economists!) are illustrated by, rather than limited to, the health territory; while his great contribution to outcome measurement – the QALY and the EuroQol – is a theoretical concept extensible *mutatis mutandis* to a far wider range of applications than health and health care.

Alan Williams is, however, not just an economist's economist. He has, for the greater part of his career, sought to understand and enlist the ideas of people from other disciplines, to communicate economics to their members and to understand and meet the needs of policy makers at all levels of decision making. In this there can be no doubt of his success. He has become a household name to many well beyond the professional economics community.

Not all professions honour their great men as they ought, though economists have often gathered their works together so that they may be treasured and studied as an evolutionary *corpus*. We seek to honour Alan Williams in this way as one for whom humane ideas, rigorously pursued to their logical conclusions and applied within practical contexts, are *de rigeur*. The combination of exemplary explicit views as to the moral endeavour on which we are engaged, theoretical and conceptual clarity, and careful empirical application, is rare amongst the economists of today, though not unusual in the nineteenth century and the first half of this: drawing ideas from many disciplines, engaging with practitioners and their clients, forging new theory and analytical tools, applying them rigorously to address important practical problems and then feeding back the results to fellow academics across many disciplines, to the practitioners and to the wider public. He is therefore also the very model of what a modern economist might aspire to be.

PART ONE

Economic Approaches to Health and Social Welfare

1. Cost–Benefit Analysis: Bastard Science and/or Insidious Poison in the Body Politick?

In the booming business of management consultancy there is no shortage of glossily packaged analytical techniques of varying degrees of power and reliability being peddled to ill-informed clients by pushful salesmen. My own early activities in this field produced a *cri du coeur* which deserves wider circulation than it has hitherto received, not only for its literary merit in the field of satirical doggerel (well up to the standards of W S Gilbert), but also for its aptness to a situation that is likely to be with us for some little time yet. By reproducing it here I hope that I may inspire some latter–day Arthur Sullivan to set it to music!

The management consultant's lament

We showed them how to maximise the Branch's optimality,
How microeconomics had become a stern reality;
They listened with politeness but preferred their old autonomy,
Their faulty input–budgeting and massive diseconomy.
Evaluative charting could not penetrate their fallacies;
They could not see the logic of cost–benefit analysis;
Their data-handling system was in urgent need of focusing;
They had not learned the rudiments of information-processing.
Their output was unprogrammed, their retrieval rate was minimal;
Their information-network was so crude as to be criminal;
We gave them multi-access on a scheme multi–dimensional:
Their coldness was so cutting that it might have been intentional.
Our trees and algorithms were rejected with acidity;
It really is heartbreaking to encounter such stupidity.
If cybernetic doctrine cannot change their methodology,
Can't they at least be men enough to learn the terminology?
Ross (1968)

Cost–benefit analysis is one of the techniques most prone to misunderstanding and misapplication in the hands of the uninitiated (not to mention the unscrupulous!). Machlup (1965) has identified the following five connotations of the term 'benefit cost comparison':

Type No. 1. is the implicit sentiment, sans analysis, in favour of the particular project .... Everyone with a favourite project will, when asked, assert that 'of course' the benefits from it are greater than the costs ... even if they have neither theoretical arguments nor empirical data to support such contentions.

Type No. 2. may be seen in explicit propositions about the *value* of the program, in which you impose your own value judgements upon the people of the commonwealth ... though you do not care to say why, or to defer or refer to the opinions of others.

Type No. 3. is ... an assumption that you and others know what society wants, what kinds of benefits it may derive from the program, and how big these benefits are. There is still no analysis, but the assumption appears to be moving closer to matters that deserve to be taken seriously.

Type No. 4. is based on theoretical analysis, where you at least state what kind of empirical information you would need to make a calculation or an estimate of the benefits and costs ....

Type No.5. is the actual calculation of benefits and costs in which all or much of the required information is available and analysed .... (pp. 150–151).

It needs to be made clear from the outset that I reserve use of the term 'cost–benefit analysis' (or CBA for brevity) to exercises of types 4 and 5, not because I object to the authoritarian/paternalistic aspects of type 3, but because I regard the crucial feature as being *explicit and formal analysis*[1] , which is still absent from type 3.

In this chapter I shall argue that CBA is a natural and logical extension of systems analysis (SA), operations research (OR) and cost-effectiveness analysis (CEA), but more ambitious than they in evaluative scope and technique and hence rather more vulnerable at certain (well-recognised) points. I shall then review some of the criticisms of CBA emanating from political scientists and argue that, although some of them are based on misconceptions, others are perfectly valid. Unfortunately, the valid criticisms are applicable *a fortiori* to the only alternative techniques of analysis and all too often the argument appears to degenerate from attempts to replace poor analysis by better, to assertions that since analysis is difficult, costly and troublesome we should abandon it! It is not a conclusion I can share.

BASTARD SCIENCE?

Is CBA different?

I take the objective of CBA to be to assist choice (not to *make* choice, nor to *justify* past choice, nor yet to *delay* matters so that some previously chosen course of action has a greater chance of adoption, although I recognise that each of these purposes *may* also be served by the skilful employment of CBA).

There are many other techniques which also serve this same purpose, so it is necessary here to identify what (if anything) distinguishes CBA from them. Some of them are mainly concerned with the systematic exploration of what might be termed 'technical possibilities', where 'systematic' connotes being orderly and comprehensive in search, explicit in procedures and quantitative rather than qualitative in mode of measurement. The system which is the focus of interest is 'described' or 'simulated' or 'modelled' at some level of abstraction and the effects of various feasible changes in any controllable variable (or input) are then explored with respect to the behaviour (or output) of the system. The inputs and outputs represent the points of contact between the system and its environment, which implies that we are dealing with open systems, which in turn raises problems about defining the boundaries of the system (or of 'the problem space'). Inputs need not be commensurable with each other nor need outputs. Nor, *a fortiori*, need inputs be commensurable with outputs. Much of the work conducted within the natural sciences and engineering, much systems analysis and even some operations research fall within this category, which economists would regard as logically prior (and essential) to their own role in the analytical process.

What separates this group of techniques *in principle* from the next group is that the former are positive in orientation rather than prescriptive. Such analyses are designed to facilitate understanding, proceeding via prediction or hypothesis tested by experiment and/or observation, with variables chosen and measured primarily so as to facilitate monitoring and replication, with suitability for manipulation for policy purposes playing a subsidiary role. When this policy manipulation becomes the central purpose of the analysis, we move more firmly into the acknowledged territory of operations research and, although in practice the differences in substance between the two kinds of exercise tend to shade off into each other in almost imperceptible gradations, this 'consultative' role needs to be regarded as something different from the 'purely scientific' role. But still it will be common to find little commensurability within or between inputs and outputs though in applications in the field of business and governmental activities OR tends to accept as a limited basis for establishing commensurability the money values of any inputs or outputs which are conventionally so valued. But since other significant inputs and/or outputs will typically remain outside the realm of conventional financial evaluation, the result will be what is commonly called 'cost-effectiveness' analysis (CEA), 'cost' embracing all those items valued in money terms (usually inputs) and 'effectiveness' some multi–dimensional pay-off exhibited without any attempt at further evaluation in commensurable terms.

The characteristic of the third group of techniques (to which cost-effectiveness studies already partly belong) is that while they too are prescriptively oriented (that is, designed to help a client improve his situation) they try hard to render inputs and outputs commensurable. Thus the 'ideal' CBA will have all inputs and outputs evaluated in money terms (which, it should be stressed, is *not* the same as evaluating them all at market prices), though in principle any other common unit would serve equally well (for example years of life-expectation, utils, cowrie shells). But since it is unlikely that any but the most low-level CBA will, in the foreseeable future, succeed in evaluating all inputs and outputs in commensurable terms, the distinction between *actual* CBA and *actual* CEA will only be a matter of *degree* (and, on occasion, perhaps only a matter of *intent*).

I therefore conclude that CBA accepts and pushes further the scientists' preoccupation with quantification already central to OR – and pushes much further down the path already trodden by CEA towards evaluation of inputs and outputs in money terms, tending at the same time to enlarge the scope of the inputs and outputs that are brought within the area of systematic comparison.

Is CBA scientific?

This question is better dealt with by breaking it down into three related questions: (a) Is CBA systematic? (b) Is CBA objective? and (c) Is CBA precise? Brief answers can be offered both at the level of principle (that is, what could conceivably be true one day) and at the level of present practice. In principle CBA *is* systematic (that is, assumptions are required to be explicit, evidence presented, results communicable and replicable, etc.). In practice, of course, one may find studies which violate these norms, for good reasons or bad. As regards *objectivity,* CBA is no more exempt from the general methodological critique emanating from the philosophical debate surrounding 'welfare economics' than is any other branch of economics, nor for that matter, than any other social science. Since these criticisms apply to all other analytical aids, they are not peculiar to CBA and should not be so regarded. Indeed, we may be on safer ground in this field than elsewhere simply because the methodological foundations have been laid bare, so we are more keenly aware of the risks we run that are those who advocate substitutes which have so far escaped such scrutiny. As regards *precision*, there is no reason in principle why CBA should be more suspect than any other forward-looking analysis but in practice the data it requires may well be somewhat more inaccurate than some more conventional method, since we have not yet accumulated much experience or expertise in collecting and analysing it. Whether this leads one to conclude that CBA should be rejected as unsound, or pursued by every possible means to make it more accurate, rests partly on one's psychological predispositions concerning the role of analysis and quantification in human affairs, of which more anon.

The Modus Operandi of CBA

In order to explore further the strength and weaknesses of CBA let us look more closely at its *modus operandi* and compare it with other systematic techniques of analysis, (SA) and OR.

In order to structure this discussion along lines which facilitate that comparison, I shall start with an extended quotation from Dorfman (1966) where he writes of the role of OR in business and economic administration:

> If an experienced operations analyst were asked to describe the problem he would most like to meet, I suspect that he would mention four characteristics: first, the objective of the operation should be clearly defined, second, the operation should be describable by a reasonably manageable model, third, the data required for estimating the parameters of the model should be readily obtainable, and fourth, current operating practice should leave plenty of room for improvement ... (pp. 64–5).

> The foregoing summarises the circumstances in which operations research is likely to be successful in the sense of disclosing significantly improved policies and practices. But operations research can be successful in other senses also. We have the educative value of collaborating with an analyst and looking at the problem from his viewpoint. We have seen that operations research often suggests and sometimes accomplishes valuable substantive research. In many instances the contribution ... is to improve the planning process itself, without improving the quality of the plans ... (pp. 69–70).

> Finally, consider the really tough and important problems where there is no objective basis for making a usefully precise evaluation of the consequences of possible actions or policies. Contending with such imponderables is the primary responsibility of the high executive, a responsibility that cannot be delegated ....

Nevertheless, an operations research study of such a problem can help the executive by organising the data, focusing the issues, bringing out the implications or possible assumptions and hunches, delimiting the range of the inevitable uncertainty. Any detached, analytic, sceptical viewpoint can help clarify such problems and the analyst has such a viewpoint to contribute. (pp. 70–1).

Most of those observations are equally apt as comments on the role of CBA but because of the drive to push valuations further and the tendency to broaden the scope of the analysis beyond the confines regarded as optimal within the OR culture, CBA finds itself persistently in the terrain which Dorfman regards as the more difficult. There is one other important feature of OR work which Dorfman surprisingly omits to mention and which is even more characteristic of CBA: namely that 'the problem' is not usually formulated at all clearly in the analyst's brief and that he may well need to expend a sizeable chunk of time and effort discovering what it actually is and getting it into an analysable shape. Over and over again one hears of cases where the problem the clients thought they had turned out not to be the problem they actually had. Any OR or CBA practitioner who accepts the client's initial formulation of the problem uncritically is heading for disaster.

This leads naturally into the vexed issue of 'objectives' and 'constraints', to which I cannot possibly do justice here. Although expressed in a form which suggests that they are concerned only with the limits of 'feasibility', 'constraints' are in principle to be interpreted as objectives cast in a form which gives them overriding priority over the variables that are actually *called* 'objectives'. 'Objectives' cast in the form of minimum target levels of performance are of precisely the same ilk as 'constraints', except that analysis is permitted to proceed with a view to discovering whether or not they can be satisfied (and if so, in what ways), whereas it is implied that no solution falling foul of a feasibility constraint (be it technological, legal, administrative, moral, political or budgetary) is worth analysing further at all. If one had faith in the capacity of the client to think through the full implications of these ostensible constraints, there would be no need to be sceptical about accepting them without discussion but, if a client had such a capacity, it is unlikely that he would need an analyst anyway, so it seems to me that part of the dialogue about the nature of the problem (and about the objectives of the decision makers) must give comparable attention to the reasoning behind (and purpose served by) the postulated constraints. Even then, a counsel of perfection would be that all such constraints should be varied parametrically in order to indicate the effect on the various outcomes of shifting the boundaries of the problem, for this might well indicate that the next round of effort might more profitably be expended on extending one of the supposed 'feasibility' limits, or that others are so trivial in their impact that there is no point in expending resources to prevent them becoming more stringent.

An important and often neglected dimension in problem definition is determining who constitutes the relevant community with respect to the appraisal in question. If limited narrowly to the direct impact on the client (or client organisation) the analyst is likely to miss 'externalities' which may feed back through the system at large and affect the client indirectly (but still significantly). For example, the generation of noise or other forms of pollution may generate countermeasures which could prove more costly than reducing the pollution in the first place but which in any event should be included in the initial analysis. More difficult still is the situation in which the client is a governmental agency which is ostensibly pursuing the interests of the community at large and which may be confronted with difficult decisions like whether foreigners are part of the 'community at large' when determining airport location policy, or what weight consumers' interests should be given *vis–à–vis* those of producers (an acute problem for departments such as Technology and Agriculture. Because I can see little scope for incorporating redistributive or 'grand efficiency' considerations *à la* Weisbrod (1968), into the analysis in any acceptable systematic manner at present, I would be content to see incidence calculations merely appended as exhibits to each appraisal, leaving the client to incorporate these as best he can, though I am sure there is plenty of scope for fruitful dialogue as to the principles which underlie his judgements and eventually we should be able to discuss and analyse these too.

This redistributive aspect is one of the incommensurables I am willing to accept at present but I would like to pursue the general problem of incommensurability a bit further, still in the context of problem definition, through a brief excursion into the territory of SA. My qualifications in this field are so small as to be derisory but in my dilettante way I have picked up the following smattering of knowledge which seems relevant to my present purpose. General systems theory distinguishes between open and closed systems, the former being those in which there is import or export and the latter being those in which there is none. Social systems may, in principle, be of either kind but open systems seem typically to be more relevant for our purpose than closed ones. This says no more than that we are usually trying to assist in the optimisation of some subsystem and we therefore need to be clear about (i) the boundaries of this subsystem, (ii) the relationship between this subsystem and its 'environment' (that is, the rest of the system) and (iii) which of the variables are 'decision–variables' (that is, those which can be directly manipulated).

Judging by the comments of Emery (1969) there appears to be some dispute among systems theorists as to whether it is sufficient 'to cope with the problem by simply adding some exchange equation to the equations defining the inner system' or whether it is true (as he believes) that:

> ... open systems analysis cannot hope to stop with a specification of an exchange equation; it can hope to approach adequacy only when there is some characterisation of the environment. The reluctance to tackle environmental analysis appears to have arisen from the forbidding nature of two problems – (a) the sheer complexity of most environments and (b) the incommensurateness of the many heterogeneous processes that make up the system and its environment (for example psychological, economic, technical, meteorological).
>
> Sommerhoff had, as far back as 1950, provided a perfectly satisfactory answer to the second problem. The first problem should not have been taken so literally. Not all aspects of an environment are equally relevant to any particular system or class of systems. In economics, military theory, ecology, psychology, to name but a few, there has been a long and successful tradition of identifying and characterising global aspects of environment and relating these to adaption.
> (pp. 203–4)

On this 'first problem' (the choice of environmental characteristics for explicit analysis) the characteristic approach is either conventional (ie 'traditional') or else subject to largely intuitive and scientific processes of search for (and selection of) promising modifications. Only the subsequent results 'justify' one selection as against another. Thus, whether the economists' conventional selection of variables is better or worse than the sociologists' or the psychologists', will depend on whether they generate results which prove more useful for the problem in hand. Part of the dispute about cost–benefit analysis and its relationship with other techniques concerns precisely this dimension, that is, the width of the range of environmental characteristics which it attempts to encompass. Nonetheless, some people argue that it is not yet wide enough and I will return to this issue later.

The second problem, that of incommensurability of variables, is one that continues to plague cost–benefit analysis, as indeed many other evaluative techniques, so I was elated to read that it had been solved by Sommerhoff (1950, 1969). As far as I can see the solution boils down to the following statements: (1) if a set of activities, or states, are so related (whether purposively or not), to uncontrolled variations in the environment that, despite the latter, some specified terminal configuration can be attained, then those activities or states are said to be 'directively correlated' with the environmental changes; (2) since the specified terminal configuration (called the 'focal condition') could be a very particular, proximate goal (like picking up a stone) or a very general ultimate one (like ensuring survival of the species) these focal conditions can be arranged hierarchically and hence so can the 'directively correlated' activities or states which bring them about; (3) from which it follows that any pair of such activities or states can be made commensurate by finding the lowest level in the hierarchy of goals to which they can both be related. This leads Sommerhoff to observe that:

> Apart from any other features they may have in common, any set of objectives whose states or activities are directively correlated may be thought of as a unit by virtue of their joint partnership in this single system properly and particularly on the strength of their joint relationship to a single focal condition. We thus get the conception of a set of objects which are united by the common goal of their activities. (p. 136)

Whether or not this is a fruitful way to deal with the problem of incommensurability in biology I cannot judge but I am sure that it is rather empty, though formally correct, procedure in the social sciences. In economics, it is tantamount to saying at individual level that, since individuals have preference functions and manage to make choices, everything must be commensurable in terms of individual 'utility'. Likewise, for a society, one needs only to consult the social welfare function and all incommensurability problems disappear. True, but unhelpful. It is always part of the analyst's job during the exploratory phases of a study to attempt to identify objectives, relate them hierarchically or otherwise as appropriate and to find other 'directively correlated' activities (that is, alternative courses of action) which serve the same ends and may therefore be regarded (albeit partially) as substitutes. It is frequently the case that, when this process has been exhausted, there are still aspects of a problem which cannot be rendered commensurable in this way (that is, by finding some higher level index of 'effectiveness' to which the lower level indices can be systematically related; the ultimate in such indices being the social welfare function).

The economists' approach (I hesitate to call it a 'solution') is to use money values as the dimension by which as many features of a situation as possible are measured. The price mechanism provides a wealth of information which can be drawn upon for this purpose, although it needs very careful handling. The rather conservative conclusion reached by McKean (1968),[2] is worth repeating here:

> ... What we really face ... is an amorphous and shifting preference function emerging (implicitly) from the political process and a complex economy dominated by uncertainty. I have to conclude: Either market prices or imputed prices are likely to be off the mark. Where market prices do not exist, one often

prefers imputed prices to no prices at all. But in what specific instances to impute prices, how far to go in refining those imputed prices and whether to adjust market prices where they exist depends on one's judgement about the costs and worth of the information. For external effects and public goods, where no market exists at all, I would often be confident that some effort to impute prices is worthwhile. Where market prices exist, even though they are off the mark in terms of the appropriate (but dimly perceived) preferences function, my personal judgement is that they are not completely unrelated to that objective function, and I would be reasonably confident only in a few instances that derived or adjusted prices would be an improvement over observed prices. (pp. 57–58).

Even this restrictive view leaves a great deal of scope for evaluative work but it needs to be clearly understood that attempting to place a money value on non-traded 'goods' does not imply *either* that one is advocating the establishment of 'markets' in such goods *or* that one is restricting one's attention to their 'economic' attributes. Thus, if I try to place a 'value' on the reduction of road accidents, I am not advocating that the victims should be required to pay for medical treatment or that they should be compensated, nor am I solely concerned with the effects on GNP.[3] The use of money as a common measuring rod in making diverse values commensurable is not to be confused with an obsession with the more sordid aspects of profit–maximisation!

I am also willing to accept that there is a role in CBA for the 'postulated' price, that is for a client (say a Minister of Transport) to say that for the purposes of this analysis I want variable X valued at £Y per unit (where X might be the intrinsic value of life), for then £Y represents a precise statement about the client's policy trade-off between X and the other variables in the analysis. It is 'arbitrary' in the sense that it cannot be supported by empirical evidence, but it does imply a commitment to allocate resources in a particular way for which the client will have to answer, hence it is not irresponsible and it provides a precise focus for discussion by others (say Cabinet colleagues) who might wish to influence his judgements on this matter or modify their own. It is thus systematic, quantitative and communicable and hence an improvement upon the casual, implicit and possibly capricious evaluation that can so easily be concealed if such matters are left to piecemeal informal judgement 'on their merits'.

INSIDIOUS POISON

I earlier asserted my belief that CBA was intended to assist choice, without qualifying in any way the kind of choices I had in mind. I will now add that it will by now be obvious that it is with resource allocation decisions that CBA has been concerned. The traditional connotation of an 'allocation decision' in economics is a budgetary one and usually also a financial one, although more recently 'manpower budgets' have been much discussed too. But some economists have recently begun to extend the application of economic models of resource–allocation to decisions which at first sight are far removed from the traditional area of discourse, as when political behaviour is seen as a 'market' for votes or for the accumulation and enjoyment of political power rather than 'wealth'.[4] This broadening of the notion of an 'allocation decision' has led economists to make ever-deeper incursions into the territories of neighbouring disciplines, not simply in the role of tourists seeking information and enlightenment, but with a view to permanent colonisation. This has led to a further blurring in the delimitation of the area of applicability of resource allocation techniques like cost–benefit analysis. It has also engendered plenty of anti-imperialistic resistance from those occupying these neighbouring territories, as they see with horror the brash and uncompromising behaviour of the parvenus!

Wildavsky's disease: the diagnosis

In the vanguard of the political scientists' counter–attack is Wildavsky (1966) (whose article should be read by all practitioners of CBA). In that article, he argues that:

> The economic model on which cost–benefit analysis depends for its validity is based on political theory. The idea that in a free society the economy is to serve the individual's consistent preferences revealed and rationally pursued in the market place. Governments are not supposed to dictate preferences nor make decisions. (p. 294)

> The dependence of cost–benefit analysis on a prior political framework does not mean that it is a useless or trivial exercise. Decisions must be made. If quantifiable economic costs and benefits are not everything, neither would a decision-maker wish to ignore them entirely. The great advantage of cost–benefit analysis, when pursued with integrity, is that some implicit judgements are made explicit and subject to analysis. Yet, for many, the omission of explicit considerations of political factors is a serious deficiency. (p. 297)

He quotes with approval the philosopher Diesing (1962) who wrote:

> Political rationality is the fundamental kind of reason, because it deals with the preservation and improvement of decision–structures, and decision structures are the source of all decision ... (p. 203)

Wildavsky then argues that:

> Once the political process becomes a focus of attention, it is evident that the principal participants may not be clear about their goals .... The mixtures of values found in complex policies may have to be taken as packages .... In a political structure, then, the need for support assumes central importance. Not simply the economic, but the political costs and benefits turn out to be critical. (p. 308)

This leads him to sketch out a political cost–benefit calculus to complement (and by implication to override) the economists' efforts, leading him to conclude that although:

> Studies based on efficiency criteria are much needed and increasingly useful ... I would feel much better if political rationality were being pursued with the same vigour and capability as is economic efficiency. In that case I would have fewer qualms about extending efficiency studies into the decision-making apparatus.

Thus, although by implication he accepts that a broader based cost–benefit analysis would meet his objections, he sees this as a task for political scientists because 'the qualifications of efficiency experts for political systems analysis are not evident' (p. 305).

> ... policy analysis, with its emphasis upon originality, imagination and foresight, cannot be simply described. It is equivalent to what Robert N Anthony (1965) has called strategic planning; '... the process of deciding on objectives of the organisation, on changes in these objectives, on the resources used to attain these objectives .... It connotes big plans, important plans, plans with major consequences'. While policy analysis is similar to a broadly conceived version of systems analysis, Yehezkel Dror (1967) has pointed up the boundaries that separate a narrow study from one with larger policy concerns. In policy analysis:
> (1) Much attention would be paid to the political aspects of decision–making and public policy–making (instead of ignoring or condescendingly disregarding political aspects) ...
> (2) A broad conception of decision–making and policy-making would be involved (instead of viewing all decision-making as mainly a resources allocation) ...
> (3) A main emphasis would be on creativity and search for new policy alternatives, with explicit attention to encouragement of innovative thinking ...
> (4) There would be extensive reliance on ... qualitative methods ...
> (5) There would be much more emphasis on futuristic thinking ...
> (6) The approach would be looser and less rigid, but nevertheless systematic, one which would recognise the complexity of means–ends interdependence, the multiplicity of relevant criteria of decision, and the partial and tentative nature of every analysis .... (p. 190)

Apart from item 4, this is pretty much the agenda of those who wish to broaden the notion of an allocation decision beyond what is implied by those in item 2 and to use economic, statistical and mathematical analysis in the political field. Hence I suspect that 'policy analysis' is another bit of terminology that we need to take on board as meaning any use of evaluation techniques which attempts to incorporate the political aspect of a problem into the analysis. Or to put it more provocatively, it is 'cost–benefit analysis' writ large. Who is best qualified to conduct it is an issue I shall return to later.

Wildavsky's disease: the cure

This is a convenient point at which to return to the role of 'constraints' in problem definition. Every analyst quickly becomes aware of the delicate nature of the dialogue which has to be pursued in order to clarify the nature, purpose and rigidity of these 'constraints' for, while it is tempting to accept such constraints readily in order to keep the problem within manageable bounds (indeed it is not uncommon for the analysts to add a few of their own with this very consideration in mind!), the penalty to be paid is the danger that the problem becomes trivialised (or at any rate circumscribed) to such a low level of suboptimisation that it may offer only a very small pay-off to the client.

But my interest here is with a different aspect of 'feasibility', for one analyst's 'constraints' are another analyst's field of optimisation and hence, in defining and accepting such constraints, one is defining and accepting the borderlines of one's discipline. I do not mean that statement to be interpreted as meaning that, as an economist, I would want every economic variable left unconstrained in every problem I tackled but, rather that, in a run of problems, I would expect to have an opportunity of tackling every one of them as variables at one time or another. I would not, however, have the same expectation about legal or organisational features of a problem. Here I would expect to be told (preferably after some dialogue) what assumptions to make about these aspects of the problem. Again, I must explain carefully what I do and do not mean by that statement. I do not mean that the investigation must necessarily be constrained within one specified framework of such 'feasibility' constraints. Not only would I expect alternative assumptions to be permissible but they might well be the central feature of the analysis (for instance, in a cost–benefit analysis of alternative safety regulations).

It is at this point that I would take issue with some of Wildavsky's propositions quoted earlier about the scope of cost–benefit analysis, for he stated that it is limited in applicability to political systems which assume that 'the economy is to serve the individual's consistent preferences revealed and rationally pursued in the market place. Governments are not supposed to dictate preferences nor make decisions.'

This is so patently false that if I had not had an opportunity to read and re–read this statement carefully I would not have believed it possible that such a wellinformed and normally judicious observer could have made it. For while it is undoubtedly true that much of the analysis that goes into most cost–benefit studies *is* based on the assumption that the market does constitute a good guide to the value of costs and benefits, it is precisely because cost–benefit analysis may require (a) non-market valuations to be substituted for market valuations and/or (b) valuations inferred from market behaviour to be placed on items which have no market values and/or (c) the use of valuations postulated by the policy-makers or decision-makers themselves, that it is frequently criticised for precisely the opposite reasons to Wildavsky's. In other words, it is too often used by governments as a (soft) justification for over-riding the (hard) realities of market![5]

But I think that a bridge can be built over this particular chasm by a careful formulation of the 'constraints' in a problem. If I am asked to conduct an optimisation exercise which involves minimising the cost of meeting a specified minimum bill of requirements, it is well known that the conventional linear programming formulation of such a problem will yield a notional (or 'shadow') price for the constrained variables (that is, those specified in the minimum bill of requirements). Interpreted more generally, this shadow price is the value implicitly placed on the marginal unit of the constrained variable so if one of the political 'constraints' in a problem is that group X must not be disadvantaged, it is, in principle, possible to compute the cost (= implicit value) of meeting that constraint. If Wildavsky's political analysts don't like that implicit valuation, they are free to replace it with a postulated value of their own, to promulgate it, to argue about it and to justify it, etc., by whatever structure of political rationality he approves of for the purpose. The cost–benefit analyst can then insert the postulated figure in his analysis and grind out a new set of results accordingly. The effective limit on all such shadow pricing, as McKean (1968) has pointed out, is the cost involved in getting the information (and judgements about how much it is worth spending for an uncertain payoff in additional information are themselves tricky optimisation exercises!). My solution, then, would be to use shadow prices, if necessary, as the vehicle by which information is carried from other analysts' territory into the realm of cost–benefit analysis as understood by economists.

This does still leave us with a tricky problem of co-ordination, however, because we are now in the classic danger of assuming that the results of a set of suboptimisation exercises constitute the grand optimum solution! To spell out the difficulty more fully, when I accept a restriction (based, say, on ground of 'political unacceptability') upon the set of alternatives to be appraised, I have to assume that some political analyst has done his homework properly. But what if he reached his conclusions about what is or is not 'politically acceptable' by making some prior assumptions about the effects of various economic measures which an economist might well have been able to show to be false if asked to do so? For instance, in conducting an appraisal of alternative means of financing university education, loans to students might be excluded from consideration on ground of 'political unacceptability' because it is implicitly assumed by the political analysts that these must have adverse effects on access to university education by women or by the children of poor families. Now the cost–benefit analyst may know this to be a false assumption, yet accept without question the restriction ruling out student loans from consideration on grounds of political unacceptability, because he may think that the declared 'political unacceptability' of student loans was based on a strong ethical aversion to usury which, as an economist, he has no right to challenge. It is the danger that a hiatus of this kind may occur that has led many cost–benefit analysts to move into the role of general 'policy analysts'. It is very tempting to do so when you see pieces of analysis which seem (a) wrongly conceived, (b) poorly understood, and/or (c) misguidedly implemented (or misguidedly *not* implemented as the case may be).

Nevertheless, I wish to emphasise the dangers of succumbing to this temptation. The progress of human society has depended in a very large part upon specialisation and division of labour. We all know that this has not been a costless process, in terms of growing impersonality and rigidity in the system and its greater vulnerability to major breakdowns. But if every cost–benefit analyst has to become his client's general policy adviser, then his capacity to do technical economic or statistical work must inevitably suffer as a result. While I recognise the dangers of compartmentalisation, I cannot accept as a solution the notion of total commitment and universal expertise (that is, no disciplinary constraints). I would instead propagate an old-fashioned and currently unfashionable, doctrine that there is a valid and useful role both for specialists and generalists (which distinction is not the same as that between 'professionals' and 'amateurs' respectively). It is the professional generalist who should be the general policy adviser and whether his original discipline was economics, political science, mathematics, engineering, history or classics is not of prime importance. What is important is that he understands the nature of all the disciplines relevant to his problems so that without necessarily being familiar with or even understanding their more intricate and esoteric parts, he can marshal and deploy them intelligently. Just as the general in charge of an army does not necessarily have to know how to fire a gun or drive a tank, so a generalist policy adviser does not necessarily have to know how to invert a matrix or how to calculate the internal rate of return on an investment project. What each does need to know are the capabilities and limitations of the forces at his disposal, their complementarity, substitutability or more complex patterns of interdependence, their relevance to his problem and how to give instructions that cannot be misinterpreted and which will produce answers that are relevant and comprehensible. His is thus the responsibility for ensuring that no hiatus occurs and that the constraints on 'feasibility' imposed on each specialist's suboptimisation exercise are not unduly restrictive. His also is the role of 'policy analyst' as conceived by Dror (1967) and Wildavsky (1969) (mistakenly identified by them as being concerned with 'decision structures', for I see this narrower role of political analyst as just another specialism like that of the cost–benefit analyst).

This seems to accord well with the views of Schultze (1968) when he writes:

> There are political opportunity costs to any decision, just as real as the economic opportunity costs. Because values conflict among different groups and among themselves, securing the agreement necessary to pursue one line of action must often reduce the opportunity to pursue other lines of action .... A civil rights bill may 'cost' a housing bill .... These costs dictate a set of efficiency criteria for political decisions equally as real and valid as the resource costs which lead to the efficiency criteria of systematic analysis (pp. 45–47). ... the purpose of advocacy process and political bargaining is to reach decisions about specific programs in the context of conflicting and vaguely known values. Systematic analysis makes a major and essential contribution to this process by forging links between general values and specific program characteristics – links that are immediately and directly evident only in the simplest cases .... Proceeding by trial and error is of no use unless we design the trials meaningfully, recognise the errors promptly, and make the necessary corrections. Where complex issues are involved, we must rely on analysis to help. Intuition and goodwill alone will not suffice. It is not really important that the analysis will be accepted by all the participants in the bargaining process. We can hardly expect that information systems will be so complete, necessary assumptions so obviously true, or constraints so universally accepted, that a good analysis can be equated with a generally accepted one. But analysis can help focus debate upon matters about which there are real differences of value, where political judgements are necessary. It can suggest superior alternatives, eliminating, or at least minimising, the number of inferior solutions. Thus, by sharpening the debate, systematic analysis can enormously improve it. (pp. 74–75).

Analysts ... can and should play the role of ... 'partisan efficiency advocates' ÷ the champions of analysis and efficiency. They are indeed partisans, and the ultimate decision-maker has to balance their voice against the political, tactical and other considerations ... (p. 96)

How to alleviate the Lindblom–Simon disease

Some sharp things have also been written about the primitive naive notions entertained by economists concerning the nature of means–ends relationships, which may be represented crudely as being that the ends are 'given' and that the economist's problem is simply to find the most efficient means of attaining them.

The first common objection (for example, Lindblom 1959) to the economist's position is that people do not have clearly articulated ends, they just have vague ideas about what it would be nice to be able to do or have it and that these vague ideas get crystallised only through the process of seeking means to attain them. This I regard as a semantic problem within the present context, or at any rate one mainly requiring more careful role differentiation. It seems to me to be saying no more than that there is a feedback and what the economist is saying is that his professional interest and expertise is relevant to one part of the loop, in which the assumption of 'given' ends is valid. At the next 'iteration' a different set of 'given' ends may need to be considered but how the ends get adjusted in the light of experience is not his bailiwick.

The second objection to the economist's position is that people are not 'optimising' anything at all, but simply seeking a 'satisfactory' or 'acceptable' outcome. This position has two strands, the first being the concern with costs of search and if that is all that is being said it amounts to saying that economists are very prone to ignore certain important costs and hence are likely to reach misleading results. That is a fair criticism, but it is not an argument against the postulate that people are optimisers. It is an argument against the practice of inadequate optimisation. The second strand (for example Simon 1955, 1956) is more subtle, depending on the assertion that people have limited aspirations and, once these are satisfied, no further effort is devoted to them until the individual is given cause to be dissatisfied once more (that is until his level of aspiration is raised). Again, I reply that, as an economist, I am not interested in how levels of aspiration are raised (though if I were in the advertising business I could make a fortune if I knew!) but what the consequences are if they are whatever they are or might be. I expect the sociologists and psychologists to produce information and ideas on the value–systems of (say) corporate enterprises and it is then up to us economists to listen attentively and attempt to work out the consequences for their economic behaviour.[6] If some systematic relationship can be established between attainment and aspirations and then if it can be cast in terms of the variables of the model, then by all means let us incorporate it. But if it requires a very complex apparatus of sociological and/or psychological theory to establish the linkage, then the economist may surely be permitted to say (as I did earlier) that what goes on in that part of the loop is not his concern and he will pick up the signals again when they emerge as a fresh set of assumptions for him to make about motivation at the beginning of his segment of the loop.

Self on Roskillitis[7]

This brings me finally to the objections raised by Self (1970) in a recent attack on the role and activities of the Roskill Commission on the Third London Airport. Self distinguishes two roles which the Commission is supposed to play in this particular instance, first 'to achieve a full, fair and judicialised treatment of all objections' and second to supply a 'scientific basis for an eventual decision ... through the elaborate cost–benefit exercise of the commission's research team'. Although the former role of CBA is not without interest and significance, I shall here concentrate on Self's criticisms of the latter aspect of the Roskill Commission's activities (see Roskill 1970, 1971).

In this context, the essential points made by Self are the following:

(a) ... the kind of aggregate monetary figures used ... could not be made even tolerably credible or objective *upon any basis whatever*. Cost–benefit analysis gets its plausibility from the use of a common monetary standard, but the common value of the £ derives from exchange structures. Outside such situations, common values cannot be presumed .... The greater part of the figures used ... represent notional values which will never adequately be tested or validated by actual exchanges, and which are highly arbitrary in the sense that a very wide range of values can plausibly be predicted, depending upon innumerable opinions and assumptions ....

(b) ... the framework of analysis becomes distorted by the grotesque attempt to place all factors on the same monetary basis ... such diverse items as 'capital construction costs' ... which can be estimated within tolerable limits ... (and passenger surface travelling costs) ... which ... far outweigh differences in capital costs ... (and are) based upon an enormous chain of speculative analysis.

(c) The cost–benefit figures are incredible, not only because of the disparate basis of the items included, but because of the important items excluded. Of course, important factors often are excluded from policy judgements, but the appropriate remedy ... is to point out the missing factor with the aid of as much supporting evidence as possible .... Instead of ... being argued upon its broad merits ... (such an element) ... has to be translated into yet more items within a cost–benefit equation .... The limits of the financial quantifiable have to be stretched past absurdity, as the logic of the exercise will collapse. And once again more limited types of judgement – which ultimately in fact cannot be avoided – become confused and displaced.

(d) The ultimate absurdity of the whole exercise is revealed, paradoxically, through efforts to increase its rationality. Following classical welfare economics, Professor Lichfield has argued ... that distributional values should be included in the exercise .... But is it in any way sensible to express such factors in monetary terms? Clearly there are almost as many views upon the precise weight that is to be attached to these 'distributional' ... values as there are people in Britain. How is the Roskill Commission to proceed? Conduct a Gallup Poll of popular valuations in circumstances in which nobody will have to pay for the figure he hits upon with actual costs? Or ask the Government for directives? One only has to pose the questions to see the absurdity of such proceedings.

Objection (a) is a peculiarly two-edged weapon for Self to be wielding. He appears to accept market valuations as 'objective' but not non-market values. Yet market values are merely the outcome of a large number of subjective valuations and disparately based decisions which are on all fours with 'non-market' decisions abut time savings, accident prevention, the allocation of leisure time, etc. So one possible inference from his position is that since *everything* (even market price) rests on subjective valuations (upon which it could be also said that 'there are ... as many views ... as there are people in Britain'), no 'objective' analysis of any sort is possible. The alternative (and opposite) inference is that the market is to be regarded as an 'objective' mechanism for reconciling conflicting valuations but no other method is to be so regarded, least of all those which use market-type concepts and terminology (for example imputing a 'price' for time, or calculating the implicit 'cost' of accidents, etc.). This would be an even more extreme position than McKean's, would undermine any attempt to get away from naive financial appraisal and surely cannot be what Self intends. If, on the other hand, he accepts the former implication, then his strictures apply as much to political science as to economics. I cannot see any way of interpreting his position which leads to a different implication from these two and of them I would accept the former and reject the latter. But in accepting the former implication, I would assert that I am doing no more than accepting the well-recognised proposition that you cannot ascribe values without making value judgements. Market prices are acceptable if the value–premises underlying market behaviour are acceptable (for example accepting the prevailing distribution of economic power and the relevance of individual market-oriented valuations to the making of social judgements). Imputed prices are acceptable if it is believed that the incidental values of x to be inferred from people's associated (market) valuations of y and z are an accurate reflection of individuals' utility-maximising behaviour and that this is the relevant consideration. Postulated prices are acceptable if it is believed that the value–premises underlying both market and imputed prices are misconceived for the purpose in hand (for example if one accepts the propriety of a paternalistic or collectivist basis for valuation). To say that any of these is 'arbitrary' is meaningless ... each follows logically from a particular (but different) set of premises. The fact that each of them may be *imprecise* in the actual valuation it yields is a different point and Self merely confuses matters by linking it with the earlier methodological points. Wide differences in actual valuations would flow from differences in methodology (for example whether 'market', 'imputed' or 'postulated' values are relevant), differences in methods of estimation, or differences in data. One of the advantages of formal and explicit procedures of evaluation is that it should be possible to identify which of these differences is present in any particular dispute.

Objection (b) appears to be saying that estimates embodying differing degrees of complexity and accuracy (which are independent dimensions!) should not be brought together. He surely does not mean it, for almost no comparisons could survive that test. If he means that likely margins of error should be indicated, one cannot disagree with him. But presumably such margins of error are needed so that they can be compared. How can they be compared unless they are in commensurable terms? What practicable alternative is there to the use of monetary units?

Objection (c) seems even more confused because he appears to be saying the CBA is more likely to lead to the omission of relevant material than other modes of analysis and, at the same time, that it prevents proper analysis of these excluded items by dissecting them and scrutinising each component so closely that the 'Gestalt' is entirely obscured. If it is to be interpreted as an attack upon *analysis* as such (which I take to mean 'the resolution of anything complex into its simple elements' OED) then the rift between CBA (and all other analytical techniques) and the Selfs of this world is irreconcilable. If it is intended to be a more narrowly directed assault on the use of money valuations as the medium by which comparative values can be expressed, I ask once more for suggestions as to a practical alternative.

This brings me to the final quoted objection concerning distributive justice and the difficulties inherent in assessing it. His initial assertions are well-taken, but his rather lame rhetoric at the end leaves me unmoved. As I see it we are once more in the realm of choosing between 'the market' (that is, direct observation of individual behaviour), 'imputation' (that is, inference from associated behaviour) and 'revelation' (that is, guidance from on high) as a source of 'weights'. I see none of these as inherently 'absurd'. The more important issue is which of them is relevant in the particular decision-making context. From the analyst's viewpoint, the cardinal sin is obscurantism ... virtue lies in being explicit.

Is 'Roskillitis' really curable?

Even if all Self's strictures on CBA were valid, however, the sad thing is that he has nothing substantive to offer in its place. He says we must return to 'planning' which he describes as offering 'multiple and interlocking perspectives which can be related in various ways'. Each of these perspectives has its own criteria of desirability and putting them together 'cannot be free of opinion and argument'. 'But the extent of these disputes cannot be known or arbitrated unless each principal set of desiderata is investigated and classified'. In other words, objectiveness need to be clarified and rendered commensurable! 'In so far as these differences cannot be bridged, some ultimate valuations have inevitably to be made by somebody'.

At this point it sounds like a recipe for CBA but he shies away from this and writes 'In so far as the exercise involves the valuation of questions of opinion, one might suppose that truer indications could be secured through the direct collection and confrontation of the views of the interested or expert bodies. Such a process may seem to have less scientific precision than hypothesised monetary equations but it is at least grounded in reality'. This hardly squares with his own earlier proposition that only judgements backed with cash were to be regarded as so grounded. It also elides the crucial point that the intensity of benefit or loss by each of the 'interested bodies' has still to be weighed according to *some* criterion and he implies that 'weight of public opinion' would be the relevant consideration. He does not tell us how it is to be weighed, however, nor what sort of information the public needs in order to form an opinion, who provides it and whether this can be done without engaging in a process essentially indistinguishable from CBA. Indeed, in his concluding paragraphs. Self seems paradoxically to be doing no more than invoking 'the planning process' as a means of ensuring that non-market values are brought into the picture more explicitly so as to override the market values to which he ascribes such 'objectivity'! 'The claims of interested bodies must not only be checked, but sometimes overridden. However, the basis for actually doing this ought either to be the demonstrated superiority of a broader perspective – which is the real meaning of planning – or alternatively the use of some explicit valuation, such as ... the interest of future generations in environmental protection ...'

What Self does not seem to realise, or will not accept, is that unless such valuations are expressed directly (or indirectly)[8] in money terms they are bound to be vacuous (and therefore useless operationally). Instead he ends his polemic associating 'notional' values with 'imprecision' and castigating evaluation in money terms as a facade surreptitiously concealing value judgements rather than recognising it as a means of making them more explicit and precise. 'Greater rationality' he concludes 'is not helped, but hindered by the use of notional monetary figures which either conceal relevant policy judgements or else simply involve unrealistic and even artificial degrees of precision. Those who suppose otherwise are heading for a particularly dreary version of 1984'. One can only respond despairingly, that greater rationality is certainly not helped either by vacuously broad policy judgements cast in terms which defy analysis. Nor does it inspire confidence when their consequences are protected from systematic scrutiny by resort to obscurantist assertions that any attempt to turn them into empirically researchable propositions reduces them to the sordid status of terms in an equation and hence is to be resisted on some unstated (quasi-religious?) principle such as the belief that there is but one way to the truth (and although we do not know what it is, it is not CBA!).

CONCLUSION

In disputes of this kind two great dangers are always present. One is that the advocates (of CBA, for instance) will oversell their product, especially as they encounter what they may (perhaps quite rightly) consider to be ill-informed resistance, both from potential clients and from those peddling rival products. This has undoubtedly occurred with CBA and it is salutary to be reminded of the limitations of the current state of our knowledge, both in principle and in practice. The other greater danger is that the perfect becomes the enemy of the good, that acknowledged limitations are made the excuse for not abandoning practices which have even more defects, on the curious notion that we should change over only if a *perfect* product is offered in place of the imperfect one we are already using. CBA is not the way to perfect truth, but the world is not a perfect place and I regard it as the height of folly to react to the greater (though still incomplete) rigour which CBA requires of us by shrieking '1984' and putting out heads hopefully back into the sand (or the clouds) hoping that things will look better when we look around again in 15 years' time. In contemplating CBA, I prefer the philosophy embodied in the answer Maurice Chevalier is alleged to have given to an interviewer who asked him how he viewed old age: 'Well, there is quite a lot wrong with it, but it isn't so bad when you consider the alternative'.

NOTES

1. The qualification 'explicit and formal' is necessary here to establish a distinction from implicit and informal analysis by largely intuitive judgmental processes.
2. Margolis's comments (pp. 71–77 of the same volume) on McKean's paper present a more radical view, more in accord with my own.
3. That some cost–benefit analyses have given this impression is regrettable and it has led at least one practitioner to conclude that 'cost–benefit' studies deal solely with elements that can be measured in economic terms; a cost-effectiveness study considers other 'benefits' in addition to those that are purely economic' Packer (1968). This led him to conclude that 'The cost–benefit approach distorts the conclusions by over-emphasising morbidity among males in their maximum earning years; by under-estimating prevalent problems such as neurotic conditions, common colds and certain allergenic conditions; and by ignoring intangible costs, etc. Actual and implied costs are important dimensions of any resource allocation decision; however, the cost-effectiveness framework points (and demands) explicit consideration of variables that are measured not in dollars and cents but in terms of basic human values' (ibid. p. 251). Although I cannot deny that there is a danger that casting evaluation in money terms carries the danger Packer mentions (just as attempting any precise measure of normally unmeasured phenomena leaves the analyst open to the charge Wildavsky makes of obsessive and mindless quantification), this is not inherent in the method of analysis but only in the way in which it may be used by some practitioners.
4. This literature stems directly from early work by Bowen (1948) and Black (1958), which itself owed much to the Voluntary Exchange Theorists (see, for instance, Knut Wicksell *A new principle of just taxation* and Erik Lindahl *Just taxation – a positive solution* in Musgrave and Peacock (1958)). Among the better known and influential recent works of this genre are Downs (1957), Olson (1965) and Buchanan and Tullock (1962). The latest example to come to my notice is Curry and Wade (1968) which contains a select bibliography of relevant works. An interesting review of the methodological differences implicit in the various approaches is contained in Barry (1970) which contains a much longer bibliography.
5. In a paper on cost–benefit analysis present to the British Association for the Advancement of Science at Exeter in September 1969, I have rehearsed the case for overriding the market in cases where the results of an orthodox financial appraisal (for example via DCF techniques) conflict with the results of a cost–benefit analysis: see Williams (1970).
6. See, for instance, the work of Baumol (1967) chapter 6; Williamson (1964) and Cyert and March (1963).
7. In Williams' (unpublished) Medical Encyclopaedia, defined as 'the state of being inflamed by Roskill'.
8. By 'indirectly' I have in mind expressions such as the marginal rate of social time preference which, although a pure number, can be used in conjunction with variables cast in monetary terms.

2. The Cost–Benefit Approach

There are not, and probably never will be, enough resources to satisfy the community's desires for things that improve the quality of life. This poses the necessity for choice and hence the consideration of priorities. In the medical field two distinct kinds of choice arise: one at the clinical level and the other in the planning process. At the point of contact with a particular patient, a doctor's duty is to do the best he can for that patient within the confines of existing knowledge and facilities. In the planning process, on the other hand, the concern is with large groups of potential patients at some future date and with decisions that will to some extent determine what the confines of existing knowledge and facilities will be at that date. The planning process is my concern here and, in particular, one important and increasingly wide-spread approach to the problems that arise within it, namely, cost–benefit analysis.

WHAT IS THE ESSENCE OF THE COST–BENEFIT APPROACH?

Cost–benefit analysis rests on the proposition that we should provide services only if their benefits outweigh their costs. In subscribing to that view, however, we (perhaps unwittingly) commit ourselves to the following set of propositions: (i) it is possible to separate one service from another service in a sensible way; (ii) there is the possibility of choice between them; (iii) it is possible to estimate the outcomes associated with each alternative service; (iv) it is possible to value these outcomes; (v) it is possible to estimate the cost of providing each service; (vi) these costs and benefits can be weighed against each other; and (vii) we should cease providing those services the costs of which outweigh the benefits.

The technical problems of cost–benefit analysis lie in establishing propositions (i) to (vi). If these are resolved satisfactorily, item (vii) is merely a test of society's belief in the basic principle. In this section I shall consider items (i) to (vi) in turn.

i. The separation of one service from another is the essence of analysis and there can be no single correct way of doing it. Differing systems of separation will give different insights and in the present context they are to be judged by their usefulness for health services policy. Thus, whether it is more useful to separate patients by disease, speciality, location, age, sex, weight and so on, depends on the problem in hand. Often it will not be possible to know in advance which classification scheme is going to be the most fruitful. Moreover, abortive analyses may add to our knowledge just as much as successful ones. Also involved is the general problem of 'project definition', that is establishing the boundaries of the analysis.

ii. People frequently adduce the alleged impossibility of choice as a reason for not conducting analysis. However, what they are usually saying is that, if certain assumptions are accepted, there is only one logical outcome and hence no possibility of choice. In my experience it has yet to be the case that the logic is so restrictive that it tells you precisely, for example, how much of a particular service should be given. Usually, there is room for manoeuvre over 'how much'?, or 'how soon?', hence the necessity for choice through the setting of priorities. But the analyst usually needs to go beyond this kind of exploration and challenge all or some of the assumptions themselves; one of his objectives should be the creative one of broadening people's horizons as to what is possible and prima facie worth considering.

iii. The cost–benefit analyst (usually an economist) has to rely on the expertise of other groups. In the medical field, he needs to know what will be the likely consequences of one intervention rather than another (where the latter can mean doing nothing). Here cost–benefit analysis can encompass what is known to medical science but it cannot be expected to make good what is not known. It may be possible to deal with ignorance here by making alternative assumptions about efficacy in order to see whether intervention is justified. It can also play a useful role, by indicating more clearly what needs to be known and in what form, if particular policy issues are to be better illuminated. It may even be possible to establish some rough order of relative social values for some kinds of new knowledge compared with others. But cost–benefit analysis cannot supply answers to technical medical questions, since that is the role of medical science. What it can, and does, stress is that, in matters of health services policy, medical knowledge alone is not enough.

iv. Whether a cost–benefit approach is adopted or not, valuation is inescapable. Much of the valuation of benefits is implicit, concealed in some broad judgement about priorities, or deeply embedded in some index or scaling mechanism designed to act as a measure of the seriousness of a condition. Since cost–benefit analysis is concerned with medical or humanitarian benefits (in terms of improved social function, absence of pain and distress and a sense of being cared for) as well as with the more material benefits (the productive capacity of working individuals, now patients), we also need to find some way of bringing the two together.

v. Here cost–benefit analysts have a highly distinctive stance, derived from one of the fundamental principles of economics, namely that cost means 'forgone benefit'; in common-sense terms, the cost of any service is what you sacrifice in order to obtain it. Hence in economics it is known as 'the doctrine of opportunity cost'. Thus the cost of any kind of medical intervention is represented by the value (in the best alternative use) of all the resources so employed. This may or may not be accurately measured by what (if anything) is actually paid for them, so it must not be assumed that 'opportunity cost' and 'cash expenditure' are interchangeable. Herein lies a great deal of the distinctive contribution of cost–benefit studies to public policy, resting on some very complex technical analysis to which I cannot do justice here (see Dasgupta and Pearce 1972, Layard 1972), but some glimpses of which will be provided later in this paper.

vi. The final hurdle is the proposition that if costs and benefits are to be weighed one against the other, they must be made commensurable. This implies that, if costs are to be measured in money terms, then benefits must be measured in money terms too. This prospect stirs deep emotions in many people, who seem to prefer matters of relative valuation to be left obscure (Williams 1973). But decisions that are never systematically analysed, never confronted and compared with evidence from other related fields, are likely to become haphazard and hence the strong drive in cost–benefit studies to find suitable bases for comparison across different fields of public policy (for example between improved health and road safety or the reduction of occupational hazards) as well as between apparently similar decisions within any one field.

Another aspect of the cost–benefit comparison that needs stressing is that cost–benefit studies stress the simple truth that the decision whether or not to pursue a particular course of action depends on both costs and benefits. Yet we see far too many recommendations based on assertions that x is cheaper than y (without adequate consideration of relative benefits) or that x is more effective than y (without adequate consideration of relative costs). Accountants are prone to perpetrate the former fallacy and medical men the latter and if the cost–benefit approach did no more than keep these errors in check it would have made a valuable contribution to clearer thinking!

IS COST–BENEFIT ANALYSIS APPLICABLE TO HEALTH SERVICES POLICY?

In principle, cost–benefit studies are appropriate wherever resource-allocation decisions have to be made; this leaves most of the field of human choice susceptible to cost–benefit analysis. In practical terms (since such studies are costly) the potential benefits of analysis are not likely, in many cases, to outweigh the costs; hence cost–benefit studies should be concentrated where the reward is likely to be greatest. Among the ingredients by which I would seek to identify such areas of choice would be that (i) sizeable amounts of scarce resources are at stake; (ii) responsibility is fragmented; (iii) the objectives of the respective parties are at variance or unclear; (iv) there exist acceptable alternatives of a radically different kind; (v) the technology underlying each alternative is well understood and (vi) the results of the analysis are not wanted in an impossibly short time. Items (i), (ii) and (iii) on this list specify situations in which the potential benefits from a cost–benefit study would be great. Items (iv), (v) and (vi) ensure that the analyst would have something worth while to consider.

In reality one never expects to see all these desiderata fulfilled simultaneously, so one has to decide whether the scope for analysis and the available material are promising enough to justify the time and effort involved – compared, of course, with the alternative uses of scarce analytical talents. Some problems will be worth only a short and simple exercise, others will warrant million-pound exercises of the most elaborate and sophisticated kind and both may legitimately be termed 'cost–benefit studies' if they conform to the general rubric set out at the beginning. But it will be important here not to place greater weight on the results of any particular study than it can legitimately sustain. As with every systematic investigation, interpretation and generalisation remain a skilled and dangerous business which the analyst, writing up his results, cannot fully anticipate. Hence everyone needs to develop a basic critical faculty with respect to such studies.

There are plenty of problems arising in the health services which are prima facie susceptible to cost–benefit analysis, even within my more restrictive version of the terms of reference. Leaving aside the many choices that have to be made which are not peculiar to health services as such (e.g. in the general fields of catering, domestic services, engineering and building), there are important decisions concerning different types, places and times of treatment for a particular condition and priorities for treatment within a particular condition and between conditions (or patients). Each of these generates needs for similar types of data and, even conceptually, they are not so dissimilar as they appear at first sight. They give rise to the need for more fundamental studies, concerned to elucidate certain common problems, such as the notion of cost that is appropriate for a particular context of choice and how the effectiveness of health care systems can be measured (see section 3). Initially, a few examples will be given of actual studies that have been published.

The choice between different types of treatment is, I suppose, one of the classic problems of medicine. It has been studied recently, for instance, in a cost–benefit framework, with respect to such problems as those posed by varicose veins and chronic renal failure. In the case of varicose veins, the medical benefits for the population selected were found to be much the same from two kinds of treatment – injection-compression sclerotherapy and surgery hence the choice turned on their respective costs (Piachaud and Weddell 1972). It was assumed that all the costs of injection-compression sclerotherapy were avoidable (and hence relevant) and were accurately measured by their money costs to the health services. Since the alternative, surgery, involved the use of a fair proportion of common facilities, cost allocation was based on average (non-medical) costs for each day in hospital plus a specific assignment of medical, surgical and nursing costs, again based on what the health services pay for these resources. No estimate was made of the cost to patients but it was argued that, since this would be higher for surgery than for injection-compression sclerotherapy, and since the latter was in any case the cheaper form of treatment, this omission reinforces the conclusion that it would be to everyone's benefit if the majority of patients were treated in outpatient clinics by injection-compression therapy. The standard analytical device of holding benefits constant, thus converting a cost–benefit study into a cost-effectiveness study, is acceptable provided that (i) the alternatives really do generate the same benefits and (ii) one is not interested in finding out whether anything at all should be done, since it is quite possible that, even for the least costly alternative examined, the costs exceed the benefits.

A study on chronic renal failure (Klarman 1968) in the USA goes one step further in terms of benefit evaluation by taking years of life expectation as the relevant unit and measuring the effects of various treatments (kidney transplants versus renal dialysis) upon it, recommending that treatment which provides a year of additional life expectation at the least cost. Benefit (in extra years of life) was roughly adjusted for 'quality', transplantation yielding a higher quality of life than dialysis, so the latter was regarded as worth 0.75 of the former. As regards costs, we still have preponderant attention given to actual expenditures on medical services and relatively little to other costs (such as loss of output caused by patients being unable to work, additional family expenditures and loss of leisure time). The study also worked with average data for a hypothetical cohort of 1000 people with chronic renal failure, whereas it is likely that the medical effectiveness of the respective treatments would be worse than average at the margin (if the patients with the best prognosis are taken first) and the costs of providing more or less treatment would differ from the average for the group as a whole (depending on whether there are economies of scale). The conclusion was the '... transplantation is economically the most effective way to increase the life expectancy of persons with chronic kidney disease.' The method adopted does not, however, enable us to tell whether any transplantation at all is justified, compared with other possible uses of these resources.

Both these examples involved differences in the place of treatment as well as in its type. Indeed, there is some discussion in the study by Klarman *et al.* (1968) of the relative merits of hospital dialysis versus home dialysis, with the latter showing up much better on grounds of cost. That this was so was probably partly owing to the non-inclusion of some relevant costs, as is hinted at in the conclusion:

> ... that one limitation on the performance of dialysis at home is the condition of housing. Accordingly, it may be desirable to consider alternative ways to improve the housing of patients with chronic renal disease and their families.

A study which tackles institutional care vs domiciliary care more directly is that on the care of the elderly, commissioned by Essex County Council (Wager 1972). Here a key concept was an 'index of incapacity' based on 14 different components which cluster into five groups relating to sensory perception, intellectual processes, personal care, physical mobility and domestic duties. By scoring points for these components, 'clients' are classified according to five grades of incapacity (none, slight, moderate, substantial and severe). The study shows that those with moderate incapacity, or less, account for about half of all applicants for places in the local authority's welfare homes and for them domiciliary support is, in principle, a viable alternative. The study makes the important observation that the accommodation occupied when domiciliary care is being offered is usually not without cost to the community (quite apart from the costs – material and psychological – falling on the family itself). Estimates of the cost of such accommodation showed that it was likely to be higher for old people living alone in ordinary houses than for an old person living with others, since only the marginal (additional) resources used directly by that person would be released for other purposes. The study suggests that there are significant potential economic savings to the community through making potential economic savings to the community through making small purpose-built dwellings available to elderly people at present living in larger dwellings which have become too big for their needs.

> The benefit to the community's housing problem arises irrespective of whether the accommodation released is local authority or privately owned; this suggests that the optimum use of the community's housing resources is not served by the practice ... whereby some local authorities do not consider applications for accommodation from owner–occupiers (Wager 1972, p. 59).

Turning next to alternative times of treatment brings us into the field of preventive medicine. Following a general survey of the field (Pole 1968), a particular example pursued in greater depth was mass radiography (Pole 1971, 1972). As regards the benefits of early detection, it is observed that:

> ... the whole relationship between the existence of a primary active case and the existence of a secondary case requiring treatment, though complex, can be described by a series of interrelated but unknown probability distributions (Pole 1971).

In the absence of a better understanding of the underlying epidemiology, a range of assumptions is made. The costs of a case are then estimated and the saving of such costs is taken to be the benefit derived from preventing the occurrence of that case. It is found that, even on the most favourable set of assumptions, the costs of mass miniature radiography are twice the amount of the expenditure needed for treatment of the patients. From this it is guardedly concluded '... that this analysis does not tend to show that the decision to abandon mass screening for pulmonary tuberculosis was mistaken' (Pole 1971). Prevention is not always better than cure!

I have cited two final groups of alternatives – between different subjects for treatment, either within or between conditions – because I have come across few cost–benefit studies in the health field which tackle them directly. As regards road safety, there have been attempts to rank widely different measures in the order in which they would contribute to casualty reduction, but I believe that nothing comparable has been done for the closely allied field of preventive medicine, or for general medicine. Yet it seems to me to be one of the central problems for the setting of priorities in medical care and frequently has economic undertones. The death or incapacity of a young productive individual costs the community more than that of an inactive person with short life expectancy and one of the excruciating, but inescapable, issues of health services policy concerns the relative values to be attached to the 'humanitarian' and the 'material' benefits and costs involved in medical care. On the resolution of that issue will depend the operational priorities of the system.

HOW ARE BENEFITS AND COSTS TO BE MEASURED?

A variety of benefit measures have been used in the studies cited. In debates on health services policy, indicators such as doctors/1000 population, beds/1000 population, costs/day in hospital, costs/patient, perinatal mortality and life expectation at birth are all used as criteria of success (or failure). If it is accepted that the objective of health services is to improve health, then only the last pair of indicators is really a measure of the output (that is, effect on health) of the system, the first pair relating to inputs (resources used) and the middle pair to throughput (amount of work done). Unless one is convinced that more input of doctors and/or hospital beds automatically generates more good health, or more patients, then one had better stick firmly to indicators that relate directly to health conditions of people.

Morbidity statistics are of some help here, but only indirectly, since it is well known that the degree to which people suffer from any condition varies widely. It seems therefore advisable to concentrate on indicators of social functioning as the key to benefit measurement. This means going further down the road pursued tentatively in the Essex study cited above (Wager 1972) and building up some generally acceptable set of criteria, perhaps incorporated in an index of health, which becomes the touchstone for comparative evaluative studies in the health field.

Even then it will still be necessary to confront the issues of relative valuation of outcomes. At present these tend to be resolved by experts exercising their judgement and it is likely that this will continue to be the prop on which we lean for a good while yet. But ultimately these relative valuations, be they weights in an index, scores in an assessment schedule, or operational decisions that x shall and y shall not qualify for a particular 'treatment', must be recognised as statements about health services policy and be accepted according to criteria that society has approved. They must first be systematised and made explicit.

Routine ways of measuring the extent of functional disability and adverse psychological effects of various conditions and the changes that occur with and without treatment, are already being devised by people in the field responding to the need to overcome particular clinical problems concerning assignment of treatment.   An assessment schedule, designed to be administered once only at some stage of a diagnostic process, could be adapted for routine monitoring or follow-up and be transmuted into a measure of effectiveness. It is but a short (though difficult) step further to get the various health conditions of people ranked in order of goodness to badness.   At that stage we have the problem of benefit measurement essentially solved (see Culyer *et al.*1971). The remaining problem will be that of finding enough common elements in the scales used in different places and for different conditions to enable one to synthesise a more general health index.

All this is necessary and valuable whether or not we go to the final step of attaching explicit monetary values to health states. The acceptance of the fact that an additional year of healthy life is intrinsically worth, say, £1000, no matter to whom it accrues, would probably lead to a much finer, humanitarian and egalitarian health service than we have at present, and does not imply that the pursuit of profit rules all, since my supposed figure of £1000 is merely a precise expression of what society should be prepared to devote to the cause of increasing expectation of healthy life, irrespective of these financial considerations! I fear that the reluctance of well-meaning people to accept this part of the logic of cost–benefit analysis may be deleterious to their cause, because it may lead to an excessive amount of attention being devoted to the other items in the calculus.

The measurement of costs is the other key issue that generates much misunderstanding. 'Cost to the community' does not mean cost in terms of 'public expenditure', or even 'private and public expenditure', since neither of these notions may properly represent the value of what is sacrificed if resources are used in one way rather than another. For instance, the use of an existing capital asset (for example, a piece of land) may involve no additional expenditure if it is land belonging to the Crown or is held freehold, yet the use of that plot is being denied to other people and involved a sacrifice which needs to be taken into account. Conversely, if some rearrangement of work is envisaged that releases labour which is otherwise unemployable, then the gain to the community will be minimal and will certainly be overstated by the savings in expenditure on wages.

This brings me to an important point concerning the time context and general scope of the decision under analysis. Costs are not immutable 'facts' scattered about waiting to be gathered and processed. What is a relevant cost in one context is not so in another. Suppose we are asked the question: What does it cost each day to keep someone on renal dialysis at home? At one extreme the question could be answered with respect to an individual who already has his home adapted, equipment installed and is himself fully skilled in what is required of him, so that the costs of keeping his system going for another day may be very small and within a day it would be impossible to reallocate the equipment if he were not using it. So we shall get a very low cost figure in that context, because most of the sacrifice to be made by the community has already been made and is immutable within the time context specified. At the other extreme we could posit a context in which no such dialysis exists, in which development work on equipment has to be undertaken, staff trained and patients selected and established in the new routine. Here all the sacrifices are prospective, hence relevant, and we are thinking in terms of a long planning horizon in which there is ample scope to redeploy resources. Thus the cost/day of home dialysis is the present value of a long stream of prospective costs divided by the number of days of home dialysis it provides for prospective patients generally. These costs may well be sensitive to the scale of the undertaking and hence need to be so calculated and presented.

Between these two extremes there is an array of intermediate stances, each of which could give rise to such typical questions as: Assuming that we have the equipment and staff, how can they best be deployed on a continuing basis with replacement as necessary, at the present volume of activity? What would be the long-term consequences of varying the present volume of activity up or down by x per cent (where x is less than 25) on a continuing basis and what is the optimum rate of transition from the present to the new level? Each of these is a perfectly legitimate formulation of a possible context in which the original question about costs might be answered and each yields a different notion of which costs are relevant and how they are to be measured.

One final time-related element has to be taken into account, namely the fact that we are considering streams of benefits and costs stretching into the future. It is evident that the community generally prefers benefits sooner rather than later and, conversely, prefers to defer costs rather than to incur them early on. To take account of this we usually apply a discount rate (say 10 per cent per annum) to the respective streams, which has the effect of reducing the value of more distant costs and benefits relative to those close to the present time. There is much dispute as to what should be the proper basis for calculating such a rate and as to the appropriate rate itself – though for most decisions on public expenditure this discussion is effectively short-circuited by a Treasury ruling saying what rate shall be used. However, since the balance between benefits and costs can be very sensitive to the discount rate used, this deserves more consideration than it frequently gets in cost–benefit studies.

IS IT ALL TOO DIFFICULT?

Enough has been said to indicate that there is more to cost–benefit analysis than generalised discussion of advantage and disadvantages. It represents the outcome of much hard thinking about what is involved when we seek a systematic answer to the question: Do the benefits of a certain action outweigh the costs? It proves to be a very difficult problem even to formulate the question satisfactorily and attempting to answer it frequently impresses one with the vast area of ignorance that pervades even quite commonplace activities. At this stage, therefore, the weak-spirited usually abandon the cost–benefit approach as too demanding and return with relief to more comfortable ways.

The trouble with the more comfortable ways is that they foster the illusion that, if cost–benefit analysis is not done, the issues which it poses can be avoided, whereas the reality is that these issues are all still present and they all still have to be resolved. If health services planning is not to be based on the principle that unwitting decisions are likely to be better than witting decisions, then the cost–benefit approach must become a part of every decision maker's intellectual equipment. As a homely contribution to the furtherance of that worthy cause, I offer the following 'check-list' of basic questions that should be asked every time anyone makes a studied recommendation about use of resources. If it is impossible even to discern any material relevant to the questions, be especially on your guard, since the questions will have been answered by making assumptions which may be unrealistic and/or unsupportable and/or unacceptable.

A basic check-list of questions runs as follows:

1. What precisely is the question which the study was trying to answer?
2. What is the question that it has actually answered?
3. What are the assumed objectives of the activity studied?
4. By what measures are these represented?
5. How are they weighted?
6. Do they enable us to tell whether the objectives are being attained?
7. What range of options was considered?
8. What other options might there have been?
9. Were they rejected, or not considered, for good reasons?
10. Would their inclusion have been likely to change the results?

11. Is anyone likely to be affected who has not been considered in the analysis?
12. If so, why are they excluded?
13. Does the notion of cost go wider or deeper than the expenditure of the agency concerned?
14. If not, is it clear that these expenditures cover all the resources used and accurately represent their value if released for other uses?
15. If so, is the line drawn so as to include all potential beneficiaries and losers and are resources costed at their value in their best alternative use?
16. Is the differential timing of the items in the streams of benefits and costs suitably taken care of (for example by discounting, and, if so, at what rate)?
17. Where there is uncertainty, or there are known margins of error, is it made clear how sensitive the outcome is to these elements?
18. Are the results, on balance, good enough for the job in hand?
19. Has anyone else done better?

The last two have been added because I do not want to be accused of advocating a counsel of perfection. Decisions do have to be made and will continue to be made, on the basis of imperfect knowledge. But I am anxious to ensure that we know how little we know when we do what we have to do.

3. One Economist's View of Social Medicine

THE BUREAUCRACY OF SCIENCE

In principle knowledge is doubtless indivisible but, in practice, we have to divide it into small finite tracts, with plenty of overlap, in which each of us cultivates his respective plot with whatever tools he is most adept at using. It is inevitably a messy business but, as with all bureaucratic arrangements for the division of labour, it is occasionally rewarding to try to think through (and, if necessary, invent) the rationale of the system as we observe it, if only to reassure ourselves that there is no obviously better way of organising our activities.   Sometimes such a review will prove more constructive than this and indicate some rearrangement which offers the promise of more productive collaboration.

In thinking about my own subject I have found it useful to distinguish economics as an area of study from economics as a mode of thinking. By 'economics as an area of study' (or *topic*, for short) I mean essentially taking the economic system as the subject to be investigated, the presumption being that economists have special expertise on this particular topic. This 'special expertise' accumulates because, by sustained thinking over several centuries about how different economic systems work and develop, it proves useful to use certain concepts and structural relationships, to ask certain questions and to collect certain data, all of which come to constitute the corpus of knowledge transmitted from one generation of investigators to the next, with increasing specialisation and internal subdivision of expertise as the corpus grows in volume and complexity.   The 'special expertise' itself is what I call the 'mode of thinking', or *discipline* characteristic of the subject.

But although the relationship of the discipline of economics to the topic of economics may be a special and dominant one, it is not an exclusive one. It is not exclusive in two respects: firstly, the topic may be investigated by other disciplines and, secondly, this discipline may be used to investigate other topics. Let me illustrate each case in turn.

The topic of inflation is clearly within the ambit of economics and it is one in which the discipline of economics is much utilised. But it is also very enlightening and fruitful to see inflation as a political or sociological or moral problem and to apply the special expertise of those subjects to its analysis and clarification. In other words, economic *topics* are not the exclusive preserve of one discipline, not even of the discipline with a special and perhaps dominant relationship to them.

Conversely, the discipline of economics will have something to contribute to topics which are not conventionally classified as 'economic' problems, for example, whether or not particular types of crime should attract the penalty of imprisonment or be tried by jury, how stringent fire and other safety regulations should be, or how many doctors we need. In other words, the *discipline* of economics is not exclusively focused on economic *topics*.

Turning to the topic of social medicine, I was initially intimidated (and I am not easily intimidated!) by the confusing array of elements which it appeared to embrace: epidemiology, public health, sociology, social administration, public administration, management, operational research, planning and behavioural studies.

After considerable cogitation and several false starts, my professional conclusion is that 'social medicine' is a topic but not a discipline. I say this with some trepidation. The trepidation is due partly to the fact that this may be interpreted as hostile comment, which it is not intended to be and partly to the fact that at least one distinguished writer on the subject (Martin 1977), who knows a lot more about this than I do, clearly believes that it certainly is a discipline and so terms it. I will nevertheless develop the argument that led me to my conclusion, for we shall then have a sporting chance of identifying the precise source of error, if error there be.

WHAT IS SOCIAL MEDICINE?

If I am right in assuming that the labels 'social medicine' and 'community medicine' are interchangeable, then the clearest and most concise delineation of the scope of the subject appears to be 'the speciality which evolved from public health and which covers the organisation and evaluation of health care systems and the medical aspects of the administration of health services' (Holland 1977). In this formulation there can be no doubt that in my terms it is a topic, or rather a group of topics and not a discipline. This view seems implicit also in the further comment by the same author that epidemiology is 'the basic science of community medicine' and 'the study of the causes and distribution of disease in populations rather than individuals'. Moreover, since, 'health is influenced to some extent by the availability and usage of health services, epidemiology must also involve itself in the measurement of need and demand for, and use of, health care' (Holland 1977). In my terms, the crux of these assertions is that epidemiology is the *discipline* having the special (and perhaps dominant) relationship to the *topic* of social (or community) medicine.

Historically, there is little cause to doubt this statement but one could nevertheless fruitfully ask whether this is, or should be, any longer the case. What are the alternative candidates? I think we have to go through the related subjects listed earlier one by one and see what their respective claims might be.

Sociology, like economics, is both a *topic* (social structure and social relationships and the evolution of social systems) and a *discipline* (a systematic mode of thinking which employs characteristic concepts such as role, social class, stigmatisation etc. and studies their interrelationships). The topic of social medicine seems to overlap with the topic of sociology, in the sense that the health care system is itself part of the social system and may need to be studied in that context. But this is a different point from the one to which I am currently addressing myself, which is whether the *discipline* of sociology is, or should be, replacing the *discipline* of epidemiology as the 'basic science of social medicine'. If ever an innocent bystander ventured into a minefield, here is a touching instance, but – economists have never shrunk from rushing in where angels feared to tread!

As I see it, epidemiology developed out of medicine as doctors came to recognise that disease in individuals had important characteristics which could be identified only by looking at whole communities and which could be dealt with only by intervening at 'community level'. Initially these causative factors were still disease-specific ('germs', viruses, bacteria, etc.) and could be related comfortably to the concepts, if not the practice, of 'orthodox' medicine, which have an essentially biological and/or chemical basis. As mental illness came more to the fore and its underlying aetiology relied less on 'organic' phenomena as sources of explanation, and still more recently, as psychosomatic and social elements came increasingly to be recognised as at least contributory factors in much 'ordinary' illness, so the claim of sociology (as opposed to biology and chemistry) to provide the underlying conceptual framework for epidemiological work within social medicine has grown in strength. I suspect it will continue to do so. Whether epidemiology itself will change so as to accommodate this change, or whether 'traditional' epidemiological methods will prove unhelpful in this venture, I would not wish to predict. But the challenge of sociology, as a discipline, is not so much a challenge to epidemiology as a challenge to the role of medicine – and, behind it, the role of biological and chemical sciences – in the field of social medicine. This is reflected, in an intriguing linguistic way, by the differing connotations and implications of the terms 'social medicine' and 'medical sociology'.

The relationship to social medicine of social administration and, indeed, of public administration and management, is also a confusing one. Social administration is a *topic*, not a *discipline*, and in my view it is in this respect on a par with social medicine. It has tended to start from sociology and concentrate on social services, whereas social medicine has tended to start from medicine and concentrate on health services. For the same reasons as those set out in the preceding paragraph, the respective roles of health services and personal social services have become increasingly blurred and the growth of professional demarcation disputes between practitioners in the two broad fields is a significant indicator of this uncertainty. In the long run, I suspect that 'social medicine' will be absorbed by 'social administration' because I believe that a sociological perspective, rather than a medical perspective, will eventually come to predominate. This may not be as dramatic a prediction as it sounds, however, because (a) it is going to take a long time and (b) both medicine and social medicine will move (indeed already are moving) in that direction in any case, so that these sharply differentiated 'labels' will come to be attached to increasingly indistinguishable activities.

Operational research is a more awkward candidate because it is a *discipline*, not a *topic*. It is essentially about the use of mathematical models for simulation and optimisation of complex systems, so it does constitute a direct challenge to epidemiology, which claims to do much the same thing. This is especially evident if one accepts the view that epidemiology 'is based on the study of groups or populations; it implies nothing essentially medical (nor, specifically, anything to do with infections)' (Meade 1975). I see no solution but the fusion of epidemiological method and operational research, unless a convention about respective spheres of influence comes to be established.

On the rest of the candidates I will be briefer. 'Planning' seems to be an activity so diffuse that we could all be said to be interested in it and, since it is not a discipline but a topic, there is no more I wish to say about it. The term 'behavioural studies' seems to me to connote no more than an empirical orientation to a study, though I think it is also sometimes used as a portmanteau term to pull together psychology, sociology, organisation theory and, among the broad-minded, sometimes even economics! Perhaps others can see that it has greater significance to my argument which I have missed.

WHERE DOES ECONOMICS FIT?

If economics is subjected to the same treatment as that meted out to other subjects in my potted survey of the place of social medicine in the bureaucracy of knowledge, then again we must distinguish between economics the topic and economics the discipline.

Dealing first with the topic of economics, the relationship of the economic system to health and to the health care system is itself a fascinating area of study on which economics as a discipline can obviously contribute alongside other disciplines. The study of that relationship would embrace investigations into such questions as: What effects does the industrial and occupational structure have upon the level and pattern of ill health? To what extent does the health care system raise productivity – for example, by reducing absence from work through sickness? To what extent does the general state of the economy impinge on the resources available for the development of health care? Similar issues arise *within* the health care system as soon as it is seen explicitly as a resource-allocation system. For example, what are the effects of different charging systems on 'consumer' behaviour? What are the effects of different remuneration systems on 'producer' behaviour? What are the effects of various methods of financial allocation and control on real resource allocation? These are all economic *topics* related to an area of social organisation and behaviour in which we economists have a common interest with social medicine, social administration, etc. and I do not detect any great tensions in collaborative work in these particular fields, where the relevance of the economist's expertise is generally acknowledged.

But tension is much more apparent when economics as a *discipline* stakes out a claim to say something on *topics* which are not seen by others to be *economic* topics. I will take two classic and very important examples: the measurement of 'need' and the measurement of 'outcome'.

NEED

In the field of 'needology' I have, to my own satisfaction at least, sorted out once and for all everyone's confusions about need, demand, utilisation, etc. (Williams 1974b) and I do not propose here to rehearse that material yet again. The essential point is that need can be 'objective' only if we translate the assertion 'individual A needs intervention X' into 'if individual A had intervention X then, in *everybody's* opinion, individual A would be better off'. If true, this is an essentially factual statement. It does not imply that A *should* have X, however, because we do not know who else 'needs' X, how much X is immediately available, or what the priorities are between rival claimants for the resources needed to provide X now and in the future. Thus, if statements about 'need' are to get us anywhere, they must be linked to or incorporate *valuations* of some kind. Once they do this, they are ripe for analysis by economists, because the discipline of economics is essentially about valuation (and not simply valuation in markets, although as a topic that has been, and still is, our predominant interest). So what we would want to do is to move away from 'need' as quickly as possible and talk instead about *relative valuation*, or trade-offs or, in more common parlance, priorities. Since it is even more unlikely that everyone will share the same views on priorities than it is that they will *all* agree that individual A will be better off with intervention X, then we shall also have to face the questions 'whose priorities?', and 'what will be the process by which different people's preferences are accommodated, or not accommodated, as the case may be?'. Since markets are one way of accommodating such diversity, we economists tend to compare non-market solutions with market solutions, if only as an analytical device to highlight differences in outcome about which high-level value judgements will have to be made. So if it is true that 'in the middle of the 1960s the study not only of the needs of the community for medical care, but also of the demand for and the actual provision of these services were clear candidates for the second expansion of epidemiology' (Florey *et al.* 1976), then I hope any future expansionist urges in that area will be conducted more circumspectly and in harness with economists, so that they can help formulate the problem in a more policy-relevant way and also ensure that the appropriate information is collected.

OUTCOME

The measurement of outcome, or effectiveness, is another topic in which the essential element of valuation has too often been ignored. The limitations of mortality and morbidity data as measures of outcome are well known, but I do not detect a great deal of activity in social medicine in developing valid, versatile and operational health status indexes, although these seem to me to offer the only way forward in this difficult country. The most rigorous and fundamental work in this field in the United Kingdom is the product of a collaboration between a clinical psychiatrist and an operational researcher (Rosser and Watts 1972, 1975); there are also some ambitious attempts at applying the general idea in at least one department of community medicine. Be that as it may, the *valuation* of outcomes at all levels is a much neglected field and even the routine things that people do are not worked through rigorously or developed properly as evaluative tools (but see Wright 1974, Culyer 1976). Any 'index' or 'points' scheme, or implicit 'weighting' of one thing against another, contains a statement about 'trade-offs' or 'priorities' at the margin, on which the discipline of economics can often shed considerable light. I think it is fairly important, therefore, that the potential contribution of economists to the measurement of outcome – in other words, the valuation of benefits – should be recognised and acted upon more widely and urgently in the field of social medicine than it is at the moment.

COST

My final point concerns costs, which I have left until the end on purpose, because there is an unfortunate tendency to treat economists as if they were just cost-accountants and to limit their role accordingly. We must blame ourselves, in part at least, for the fact that we tend to get 'type-cast' in this somewhat restricted role, because our initial point of entry into many policy discussions has been to pose the question 'but what will it cost'? This is an important question, which still needs to be asked and to be answered more often than it is. Far too much 'evaluation' (in social medicine and elsewhere) is rendered useless for policy purposes by failure to consider costs, even in a narrow financial sense. As I have indicated, however, there is more to economics than the calculation of costs. All valuation problems are grist to our mill and there will often be alternative sources of valuation and alternative configurations of an activity which need to be incorporated in a study *at the design stage* if economic analysis is to be anything more than a last-minute cosmetic face-lift to mislead people into thinking that a real cost-effectiveness or cost–benefit study has been done. If the discipline of economics is to be a productive analytical tool when applied to the topic of social medicine, it must be allowed some influence on how problems are formulated in that field.

4. Health Economics: The Cheerful Face of the Dismal Science?

Economics is usually a rather doom-laden subject and in this respect is linked indirectly with medicine through the observation that the only two things in life that are certain are death and taxes. If you wanted to take the gloomiest possible view of the subject matter of health economics, I guess you could say that that statement sums it up perfectly ... it is all about death and taxes. Although death and taxes may be inescapable, they are, fortunately, not immutable so, if you wanted to take a more cheerful view of the subject matter of health economics, you could say that it is all about *postponing* death and *reducing* taxes. Broadly speaking, this is the view I take, though I hope to show you that there is more to life than postponing death and more to costs than shows up in taxes, so that too simple-minded a view will not in fact carry us very far in understanding what health economists are trying to do.

To clarify the nature of the potential contribution of health economics to thinking about health care I have constructed a crude schematic representation of the main elements in health economics, which is set out in Figure 4.1.

An unkind critic once said that if you taught a parrot to say 'supply and demand' it would give a passable imitation of an economist, so let me lend some credence to that caricature by starting in the middle of my chart, with two boxes *C* and *D*, labelled respectively 'DEMAND FOR HEALTH CARE' and 'SUPPLY OF HEALTH CARE'. Taking the supply side first, our interests here are in the costs of delivering health care, in comparing the costs of different ways of delivering care (for example, primary care versus hospital care, domiciliary versus institutional, surgery versus drugs), in the possibilities of substituting capital for labour, or one kind of labour for another, in the markets for these inputs (doctors, nurses, drugs, equipment, etc.) and they work and in the ways in which different remuneration systems affect the behaviour of suppliers of health care. In this kind of work one tries to see how much of what has been learned about other production systems could be used to advantage in the very complex and sensitive business of producing health care (which should be interpreted as including much of the care provided by local authority social services and voluntary organisations as well as by the health service).

Figure 4.1   Schematic presentation of the main elements in health economics

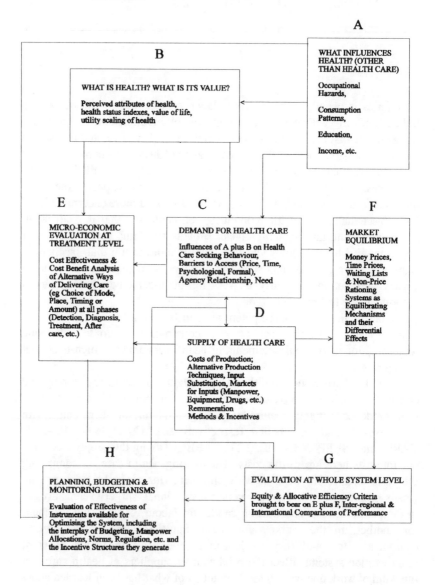

If things are complicated on the supply side, they are even more complicated on the demand side, for here we find ourselves in territory which cannot so straightforwardly be made analogous to what we do in economics generally. The first of our difficulties stems from the fact that over large tracts of the health care field we do not want 'demand' to be based on willingness and ability to pay, but on 'need' (a concept to which I shall return shortly). In order to give effect to that objective, health care is typically provided 'free' (or at a nominal price) at the point of consumption. But even very rich countries cannot provide all the health care that people would like to have if it really were 'freely' available, so in practice it never really is. Some of the non-price barriers to access are simply there 'in nature' as it were (for example, the time and trouble costs of getting to and from a consultation) whilst others are created as deliberate rationing devices (for example, rules about eligibility, or priority assessments based on 'need'). Need is a central concept in this debate but an unclear and ambiguous one, which I think essentially means some 'expert's' assessment of how much benefit someone would get from a particular 'treatment'. It will thus depend on (a) how effective the treatment is and (b) how the 'expert' perceives the value to the patient of that effect. It is therefore, in principle, partly a factual and partly a (paternalistically) evaluative statement. This brings me to the second major difficulty on the demand side, which is that patients do not usually 'demand' health care the way they 'demand' apples or pears. It is as if we go shopping knowing only that we feel hungry, with not much knowledge as to *why* we feel hungry, or whether our hunger will go away all by itself in due course, or whether there is anything that can be done to assuage it and, if so, what the most (cost-) effective way of doing that might be. For information on all these matters for health care we rely on the very same people who make a living by providing the goods in question: the doctors. Imagine what a heyday the foodsellers would have if we were in the same situation with them! In health economics this reliance on doctors is called the 'agency relationship' to denote the awkward fact that the suppliers of health care are often simultaneously the 'agents' of the demanders, with a potential for role conflict which I am sure needs no spelling out. Its analytical awkwardness for the *discipline* (as opposed to the topic) of economics is that it blurs the sharp distinction we try to draw between demand side influences and supply side influences. This happens elsewhere in economics in handling the role of advertising, where producers are clearly (despite their protestations to the contrary) trying to manipulate demand, rather than regarding 'preferences' as 'autonomous' and consumers as 'sovereign'. Could it be the same with health care?

Let me pass quickly on to where the consumers' own demand for health care comes from, which takes me to the top of Figure 4.1 and to boxes A and B. We are here concerned with the demand for *health*, which is obviously prior to the demand for *health care* and distinct from it. Clearly our health is strongly influenced by what might broadly be called our life-style and this will depend quite a lot on our own particular niche in society. If we are well-informed, prudent, moderately well–off, in a non–hazardous occupation, leading a happy and interesting life, with a balanced diet, neither eating nor drinking to excess and not smoking at all, our demand for health care is likely to be quite low, even though our demand for health may be quite high. The unfortunate people whose experiences and attitudes are the very opposite of those I just mentioned will present a very different pattern of demand for health care, mediated partly by their circumstances and partly by their own perceptions of the importance of life (that is, whether they value a good quality of life now more than extra years of life later). If society generally wishes to maintain its members in a good state of health (that is, maximise their quality-adjusted years of life-expectancy) we need some way of measuring and valuing health and economists have been active (with others) in tackling those fundamental (but very difficult) problems too.

Let us next turn to how the market for health care is likely to resolve the tensions created by balancing the demands for health care against the available supply, which is the content of box F on the right-hand side of Figure 4.1. In textbook markets money prices play the central role in bringing supply and demand into balance and in giving the appropriate signals to suppliers as to whether they should expand or contract capacity (by comparing prices with production costs). In health care we have the production costs but, for the perfectly good reasons explained earlier, we don't have any prices (or, more precisely, not any market-clearing prices). We therefore need to work with more complex notions such as time prices, waiting time and non-price priority-setting rules, all of which play the role of rationing devices which would otherwise be played by money prices. It is an interesting area of study as to what the differential effect on utilisation is of these different ways of impeding access to a nominally 'free' service.

The next logical step is to move from the 'positive' analysis of what actually happens to the 'normative' analysis of what we regard as 'better' or 'worse' ways of doing things. This is the content of box E on the left-hand side of my chart and takes us into the realm of cost-effectiveness and cost–benefit analysis. If 'effectiveness' or 'benefit' is to relate to how patients themselves perceive health and how they value it, then work of this kind must clearly draw on the material in box B of Figure 4.1. It also needs to consider the many options contained in Boxes C and D, concerning alternative ways of delivering care (choice of mode, place, timing or amount) at all phases in the process (detection, diagnosis, treatment, care, etc.). This is where the bulk of the work of British health economists seems to be located and it is the area in which we find ourselves collaborating with doctors, nurses, remedial professions, ancillary staff, managers and finance personnel, since this kind of evaluative work touches many people's interests, activities and expertise. It has great potential for the death-postponing tax-reducing aspect of health economics to which I alluded at the outset.

But the planning, budgeting and monitoring mechanisms of the health care system are worthy objects of study in themselves and constitute the content of box H in the bottom left-hand corner. Here we are not so much concerned with whether drugs or surgery are the most cost-effective ways of treating angina, as with whether the budgeting and planning systems are so designed as to give people an incentive to seek out and implement the most cost-effective way of treating angina. Would it help if clinicians became budget-holders and were allowed to retain some fraction of any efficiency savings they generated by more cost-effective practice and spend those retained savings on service developments which they thought important, without the need to get authority from higher management? At present we rely either on people's public spiritedness, or on rather arbitrary cuts (or both), to generate a drive for greater efficiency. Is there not a more constructive incentive structure to be thought out and implemented?

Finally we come to box G at the bottom right-hand corner of my chart, which is labelled 'evaluation at whole system level'. In this we try to look in a more Olympian way at the performance of health care systems, testing them against broad equity principles as well as against tests of allocative efficiency. In that kind of work it is often instructive to make interregional or international comparisons of the impact of different structures, although the more dissimilar the societies and their objectives the more difficult it is to draw relevant conclusions from such comparisons. I fear too that it is often an enterprise in which the grass seems all too frequently to grow greener on the other side, so that while some people in this country look hopefully for a 'free market' model elsewhere in the world which they fondly imagine will deal once and for all with our 'waiting list' problems, those plagued with the very *unfree* market systems, which are what one actually observes in some parts of the world, are looking to us for the secret of our success in 'cost-containment'. Without a deeper understanding of the dynamics of priority–setting and resource allocation mechanisms in different systems, it is very dangerous to think that you can graft selected bits of one system on to another and assume that the rest of the system will 'accept' the transplant and otherwise behave as before.

I think a lot of the suspicion about health economics – and about health economists – is the belief that our view of people is simply that they are (potentially) productive resources, that their value stems solely from that attribute and therefore that the most cost-effective way of caring for the elderly (or the mentally or physically handicapped) is to kill them off as quickly and cheaply as possible. If that were in fact our position, we should undoubtedly deserve any hostility or suspicion that came our way but, fortunately, it is not.

The general objective, in normative microeconomics, is to ensure that the value of what people get from an activity is greater than the value of what they have to sacrifice in order that it be pursued. Since a sacrifice is fundamentally a benefit forgone, then our objective can alternatively be seen as trying to ensure that benefits gained outweigh benefits forgone. This is also the objective of normative health economics and formulating it in that way puts benefit valuation, quite properly, at the centre of our attention. The benefits to be derived from health care are better health, by which I mean both improved quality of life and improved length of life. Some of this improved health may be used by people to earn money (that is, for work), but it may equally be used simply to enjoy life (and lucky indeed are the people – like me – who find their work enjoyable in itself, for they have the best of both worlds). So our starting point is that health is valued 'for its own sake', not just as a source of income. This fundamental aspect of health can be measured in terms of quality-adjusted-life-years, when the 'quality-adjustment' might be done as follows. Suppose that a year of healthy life expectancy is worth 1, then a year of unhealthy life expectancy will be worth less than 1 and will be rated lower the worse is the person's quality of life. Quality of life could be rated in terms of physical mobility, capacity for self-care, absence of pain and distress, ability to perform normal social roles, etc. and, if being dead is rated as 0, it is not uncommon to find that some people rate some very bad states as worse than dead (that is, on the above scale they would have *negative* values). So if we were measuring the cost-effectiveness of alternative patterns of health care provision for the elderly, we would measure effectiveness by the expected gain in quality-adjusted-life-years (or QALYs for short), not by anything as crass as their contribution to GNP, which might be nil throughout.

But you don't normally get anything for nothing in this world and it is likely that extra QALYs lead to extra costs. The notion of cost is a tricky one here, because in economics it is a much broader concept than expenditure. Expenditure simply means money spent and the raising and spending of money is the subject matter of accountancy and finance. The subject matter of economics is the deployment of real resources, whether they cost money or not. Health services do not pay for use of patients' time, yet time is an essential input into the health care system and is generally a resource with valuable alternative uses (except for the pathological cases who regard undergoing medical treatment as the most enjoyable way to spend their life!). Perhaps if health services *did* pay for the use of patients' time we should see a welcome reform in the way outpatients are organised. But let me not be deflected further from my main theme, which is that the economist's notion of cost is much broader and deeper than the accounting notion of expenditure, hence it is rather more difficult to measure and value.

Suppose we have solved all these knotty problems, however, and were in a position to say that, at the margin (for example, if we expanded some treatment activity by a small amount) the cost of a QALY (quality-adjusted life year) for a range of activities was something like this:

Activity *A*	£15000	per QALY gained
Activity *B*	£8000	per QALY gained
Activity *C*	£3500	per QALY gained
Activity *D*	£2000	per QALY gained
Activity *E*	£1000	per QALY gained
Activity *F*	£750	per QALY gained
Activity *G*	£200	per QALY gained

Does it then not seem natural to say that if we want to improve people's health as much as possible (that is, maximise the number of QALY's gained) we should first of all expand Activity *G* to the limit, then *F*, then *E* and so on and, when we run out of resources, stop. Suppose we got as far as Activity *C* before we ran out of resources, then we would *implicitly* be saying that we think a QALY is worth £3500 (because at that level we are not willing to put more resources into health care). Note that this cut-off point has nothing to do with people's productivity, for the beneficiaries of Activity *C* might be utterly 'useless' from the narrow viewpoint of their contribution to GNP.

There are, of course, some barely hidden complications in this simple-minded view of the world. The first of these is that one QALY is assumed to be of equal value to everybody. This is obviously a convenient analytical assumption but it has to be judged, not in those terms, but whether you believe that to be the appropriate ethical position for a health service to adopt. If not, what position do you think it should adopt? More complex ethical positions will of course require more complex calculations but that is a minor matter. Secondly, as you expand an activity and take on less and less promising cases for treatment, the effectiveness of treatment falls, hence cost per QALY rises, so the above table of numbers may well change as activity rates change.   Thirdly, cost-per-QALY may change if the resources used in them become more or less expensive, so the cost-effectiveness of a treatment depends on 'economic' as well as on 'medical' considerations. It may be that expanding an activity generates economies of scale which more than outweigh the reduced effectiveness of the activity on marginal patients (for example, costs fall faster than effectiveness). Fourthly, it may be that activities such as *A* and *B*, though very expensive in terms of cost-per-QALY *now*, are subject to rapid development, either of an effectiveness–improving kind or of a cost-reducing kind and need to be kept going (on a small scale at any rate) to allow this development work to proceed. Here one is essentially investing resources now in *future* cost-effectiveness. This argument has two important corollaries however, firstly that it is inapplicable to rather stable activities with little prospect of technological breakthrough and, secondly, that it implies that the activity should be conducted as a pursuit of knowledge (not as a proven therapy) and hence be subject to proper research protocols, full disclosure of data and evaluated for its cost-effectiveness by independent researchers, *not* by the protagonists of the activity.

In fact the activities I listed above are as follows:

A   hospital haemodialysis
B   heart transplantation
C   kidney transplantation
D   coronary artery bypass grafting
E   pacemaker implantation
F   hip replacement
G   GPs trying to persuade every smoker who visits them to stop smoking

The estimates are a bit rough and ready and the costs are mainly service costs.   It may be of interest to note that Activity *B* is regarded as experimental, but Activity *A* not! In looking again at that data you may also like to bear in mind that the *average* amount of GNP per head available to keep each of us going for a whole year is about £5500.

So we come back to GNP again, as of course we must, not in order to grade people in order of priority but in order to decide what we can afford. That is not a decision on which economics as a discipline can offer an answer, though obviously if we are prepared to say that improved length and quality of life are the objectives of all our activities, then in principle cost per QALY rules OK over the entire realm of society's resource allocation decisions. Fortunately health economists have enough to occupy themselves with in the terrain they already inhabit, without needing to indulge any megalomaniacal fantasies they may have in that particular direction.

A recent book on the current state of medicine, by a very well-informed medical scientist, was called *The End of the Age of Optimism*, so it may seem paradoxical that I should be swimming against the tide and arguing for the beginning of a mood of cheerfulness. My reason for so doing is that so far we have hardly begun to use the discipline of health economics for the improvement of the people's health. In a professional sense, this thought frustrates and sometimes even depresses me. However, it also means that there is great potential there waiting to be tapped. After all, we have only been in serious business as a subdiscipline for a couple of decades at most. As a professional group, we have youth and idealism on our side, though I must admit that my colleagues have an unfairly large share of the youth, which I try to compensate for by hogging an unfairly large share of the idealism. I think big gains have been made in the perceived relevance of the discipline of health economics to the problems of health and health care in this country and I am confident that the contributions of my colleagues will demonstrate the value of what we have to contribute. We are *not* defeatist prophets of gloom and doom, obsessed with death and taxes, but active, and often creative, workers for improvement, concerned to improve the quality of people's lives to the maximum feasible extent. *That* is why I think that health economics is the cheerful face of the dismal science.

5. Priority Setting in Public and Private Health Care

A Guide Through the Ideological Jungle

INTRODUCTION

Priority setting reflects ideology, so we must start by analysing the characteristic ideologies of public and private health care systems. Both systems (and their respective ideologies) have then to face the problem that the recent rapid growth of *effective* health care has led us to the point where no country (not even the richest) can afford to carry out *all* the potentially beneficial procedures that are now available, on *all* the people who might possibly benefit from them. So priority setting can no longer simply be a matter of eliminating *ineffective* activities (that is, it is now more than a matter of becoming more efficient in the *low-level* sense of getting on to the production possibility frontier). Priority setting now has to deal with the much more contentious *high-level* efficiency problem of choosing *where* to be on the production possibility frontier, that is which *mix* of efficient) activities to select from those that are open to us. This is a matter of *allocative* efficiency rather than *technical* efficiency and, inevitably, contains *equity* considerations, that is, views as to how the welfare of one person is to be weighed against the welfare of another person. That is why the ideological content has to be made explicit and given prominence.

I shall proceed from ideology to 'pure' private and 'pure' public systems and the characteristic problems they each encounter. This will lead naturally to a consideration of 'mixed' systems, in which each adopts a *little* of the other in order to reflect the pluralist ideologies of the communities they serve. This inevitably generates muddle and brings us closer to the systems we actually see operating around us and which we are struggling to understand and 'improve'. This brings us finally to the appraisal of policy making and performance in the different systems and to the question whether there is an *overall* framework within which we can

54

decide which is the best system of health care. I shall conclude that it all depends on your priorities.

IDEOLOGY AND OBJECTIVES

The ideological issues in the provision of health care have been admirably dissected by Donabedian (1971), who polarises attitudes around two viewpoints, 'A' and 'B', which may be loosely termed the 'libertarian' and the 'egalitarian' respectively. In the libertarian view, access to health care is part of society's reward system and, at the margin at least, people should be able to use their income and wealth to get more or better health care than their fellow citizens should they so wish. In the egalitarian view, access to health care is every citizen's right (like access to the ballot box or to the courts of justice) and this ought not to be influenced by income or wealth. Each of these broad viewpoints is typically associated with a distinctive configuration of views on personal responsibility, social concern, freedom and equality, which are set out in summary form in Table 5.1.

The implications of each of these ideologies for priority setting in health care are pretty obvious. Willingness and ability to pay should be the dominant ethic in the libertarian system of health care provision and this can best be accomplished in a market orientated 'private' system (provided that such markets can be kept competitive). Equal opportunity of access for those in equal need should be the dominant ethic in the egalitarian system of health care provision and, because such a system requires a social hierarchy of need[1] to be established which is independent of who is paying for the care, it dictates public provision (provided that such a system can be kept responsive to social values and changing economic circumstances). Let us look at each such system in more detail.

PURE SYSTEMS AND THEIR PROBLEMS

A simple view of these two systems would run something like this: (a) in a *private* system, access is determined by willingness and ability to pay and producers are kept responsive to consumers' demands by the profit motive with things being held in balance by price adjustments in competitive markets; whilst (b) in a *public* system access is determined by need, to which producers are kept responsive by the humanitarian motive with things being held in balance by quantity rationing based on a socially

Table 5.1   Attitudes typically associated with viewpoint A and B

	Viewpoint A (libertarian)	Viewpoint B (egalitarian)
Personal responsibility	Personal responsibility for achievement is very important and this is weakened if people are offered unearned rewards. Moreover, such unearned rewards weaken the motive force that assures economic well-being and in so doing they also undermine moral well-being because of the intimate connection between moral well-being and the personal effort to achieve.	Personal incentives to achieve are desirable, but economic failure is not equated with moral depravity or social worthlessness.
Social concern	Social Darwinism dictates a seemingly cruel indifference to the fate of those who cannot make the grade. A less extreme position is that charity, expressed and effected preferably under private auspices, is the proper vehicle but it needs to be exercised under carefully prescribed conditions, for example, such that the potential recipient must first mobilise all his own resources and, when helped, must not be in as favourable a position as those who are self-supporting (the principle of 'less legibility').	Private charitable action is not rejected but is seen as potentially dangerous morally (because it is often demeaning to the recipient and corrupting to the donor) and usually inequitable. It seems preferable to establish social mechanisms that create and sustain self-sufficiency and that are accessible according to precise rules concerning entitlement that are applied equitably and explicitly sanctioned by society at large.
Freedom	Freedom is to be sought as a supreme good in itself. Compulsion attenuates both personal responsibility and individualistic and voluntary expressions of social concern. Centralised health planning and a large governmental role in health care financing are seen as an unwarranted abridgement of the freedom of clients as well as of health professionals and private medicine is thereby viewed as a bulwark against totalitarianism.	Freedom is seen as the presence of real opportunities of choice; although economic constraints are less openly coercive than political constraints, they are nonetheless real and often the effective limits on choice. Freedom is not indivisible but may be sacrificed in one respect in order to obtain greater freedom in some other. Government is not an external threat to individuals in society but is the means by which individuals achieve greater scope for action (that is, greater real freedom).
Equality	Equality before the law is the key concept, with clear precedence being given to freedom over equality wherever the two conflict.	Since the only moral justification for using personal achievement as the basis for distributing rewards is that everyone has equal opportunities for such achievement, then the main emphasis is on equality of opportunity; where this cannot be assured the moral worth of achievement is thereby undermined. Equality is seen as an extension to the many of the freedom actually enjoyed by only the few.

Table 5.2   Idealised health care system

		Private	Public
Demand	(1)	Individuals are the best judges of their own welfare.	When ill, individuals are frequently imperfect judges of their own welfare.
	(2)	Priorities determined by own willingness and ability to pay.	Priorities determined by social judgements about need.
	(3)	Erratic and potentially catastrophic nature of demand mediated by private insurance.	Erratic and potentially catastrophic nature of demand made irrelevant by provision of free services.
	(4)	Matter of equity to be dealt with elsewhere (e.g. in the tax and social security systems).	Since the distribution of income and wealth is unlikely to be equitable in relation to the need for health care, the system must be insulated from its influence.
Supply	(1)	Profit is the proper and effective way to motivate suppliers to respond to the needs of demanders.	Professional ethics and dedication to public service are the appropriate motivation, focusing on success in curing or caring.
	(2)	Priorities determined by people's willingness and ability to pay and by the costs of meeting their wishes at the margin.	Priorities determined by where the greatest improvements in caring or curing can be effected at the margin.
	(3)	Suppliers have strong incentive to adopt least-cost methods of provision.	Predetermined limit on available resources generates a strong incentive for suppliers to adopt least-cost methods of provision.
Adjustment mechanism	(1)	Many competing suppliers ensure that offer prices are kept low and reflect costs.	Central review of activities generates efficiency audit of service provision and management pressures keep the system cost-effective.
	(2)	Well-informed consumers are able to seek out the most cost-effective form of treatment for themselves.	Well-informed clinicians are able to prescribe the most cost-effective form of treatment for each patient.
	(3)	If, at the price that clears the market medical, practice is profitable, more people will go into medicine and hence supply will be demand responsive.	If there is resulting pressure on some facilities or specialities, resources will be directed towards extending them.
	(4)	If, conversely, medical practice is unremunerative, people will leave it, or stop entering it, until the system returns to equilibrium.	Facilities or specialities on which pressure is slack will be slimmed down to release resources for other uses.
Success criteria	(1)	Consumers will judge the system by their ability to get someone to do what they demand, when, where and how they want it.	Electorate judges the system by the extent to which it improves the health status of the population at large in relation to the resources allocated to it.
	(2)	Producers will judge the system by how good a living they can make out of it.	Producers judge the system by its ability to enable them to provide the treatments they believe to be cost-effective.

approved system of rules. A rather more complex specification of the characteristics of each such idealised health care system is given in Table 5.2, which is taken from Maynard and Williams (1984).

The basic weakness of the idealised view of both of these systems is the peculiar 'agency' role which doctors play in all health care systems. The essence of this problem is that the 'consumers' rely on doctors to act as their agents, in a system which ostensibly works on the principle that the doctor's role is to give the patient all the information the patient needs in order to enable the patient to make a decision and the doctors should then implement that decision once the patient has made it. I am sure that the reader would find the above statement closer to his or her own experience if the postulated roles of patient and doctor were interchanged, so that the sentence would then read 'the *patient's* role is to give the *doctor* all the information the *doctor* needs in order to enable the *doctor* to make a decision and the *patient* should then implement that decision once the *doctor* has made it.'

Once the doctors are acknowledged *not* to be 'perfect agents' but, through the exercise of 'clinical freedom', may be pursuing interests other than those of the patient in front of them (Williams 1984), then each system manifests a characteristic bias. Private systems tend to 'oversupply' health care in areas of practice where doctors have plenty of discretionary power and where it is advantageous to them to do so, while public systems tend to 'undersupply' health care procedures where doctors have plenty of discretionary power and where it is advantageous for them to do so. More specifically, in Table 5.3 the implications for each of the points made in Table 5.2 are set out in more detail, permitting an item–by–item comparison of the *actual* versus the *idealised* characteristics of each system.

We then find ourselves in a paradoxical situation. The private system, which generates strong incentives for cost-minimisation at the micro-level, faces a severe problem of cost-containment at macro-level because of its inability to control quantities supplied. Conversely, a tax financed public with prospective budget limits has no problems over cost-containment at macro-level but severe problems in generating cost-consciousness at micro-level, due to the absence of appropriate low level financial incentives. It is, therefore, hardly surprising that a solution is sought in a mixed system which might combine the best of each 'pure' system.

Table 5.3 Actual health care systems

		Private	Public
Demand	(1)	Doctors act as agents, mediating demand on behalf of consumers.	Doctors act as agents, identifying need on behalf of patients.
	(2)	Priorities determined by the reimbursement rules of insurance funds.	Priorities determined by the doctor's own professional situation, by his assessment of the patient's condition and the expected trouble-making proclivities of the patient.
	(3)	Because private insurance coverage is itself a profit seeking activity, some risk-rating is inevitable, hence coverage is incomplete and uneven, distorting personal willingness and ability to pay.	Freedom from direct financial contributions at the point of service and absence of risk-rating, enables patients to seek treatment for trivial or inappropriate conditions.
	(4)	Attempts to change the distribution of income and wealth independently are resisted as destroying incentives (one of which is the ability to buy better or more medical care if you are rich).	Attempts to correct inequities in the social and economic system by differential compensatory access to health services leads to recourse to health care in circumstances where it is unlikely to be a cost-effective solution to the problem.
Supply	(1)	What is most profitable to suppliers may not be what is most in the interests of consumers and, since neither consumers nor suppliers may be very clear about what is in the former's interest, this gives suppliers a range of discretion.	Personal professional dedication and public spirited motivation likely to be corroded and degenerate into cynicism if others, who do not share those feelings, are seen to be doing very well for themselves through blatantly self-seeking behaviour.
	(2)	Priorities determined by the extent to which consumers can be induced to part with their money and by the costs of satisfying the pattern of 'demand'.	Priorities determined by what gives the greatest professional satisfaction.
	(3)	Profit motive generates a strong incentive towards market segmentation and price discrimination and tie-in agreements with other professionals.	Since cost-effectiveness is not accepted as a proper medical responsibility, such pressures merely generate tension between the 'professionals' and the 'managers'.
Adjustment mechanism	(1)	Professional ethical rules are used to make overt competition difficult.	Because it does not need elaborate cost data for billing purposes, it does not routinely generate much useful information on costs.

Table 5.3 Actual health care systems (continued)

		Private	Public
	(2)	Consumers denied information about quality and competence and, since insured, may collude with doctors (against the insurance carriers) in inflating costs.	Clinicians know little about costs and have no direct incentive to act on such information as they have and sometimes even quite perverse incentives (i.e. cutting costs may make life more difficult or less rewarding for them).
	(3)	Entry into the profession made difficult and numbers restricted to maintain profitability.	Very little is known about the relative cost-effectiveness of different treatments and, even where it is, doctors are wary of acting on such information until a general professional consensus emerges.
	(4)	If demand for services falls, doctors extend range of activities and push out neighbouring disciplines.	The phasing out of facilities which have become redundant is difficult because it often threatens the livelihood of some concentrated specialised group and has identifiable people dependant on it, whereas the beneficiaries are dispersed and can be identified only as 'statistics'.
Success criteria	(1)	Consumers will judge the system by their ability to get someone to do what they need done without making them 'medically indigent' and/or changing their risk-rating too adversely.	Since the easiest aspect of health status to measure is life expectancy, the discussion is dominated by mortality data and mortality risks to the detriment of treatments concerned with non-life threatening situations.
	(2)	Producers will judge the system by how good a living they can make out of it.	In the absence of accurate data on cost-effectiveness, producers judge the system by the extent to which it enables them to carry out the treatments which they find the most exciting and satisfying.

60

MIXED SYSTEMS AND THEIR PROBLEMS

In principle there are several different ways in which systems may be 'mixed'. One could have a single system with 'mixed' motives, or one could have two pure systems operating independently side by side serving the same community, with neither playing a dominant role. Or one could have one system dominant and the other playing a deliberately circumscribed role.

In a single system with mixed motives, there would be some areas of health care provision in which 'need' was the *only* acceptable way of ordering priorities and others in which 'willingness–and–ability–to–pay' was the *only* acceptable way. For instance, access to all hospital treatment might be determined by need, but access to all primary care by willingness and ability to pay (or vice versa). The problem with this sort of solution is that there is no clear line to be drawn between the two (people resort to casualty departments of hospitals in the absence of suitable primary care) and primary care usually acts as a first–line–of–investigation and as a 'filter' for access to hospital care, so it is difficult to run each on a different principle and emerge with an overall system which makes sense. The same applies if one tries to separate 'amenity' aspects of care (for example, the 'hotel' aspects of hospital care) from the 'clinical' aspects (for example, the nature of the operation or the drugs used). Clearly nursing care is partly clinical, partly 'amenity', so levels and type of nursing attention could be counted either way *at the margin*. Such a unified but dual-principle system would also have to decide whether waiting time was a matter of clinical priorities (and therefore the same for everybody) or of amenity (so that those willing and able to pay should get the dates/times for treatment that suit them best). Trying to cope with these conflicts within a single organisation would seem to be rather a horrific task.

This brings us to the type of mix which consists of two freely competing independent systems operating alongside each other. A critical issue here will be whether everyone has to contribute to the costs of the public system whether they use it or not, and the scale on which the public system is financed. Generally the better off people will opt out of the public system, unless its standards are higher than those of the private system, so the outcome depends on how good the public system is. If doctors can make more money in the private than in the public system, they will be drawn in that direction and may even have an incentive not to have the pubic system perform too well in the areas where private practice is lucrative. The same incentive applies to the consumers in the private system, especially if the richer people are forced to contribute to the public system, since they will wish to have their public system contributions kept to a minimum (by keeping standards low) if they do not intend to use the public system anyway. It is, therefore, hard to see how two such different systems could operate side by side on anything like equal terms without leaving all parties feeling somewhat disgruntled.

Perhaps this is why most actual systems gravitate towards becoming mixed systems of the third type, within which one system predominates and the other is permitted a minor, carefully circumscribed, role. Thus predominantly private systems moderate the ruthlessness of the 'if you can't pay, you can't have' rule by organising a small public system to take care of the poorer people (including the 'medically indigent', that is, those who have run out of privately-insured entitlement and have no other financial resources to draw on). Such public systems are usually inferior in standards to the dominant private system (for if they were not, who would use the private system?). Conversely, a predominantly public system will moderate the ruthlessness of the 'if you don't need, you can't have' rule by permitting a small private system to take care of the richer people (and any non-citizens who may have no entitlement to access to the public system). Such private systems may be better or worse clinically than the public system but they will certainly offer standards of convenience and amenity in excess of those offered by the public system (otherwise who would use the private system?). Both of these mixed systems could be viewed as representing an acknowledgement that the 'dominant' ideology is not held by everyone in that society and that the views of the minority must be respected and catered for, though only to limited extent.

POLICY MAKING AND APPRAISAL

As has been stressed earlier, each system has its own characteristic ideology and this generates an equally distinctive culture. In the private sector the 'culture' is that of accounting and business management, the appraisal techniques are based on financial analysis and control, with organisational forms reflecting the varying strengths and nature of the profit motive, as moderated by insurance markets. In the public sector the culture is that of economics and politics, the appraisal techniques based on cost–benefit analysis and the organisational forms turn on varying patterns of centralisation on decentralisation within a (quasi) governmental framework. The private system will be held to be successful if it is profitable (but not *too* profitable?) and if it meets the varied demands of those with the most purchasing power. The public system will be held to be successful if it meets the needs of the sick at low cost (without too great a tax burden?). This brings us back to the paradox about cost-containment and efficiency mentioned earlier and poses the question whether there is not some mixed *organisational* form which would enable each (pure) system to meet *its* objectives better.

Thus, in the private systems there are attempts to develop better insurance mechanisms to help the (poor) needy by wider risk spreading and to change the nature of the reimbursement mechanisms to give doctors an incentive not to intervene 'excessively' (Enthoven 1985). Conversely, in the public systems there is an attempt to introduce more decentralised budgeting systems which give clinicians a strong incentive to minimise costs and direct the resources under their (effective if not nominal) control towards those activities which improve people's health most per unit of resource used (Williams 1985b). In this sense one can see an *organisational* 'convergence' between the two systems which does not necessarily reflect any *ideological* convergence.

My own view is that Donabedian was right to polarise the ideological differences between libertarians and egalitarians and to emphasise the differences in view about priorities which ideological differences generate. Each of us must decide for ourselves where we stand in that particular configuration of attitudes and be honest with ourselves and with others about it. In case it is not already obvious from what I have already written in this chapter, I feel quite strongly egalitarian and would aim to make the public system stronger and the private system weaker in any community on which I depend for health care. But I also recognise the need, in a democratic country, to respect the ideological position of a minority, provided it is not actually subversive. The trouble with private systems, in my view, is that they become 'subversive' if permitted to play a *significant* role in a mixed system, because public systems rely on strong feelings of social solidarity (the rich must help the poor, the healthy the sick, the wise the foolish, the well-informed the ignorant, and so on), whereas private systems exist precisely to enable the rich, healthy, wise and well-informed to 'opt out' and look after themselves. Thus there is a dilemma for muddled-headed people like me in deciding how far such tolerance can go.

It is not my role to persuade you to adopt my view, but merely to identify the ideological dilemma as acutely as possible. Although I think that each system has something to learn from the other *managerially*, in the end each has to be judged according to its own lights, that is, according to its own ideology. If we can make changes which rate well from *both* standpoints, well and good. However, I observe that many supposed 'improvements' in 'efficiency' contain implications for priority setting in health care which seem to me to have a quite strong (though implicit) ideological component and which I would therefore feel bound to reject because of their distributional implications. It is no solution to say to an egalitarian like me that the public system would be better if it adopted the priorities of the private system. Nor would I expect to convince a libertarian that the private system would be better if it became more egalitarian (though that is what *I* believe). So, when appraising policy proposals for improving each respective system, let us state clearly whether our judgements flow from a basically libertarian or egalitarian stance.

NOTES

1. 'Needology' is itself a fascinating area of investigation on which see Chapter 11, Culyer (1976) and Sen (1985). In the present context I think it is best interpreted to mean 'capacity to benefit from treatment' and the 'social hierarchy' implies that different sorts of benefits accruing to different people are valued by a common standard (for example, an additional year of life expectancy is regarded as of equal value no matter who gets it).

6. Natural Selection, Health Economics and Human Welfare

INTRODUCTION

When I discovered that some biologists interested in evolutionary biology had found it convenient to conceptualise their central concerns using models borrowed from economics, I decided to try to find out in what ways they had had to adapt or develop 'our' models in order to solve 'their' problems, thinking that perhaps these innovations might prove useful to us if we borrowed them back! In the event this proved less interesting than mulling over the analogies that their work suggested to me between (a) the autonomous homeostatic 'rules' built into our individual physiological systems to maintain our fitness as organisms and (b) the deliberately chosen 'rules' *we* build into our collective social systems to maintain our health as communities.

I shall argue that the biological model which underlies natural selection is no longer useful in explaining human demography. The economic model is more general and more useful in according a central role to the rules we adopt as *societies* regarding the production and distribution of health. But where these societal rules conflict with our evolutionary heritage as individuals, we are likely to have great difficulty in accepting them and acting upon them: our 'instinct' or 'intuition' may pull us in a different direction. Hence we may expect great social tensions to develop as we move from innate maximisation of reproductive capacity to a more deliberative and conscious concern to improve the general quality of our lives, in which rearing offspring will play a part, but *only* a part. Dramatic examples of such tensions are the fierce debates surrounding contraception and abortion.

The substance of the chapter begins with my understanding of those bits of *evolutionary physiology* which are both accessible to me and seem to be relevant to my main theme. Evolutionary physiology deals with all organic species but I shall restrict myself to humans. It has also spawned a whole raft of mathematical models which might well repay closer attention by those more numerate than me. I can only rehearse their key features in more informal terms, running the inevitable risks of misunderstanding and unwitting misrepresentation. But these risks have to be run in the interests of interdisciplinary cross-fertilisation (in the hope that enriching our intellectual 'gene pool' might have survival value both for our species and for us as individual members of it!).

I then set out the bits of *health economics* that bear most directly on my theme. Needless to say there is a lot more to the subject–matter of the sub-discipline than I shall draw upon here[1] so the uninitiated should not be misled into thinking that I am offering an exhaustive account of what health economics has to offer to struggling humanity[2]. Indeed, this whole chapter is essentially a *personal* essay, reflecting *my own* interests and drawing heavily on *my own* earlier work. Not all health economists would agree with my views.

The next section highlights the analogies I see between the biological and the economic approaches to *the health of individuals* and, perhaps more significantly, the differences between them.

I then go on to explore the differences between the two models as regards their respective implications for *the health of whole populations* and, in concluding, will offer some *broader observations* about problems that I find of absorbing intellectual interest and which also seem to be of some considerable significance for the welfare of humankind.

EVOLUTIONARY PHYSIOLOGY[3]

Two basic concepts in this field are the *genotype* and the *phenotype*. The former is essentially a set of templates controlling the form and function of the latter (which is a complex organism – such as a human being). This set of templates (or genome) needs to include instructions to the phenotype that will maximise the probability that in the long term the individual (and thereby the species) will survive (and preferably even thrive) in whatever environment it finds itself. Errors in copying those instructions, and/or faulty execution of instructions and/or damage to the phenotype from sources external to it (for example, through disease, or accidental injury, or an impoverished or hostile environment) will, however, generate a broad spectrum of differentiated phenotypes[4]. Natural selection will favour those variants of the phenotype which prove well adapted to their environment and it will select out the others. To the extent that these favoured variations are transmittable through differential fecundity, they will affect the gene pool, the genotypes and hence the phenotypes in subsequent generations. Thus:

The study of adaptation ... asks why specific phenotype traits have evolved in association with specific ecological conditions .... In particular we shall be concerned with physiological adaptations of resource acquisition and use because we will argue that this is not only the basis on which organisms function physiologically but also the basis of their form (allocation of resources between different structures) and behaviour (allocation of resources between different activities). Patterns of resource use are controlled by enzymes and hence ultimately by genes, so ... those patterns of use will be favoured that best promote the spread of genes that code for them .... Nevertheless, one physiological process has to operate within the context of others which constitute the phenotype and this is particularly true given that the resources required for use are finite and limited .... Hence there are constraints operating within organisms on what patterns of allocation and hence phenotypic traits can evolve. Genotypes determine phenotypes, but the physiology of phenotypes constrains what association of genes (that is, genotypes) can evolve. (p. 1)

The core *neo-Darwinian hypothesis* usually adopted is that successful phenotypic traits will be those which *maximise fitness*. In principle this means the ones that maximise survival (S) and fecundity (N) and minimise generation time (T) of organisms that carry the traits. However ... it will not usually be possible in practice to maximise these components of fitness simultaneously. There will be trade-offs and constraints so that *optimisation* is often a more important element of the core hypothesis than maximisation *per se*. (Sibley and Calow 1986, pp. 5–6)

It is then suggested that the key strategic investment decisions an organism has to make are between *growth, defence* and *reproduction* (pp. 13–15) and that the contribution of each of these activities to 'fitness' (which may be broadly interpreted to mean the capacity to produce viable and fertile offspring) will vary according to the age of the organism (that is, according to its stage in the life-cycle) (pp. 19–21). For instance:

To increase fitness, it is desirable to reproduce fast, but it is also desirable to grow fast so as to be bigger (and able to reproduce faster) in the future .... Early in life, growth may be advantageous. As the end of reproductive life approaches, growth is less useful, because there will be little time to take advantage of increased size. (p. 21)

Moreover, since resource inputs may vary between organisms, so will the optima (pp. 23–24) and, since resources are neither homogeneous nor infinitely versatile, it may be necessary in any optimisation model to specify the precise vector of resources available and which particular resource constraints are binding in any particular situation (p. 25). But 'the most generally suitable currency' to use as a standard measure of resources in the case of animals (including humans) is energy (p. 25).

At this point a note of caution 'on the general validity of the economics approach' is considered appropriate:

> An economics of metabolism with energy as currency emphasises the quantitative aspects of the genotype. However, the quality of behaviour and form will often be important in determining survival and reproductive success. For example, the organisation of a nervous system is probably as important to the way it works as its size. Size nevertheless puts important constraints on form and organisation even in nervous systems .... An ... organisation and form are generated by the differential allocation of resources between structures in space. Hence the economics of resource use can make important statements about form and organisation and then about the consequences of this form and organisation. (p. 25)

One aspect of the 'form and organisation' of the human physiology that is of particular interest here is the capacity to repair damage (from whatever cause). The effective repair of damage requires accurate and prompt information about the nature and location of the damage, what the repair possibilities are and an ability to mobilise the necessary resources to carry out the repair before the damage proves irreversible. In general, local responses to local problems meet these requirements most efficiently and this includes rapid defence mechanisms that minimise damage in the first place. To the extent that effective defence and repair activities reduce the mortality risk, this 'lifts the pressure off the rate of replication and thereby permits a wider range of adaptation' though only if reducing mortality increases fitness. This will depend on the *costs* of repair. This in turn raises the question 'When is repair worth while and which types of damage should be repaired before others?'

Exponents of the 'disposable soma theory' (for example, Kirkwood 1981) argue that since an organism's life history is essentially a set of co-adapted traits that together determine its age-related patterns of growth, reproduction, senescence and death, and that this life history must have evolved because of its superior contribution to fitness, the priority rules for repair activity must themselves maximise fitness by optimising the distribution of resources between growth, defence and reproduction. Thus, up to the stage of the life cycle when offspring are born and reared to maturity, a *balanced* programme of repair will maximise fitness. A 'balanced' programme means one in which the marginal unit of any scarce resource (for example, energy) devoted to each component of fitness makes the same contribution to overall fitness no matter how it is used.[5] For instance, to try to sustain a higher than optimal growth rate will lead to suboptimal levels of defensive capability and reproductive capacity, because the energy needed to sustain them will not be available if it has gone into growth. The consequence of giving priority (in an optimal way) to the activities which contribute to fitness is that low priority is given to all other activities, so that there are relatively sparse resources available for them. Damage which does not impair fitness is likely to go unrepaired.

When this last consideration is put together with the observation that internally generated 'errors' and externally generated 'damage' are both likely to cumulate (at an increasing rate?) with increasing longevity of a particular organism, the explanation offered by the 'disposable soma' theory for senescence and finite life-spans is that the organism finds itself with too few resources available for repair activities in later life *because it has behaved optimally with respect to fitness in its earlier life* and it is this trait which is selected for transmission to the next generation. So we are not exactly 'programmed' to die at age 115 or whatever, but we are 'programmed' with *a set of priorities* which has the effect of making it very unlikely that any of us will survive much longer than that, even when living in an ecological habitat which has been made incredibly productive and in the protected environment of a peaceful 'welfare state'.

HEALTH ECONOMICS

Health economists have found it important to distinguish sharply between the demand for health and the demand for health care. The demand for health may be met in many ways, of which seeking medical care is but one. Indeed, it could be argued that in the richest countries of the world the marginal benefits from medical care are negligible (and even possibly negative), whilst in the poorest countries of the world improving nutrition and hygiene will be far more cost-effective in improving health than investing in hospitals (even though such countries may have very few hospital beds per head of population). However, it is not my immediate purpose to plunge into that territory but simply to emphasise the importance of making a distinction which is still too rarely made ... the distinction between 'health' and 'health care'.[6]

When thinking about health at the level of the individual, it is best seen, together with knowledge, as constituting that individual's human capital. Knowledge ranges from the basic skills required for everyday living (which can be quite demanding) to people's broad attitudes towards life, their worldly wisdom. In addition to these two intangible assets (health and knowledge), people also have tangible assets (such as houses, cars, personal belongings and savings). So people's paths through life could be viewed as being essentially concerned with the management (or mismanagement) of these three types of asset – health, wealth and wisdom.[7] Each can be seen as a capital stock yielding a flow of services over time. Each is subject to depreciation and capable of augmentation (within limits). The general situation is summarised in Table 6.1.

Table 6.1 Health, wealth and wisdom as capital stocks

Capital Stock	Health	Wealth	Wisdom
Measured as: The present value of the expected future stream of	Quality–adjusted–life–years	Net real income	?
The associated <u>current flow of services</u> being	Quality–adjusted time and energy	Purchasing power and the use of consumer durables	Lifestyle possibilities, adaptive skills and valuations
<u>Time-related depreciation</u> due to	Ageing (Physiological)	Inflation (and retirement conventions)	Ageing and obsolescence of knowledge/skills
<u>Use-related depreciation</u> due to	Hazardous lifestyles	Capital consumption	?
<u>Augmentation</u> possible by	Health promoting life-styles and medical care	Investment	Learning (from study or experience)

We are used to thinking about tangible (real and financial) assets as wealth, and to valuing them according to the present worth of the future flow of income (and capital gains) that we expect from them, using as the discount rate our own personal rate of time preference. We are familiar with the fact that our real assets (for example, houses, cars and other consumer durables) are subject to both time- and use-related depreciation. Our financial assets may be subject to rather different forces affecting their value over time and will not suffer 'use-related' depreciation (unless we regard the costs of asset-switching in that light). It is through labour market participation that most people acquire the resources that enable them to augment their wealth. But there are strong societal norms, not to mention formal restrictions and requirements, which influence people's access to labour markets at different stages in their life-cycle (for example, minimum school leaving ages or strict retirement rules). These constraints may prevent the building up of wealth, or force the running down of wealth, at particular times in our lives. We can 'use up' our wealth by living above our means, or add to it by net savings. We may also inherit it and give it away.

But suppose we now follow Grossman (1972) and apply this same kind of thinking to health. Health too can be seen as an asset in the form of a capital stock, which has to be maintained above a certain minimum level to avoid death (just as with financial assets in the business world you must not get into so much debt that you get declared bankrupt). Our stock of health determines the *maximum* amount of quality-adjusted time and energy available to us in any specific period, though below this maximum there will be some smaller amount of time and energy which is the most we can draw on before we start experiencing use-related depreciation (due to 'burning the candle at both ends'). The above mentioned 'quality-adjustment' refers to the extent to which the time and energy available to us during the period in question are compromised by pain, distress or disability. At any point in time our stock of health may be measured as our quality-adjusted-life-expectancy.   It is susceptible to time–related depreciation (senescence), which, as we have seen, is likely to increase noticeably in later life. This decline may be accelerated by particular lifestyles and circumstances (for example, by smoking, lack of exercise, poverty). Augmentation of the stock is possible (within limits) by health promoting lifestyles and by health care. We can inherit good or bad health and we can even 'give it away' by sacrificing it for the sake of others.

Applying the same conceptual framework to wisdom is rather more difficult, for we immediately run into difficulties over the valuation of this stock. Historically, in the economics of education such valuations have been sought by investigating the additional labour market rewards to be gained from enhanced knowledge and skills, but this seems too limited a view of the 'rewards' from 'wisdom'. What to put in its place is not clear, however, and the problem is compounded by the likelihood that increased wisdom changes people's whole value systems and, particularly, their personal rate of time preference and the relative importance they attach to material wealth. If this is so it will have a direct and pervasive effect upon the valuation of the other two assets. Thus summarising the services provided by the asset wisdom as 'lifestyle possibilities, adaptive skills and valuations' may be so wide ranging and complex as to defy measurement. Nonetheless, measurable or not, our stock of wisdom cannot be ignored. Our day-to-day living and adaptive skills have to be maintained above a certain minimum level, otherwise we cease to be viable as individuals (an increasingly common problem with the very elderly and with the mentally

ill and handicapped). Further our knowledge and skills may also be subject
to time-related obsolescence (what good now is my investment in counting
and calculating in money?). On the other hand, in the case of wisdom
there seems to be no analogue to use-related depreciation and the
augmentation possibilities are enormous (until such time as senescence
restricts them). Like love, knowledge is one of the few things you can give
generously to others and still have as much as (or even more than) you
started with!

But we now need to put this conceptual apparatus to work so as to
generate a model of health-related behaviour at an individual level. For
this purpose it will be useful to refer to Figure 6.1.

Figure 6.1 The Grossman–Williams Model of Health Behaviour

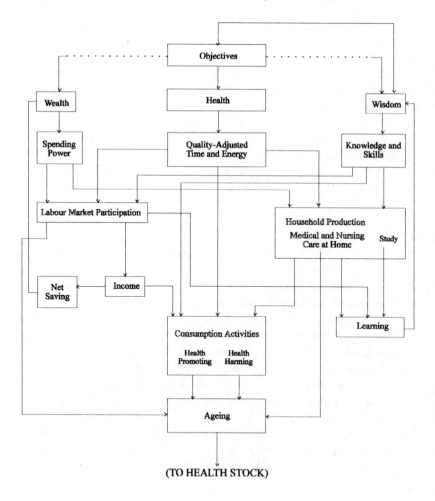

I will start by tracing out the standard simplified economic model of individual/household behaviour.[8] The individual is supposed to start each period with some accumulated wealth[9] which itself generates spending power irrespective of current earnings (top left of Figure 6.1). If the individual participates in the labour market and thereby earns income, these earnings can be distributed between consumption activities and net saving according to the (intertemporal) preferences of the individual. These preferences are derived from the individual's objectives (depicted at the top of Figure 6.1), which in turn derive from the individual's stock of wisdom. Obviously these objectives have a pervasive effect on all optimising decisions in the model. The cycle is then repeated, period after period, over the lifetime of the individual, so that it is possible in principle to trace out different (economic) life histories according to endowments, opportunities and objectives.

A significant enrichment of this simple model (in which 'firms' produce and 'households' consume) occurs when the possibility of household production is introduced into it (see the centre right of Figure 6.1). Thus certain activities (for example, domestic work) are seen as requiring both goods or services purchased in the market *and* the input of the household's own human resources (especially time and energy). For reasons already stated I shall not pursue this further at a general level but restrict myself to its implications for health related behaviour.

To do this we need to introduce health explicitly into the model as a separate asset. Thus the health stock (top centre of Figure 6.1) generates the (quality-adjusted) time and energy whose availability is largely taken for granted in the simple economic model (where money is the key resource). This time and energy can be used for labour market participation (to earn money), for consumption activities (including sleeping!) and for household production (which, besides domestic work, includes child rearing and other informal care). The appropriate allocation of time and energy then becomes the central optimising task and activities have to be appraised according to both their money prices and their time prices. Thus, deciding within a household how to treat a sick member of it will involve weighing the relative costs and benefits of health care obtained in the market as against informal care produced at home. Within each of these broad options there may be a variety of suboptions with different opportunity costs in terms of time, energy and money (and, in the context of a household, some variation in *whose* time, energy and money it will be!).

But we have to consider not only how sickness might be treated but how it gets produced in the first place. We have already examined the mechanisms underlying physiological senescence and, in simple economic models, the ageing process is usually taken as exogenous. But physiological ageing is not the same as chronological ageing and physiological ageing can be retarded by the pursuit of a healthy lifestyle (which means working in a favoured occupation, living in a favoured environment and adopting 'good' eating, drinking and leisure habits). Ageing differs from other depletions of the health stock in that (like death) it is irreversible.[10] For that reason it has been shown separately, at the foot of Figure 6.1, as a 'filter' through which all *potential* influences on the health stock have to pass before their *actual* effect on the health stock can be assessed,[11] recognising that the precise nature of this filter is itself partly determined by those same influences.

Ill health can, of course, be inherited but it can also be acquired and, although there is undoubtedly a stochastic element in its incidence, it is abundantly clear that the risks are distributed unevenly across the population in a quite systematic way. Poverty puts you at high risk and so does smoking. Certain occupations are characterised by abnormally high (or low) incidence of particular kinds of injury or disease. So an individual's economic situation is a strong influence on that individual's health (and possibly on the health of the other members of the household too).[12] An individual's economic situation contains both voluntary and involuntary elements. A child does not choose the household it is born into and many households have very restricted opportunities to acquire health, wealth or wisdom. People living in desperate straits need money badly and have to optimise over short time horizons (they may literally live from hand to mouth), so they are likely to discount heavily the long term health risks from certain employments or consumption activities. If the only way to get through the day is to smoke a cigarette from time to time, why worry about what it is doing to your life expectancy, or might do to your current health in 20 years time?

It is also possible to affect your health through ignorance and foolishness, of course, and this brings us to the role of the third asset, our stock of wisdom. The knowledge and skills which flow from that stock may also be distributed between labour market participation, consumption activities and household production. Greater wisdom may widen our opportunity set and make us more efficient at exploiting the opportunities that do come our way, in all three of these domains. Wisdom may be augmented by learning, which comes not only from 'study' (which I have treated as household production[13] but also from 'experience' (which comes from all three domains of activity). This 'learning' may contribute to wisdom not only in such a way as to increase earning or adaptive capacity but also in such a way that the individual's objectives themselves get modified (as depicted at the top right of Figure 6.1).

Looking at the system as a whole, when managing this 'portfolio' of assets, the trade-offs facing an individual at a mundane common-sense level include:

(a) how far to risk one's health to increase one's wealth or wisdom;
(b) how much wealth to sacrifice to reduce life's hazards or get the best information or advice;
(c) how seriously to set one's mind to work thinking through ways to improve one's lifestyle or to get more lucrative employment.

In these various ways each asset can be transformed into the others and (within limits) the desired balance achieved, though with an attenuated and unbalanced portfolio at the beginning of the period it may well be that the best achievable balance at the end of the period is not very good. Later I will provide an illustration of how these transformation processes work.

A COMPARISON OF THE TWO MODELS AT INDIVIDUAL LEVEL

The heroic simplifications and sweeping generalisations made so far in this chapter pale into insignificance beside those I am about to make in order to compare the main features of the two approaches to explaining the health of individuals and populations. To make this forbidding task somewhat easier I shall conduct the comparison in two stages, the first (in this section) concerned with the modelling of individual behaviour, the second (in the next section) concerned with the implications for whole populations.

The situation as I see it is summarised in Table 6.2. It highlights five key features of the models: the assumed maximand, the key input, the control mechanism, the main priority at each life-stage and the manner in which inter-generational transfers are effected. I will discuss each in turn.

The Biological Model (BM) takes 'fitness' to be the maximand, which can be interpreted as reproductive capacity (success in rearing to maturity viable and fertile offspring). The Economic Model (EM) assumes that individuals are pursuing a rather broader aim, that of leading a long and rewarding life, which is encapsulated in the notion of maximising quality-adjusted-life-years. The two maximands might broadly coincide if having sex and rearing children to adulthood[14] were by far the most rewarding experiences in people's lives, all subsequent activities being merely a self-indulgent afterglow. The BM suggests that this 'afterglow' will occur only to the extent that there *happens to be* energy left over from the reproductive

Table 6.2   A comparison of models

FEATURE	BIOLOGICAL MODEL	ECONOMIC MODEL
ASSUMED MAXIMAND	FITNESS (REPRODUCTIVE CAPACITY)	QUALITY-ADJUSTED-LIFE-YEARS (LENGTH AND QUALITY OF LIFE)
Kay Input Obtained by Measured in terms of Allocated to	Nutrients Foraging Energy Structures and activities	Real income Working Money Investment and consumption
CONTROL MECHANISM Determining priorities across	Genotype Growth Reproduction Defence/repair	Stock of wisdom Learning Sustaining own living standard Keeping healthy
MAIN PRIORITY AT EACH LIFE STAGE Pre-adulthood Early adulthood Late adulthood	 Growth Reproduction Defence/repair	 Learning Living standards Health
INTERGENERATIONAL TRANSFERS Effected through	 Genotype	 Legacies (of health, wealth *and* wisdom)

phase and only if the control mechanism's energy–allocation–rules (which will still be operating in the same way as before) happen to be conducive to it. This is most likely to happen from the 'instinct for survival' reflected in the rules governing the defence/repair mechanisms. The EM, on the other hand, suggests that people will *consciously set aside* resources for the 'afterglow', even though they know it means reducing their net reproduction rate below unity, a decision encapsulated in the view that 'we cannot *afford* to have any more children'. There is, however, a possible influence working in the other direction, namely that the successful rearing of children may be the best investment that it is possible to make for the provision of informal care (and general sustenance) in later life. Investing in the health of others is then an alternative to investing in wealth for yourself. Again I am straying into the economics of the family and of household production, so I had better stop and move on.[15]

In the BM the key input for the organism is nutrient, which is generally obtained by foraging. It is measured in terms of energy and may be allocated to sustaining either structures or activities. In the EM the key input is (real) income, generally obtained by working. It is measured in terms of money and may be allocated to sustaining either investment or consumption. Here the two models are remarkably similar and further comment seems superfluous.

This is not so with respect to the next feature, however, which is the control mechanism and its instrumental variables. In the BM the control mechanism is the genotype, which (apart from the possibility of irreparable damage) is given and invariant from conception. There are no learning possibilities, so the necessary range of adaptive rules has to be there from the word go. In the EM it is a very different story. The control mechanism is the stock of wisdom and this can be added to (within limits) throughout life. Among these limits is the physiological capacity of the system, which is set ultimately by the genotype. So the EM could be seen as seeking to explain how the (predetermined?) physiological capabilities of individuals (which include their cerebral capabilities) are consciously optimised according to those individuals' own perceived objectives. And this raises the age-old question about the extent to which our 'conscious' optimising behaviour unwittingly mimics the rules of behaviour embedded in our genotype! If it does not, will natural selection eventually bring us to heel? More on this subject anon.

We must next turn to the instrumental variables that each model envisages the control mechanism working on in order to achieve the supposed objective. In the BM at each point in time priorities have to be established at each point in time between learning (wisdom), sustaining living standards (wealth) and keeping healthy (health). In both cases inter-*temporal* optimisation is called for, as well as inter-*functional* optimisation. The odd one out here is 'learning' in the EM which, as we have just seen, has no direct counterpart in the BM. The BM does require resources to be devoted to maintaining the structures responsible for information flows in the system but this is as close as it gets to investing in 'wisdom'. Otherwise growth and reproduction can be seen as broadly analogous to the sustaining of living standards while defence and repair are obviously about keeping healthy. Both the BM and the EM envisage trade-offs having to be made between these rival claimants on scarce resources, with an implication that there is a key role for cost-effectiveness analysis in both models. In the BM the opportunity costs in any such calculations will be in terms of reductions in fitness (that is, reproductive capacity),[16] whereas in the EM they will be in reductions in quality-adjusted-life-expectancy.[17] In the BM the rules that reflect this cost-effectiveness analysis have been 'learned' in the 'school of hard knocks' through natural selection and are incorporated in the genotype. In the economic model we are supposed each to do the calculations (albeit informally and intuitively) for ourselves (except where social norms effectively restrict our behaviour to what others have already decided is cost-effective for us).

This brings us to what seems to emerge as the main priority at each life-stage distinguishing, for simplicity's sake, only three such life-stages: pre-adulthood, early adulthood and late adulthood. The distinction between the last two is loosely based on the division of adulthood between the child-rearing phase and the post-child-rearing phase. In the BM the *main* priority in the first life stage seems to be growth, in the second reproduction and in the third defence/repair. In the economic model the corresponding main priorities are learning, living standards and health respectively. This is not to imply that the variables assigned lower priority are of no importance but simply that they are not seen as the pre-eminent considerations at that particular life-stage. There are some paradoxes here. When asked 'when is it most important to be healthy?', most people say 'when bringing up young children',[18] which is in fact a period when nature has fitted us to be quite healthy, so we do not usually need to devote many additional resources to health maintenance as such at that stage. However, we do need plenty of energy, which in the EM means we need plenty of money. In the richer countries we have prolonged the period of formal education (giving priority to 'learning' in the pre-adult phase) and kept young people out of the labour market, so they will not have accumulated much wealth by the time they enter early adulthood. Thus it is in early adulthood that people typically get most heavily into debt, so that they can 'live above their means' during this phase of their lives. Moreover, before the end of early adulthood they may come to see the need to make suitable provision for old age, when they will once more find themselves out of the labour market, voluntarily or involuntarily. Both of these factors put pressure on them to earn as much money as possible in early adulthood, sometimes at the expense of their health. But it is not until old age that they typically become large consumers of health care (see Figure 6.2), trying to counteract the senescence that natural selection has bequeathed to them as their physiological heritage.

Figure 6.2 Annual NHS expenditure per head (£)

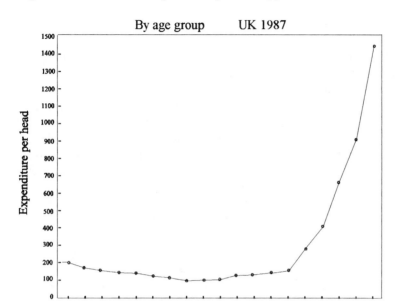

By age group   UK 1987

This brings us to the last column of Table 6.2, which is connected with what it is that we can transfer directly from one person to another within each model. In the BM what is transferred is the genotype, the set of instructions telling us what to become physiologically and how to respond as a phenotype to the various circumstances we may encounter in life. In the EM a broader range of possibilities seems open to us, since we can endow our offspring with health, wealth and wisdom. But the differences may not be so great as they appear at first sight, since the genotype may include 'instructions to parents to look after their offspring until they are viable', so that the EM is merely spelling out (at a conscious level) the way we respond to the sense of kinship that has been instilled in us from conception. But what instils in young people a sense of obligation for the welfare of their (ageing) parents? It is hard to see what value this could have in fitness terms.[19] Is this why it seems less strong than the obligations parents feel for their children? What it suggests to me is that not *all* of our conscious prioritising rules can be explained as unconscious reflections (mimickers) of the rules embodied in our genotypes and that some of our conscious rules may act counter to our biological inheritance. It is a confllict we must come back to.

A COMPARISON OF THE TWO MODELS AT POPULATION LEVEL

When we compare these two approaches with regard to their implications for human beings as whole populations, a very sharp divergence becomes evident. The BM is emphatic in stressing that evolutionary physiology works through natural selection of *individuals*, which obviously has implications for populations while the populations themselves are merely *consequences* of what is happening at individual level; they are not *independent sources of influence* upon the evolutionary process. The macro-relationship between population size and ecological habitat is obviously a determinant of demographic change, but it is mediated by the micro-responses of individuals in that habitat. Individuals are not seen as engaging in deliberative collective decisions to influence their situation. If they act in an apparently concerted way, it is because as individuals they are all programmed with similar instructions as to how to respond in that situation.

In the EM the story is quite different. The individual is not optimising in a social vacuum. Even the most atomistic models of well-functioning perfectly competitive markets require the institution of private property and the existence of collective agencies to enforce contracts based on that institution. Collective agencies (by which I mean any purposive and co-ordinated group of individual other than the family)[20] deliberately control the terms on which an individual can gain access to whatever resources the habitat offers to the population of which that individual is a member.[21] They do this by establishing and regulating contracts of employment and of sale, by taxation and social security provisions and by the provision of some benefits *in kind* such as 'free' health care or compulsory . primary education). This involves establishing rules of entitlement and operating according to some system of priorities. These entitlements and priorities may not be well articulated or consistent, nor indeed be regarded as a 'policy' at all but simply as an 'obligation'.

It will be convenient to consider the impact of a population's collective policies about health and health care at two levels.[22] Firstly, we must consider the priorities that society brings to bear on an individual's encounters with the health care system. Secondly, we must consider the priorities that society brings to bear on the higher level decisions that are made about providing health care facilities for the population at large. The higher level decisions obviously constrain the lower level decisions to some extent but in a decentralised system they do not *determine* them.

Historically the greatest influences determining health care priorities at individual level must have been the wealth of the individual (which has determined access) and the attitudes and behaviour of the medical profession. The matter of access I will set aside for a while and concentrate initially upon the key role of the medical profession in determining priorities once access has been established. Doctors play a dual role in the health care system. They are both the mediators of 'demand' (as agents of the patients) and the suppliers[23] of care. This creates various well-known tensions when what is in the patients' interests is not identical to what is in the doctors' interests. To help doctors resist the temptation to exploit patients for their own ends, the medical profession has developed a code of ethics, the six main injunctions of which are:

1. Preserve life
2. Alleviate suffering
3. Do no harm
4. Tell the truth
5. Respect the patient's autonomy
6. Deal justly with patients.

It is readily acknowledged that these principles frequently come into conflict with each other, but it is held by the medical profession that:

resolving such conflicts is central to the art of medicine (Ruark *et al.* 1988).

Thus it is seen as the role of doctors to establish priorities in the use of health care resources, bearing in mind (amongst other things) the views of patients (and, in practice, of their nearest and dearest).

It is interesting to note that doctors have tended to give strong priority to the preservation of life over all other considerations. Typically they regard the death of a patient as a defeat. They are encouraged to hold this view by the widespread use, in clinical trials, of mortality rates as the main criterion for choosing between treatments and by the common belief that it is the duty of doctors to do everything possible for the patient in front of them, no matter what the costs. It is still very difficult for a doctor to say 'we have done everything that it is reasonable to do in the circumstances and we must now let nature take its course'. In this context being 'reasonable' should mean taking into account the suffering of patients (and of their nearest and dearest) (principle 2 above), and the costs (in terms of health care forgone) being imposed upon other potential patients (principle 6 above).

The health economics model anticipates that some broad-ranging cost-effectiveness analysis will be undertaken to determine what is 'reasonable' in any specific context. To do this an estimate will be made of the benefits to be gained (in terms of quality adjusted life years) and the costs incurred (in terms of benefits forgone) of each of the alternative courses of action (including doing nothing) and the one with the best cost-effectiveness ratio will be adopted *provided the benefits are worth the costs.* Since this entails placing a finite value on (quality-adjusted) human life it excites great emotional antagonism, both from members of the general public and from members of the medical profession.[24] As far as the concerns of the medical profession are concerned, nothing is being advocated that is inconsistent with medical ethics (Williams 1988c), so I think resistance on those grounds is misguided. But this may simply be a cover for a different fear, namely a perceived threat to clinical freedom (Williams 1988d). It is interesting to note that rallying the troops to the defence of this cause has met a very mixed response (in Britain at any rate) (Hampton 1983, Hoffenburg 1987). In the case of the general public, to the extent that the rejection of the cost-effectiveness approach is not a manifestation of romantic idealism and the denial of scarcity, I think it stems from a belief that EM embodies a ruthless Darwinistic approach to the survival of the fittest, in which the emphasis on reproductive fitness in the BM has simply been replaced by an emphasis on economic productivity in the EM. As will be obvious from the earlier analysis, this is quite false, for in the EM the maximand is QALYs, not GNP per head.[25] But the QALY maximand is itself ambivalent. While at a very broad level it could be interpreted to include everything that we might consider to be an element in individual human welfare, in health economics it is usually interpreted in a narrower way to mean '*health-related* quality of life', that is, those aspects that depend on (and are affected by) the state of our health. The difference is illustrated by smoking behaviour, which is deleterious to *health-related* quality–of–life, but may improve *other* aspects of quality of life (simply by giving pleasure). Thus doctors (and health economists?) are likely to interpret the maximand in the narrower sense, whilst patients (and welfare economists?) are more likely to interpret it in the broader sense.

Turning now from the level of the doctor–patient encounter to the level of the health care system as a whole, we again encounter a complex of ethical considerations which have a pervasive, but not wholly determinate, effect on priorities at system level. This time we are in a much broader realm than medical ethics – the realm of political ideology. We must here pick up the key issue of how access to health care is to be prioritised, in a situation in which even the richest countries in the world can no longer afford to provide all the health care that might conceivably do *someone* some good, somehow, sometime.

There are two general ideological stances which are of central importance in this context, the libertarian and the egalitarian (Williams 1988f). Elsewhere (Chapter 5) I have summarised their respective features and their implications for health care, in the following terms:

In the libertarian view, access to health care is part of the society's reward system and, at the margin at least, people should be able to use their income and wealth to get more or better health care than their fellow citizens should they so wish. In the egalitarian view, access to health care is every citizen's right (like access to the ballot box or to the courts of justice) and this ought not to be influenced by income or wealth ....

Each of these broad viewpoints would generate a distinctive health care system whose characteristics would be very different.

In the libertarian system willingness and ability to pay would be the determinant of access and this would best be accomplished in a market-orientated 'private' system (provided such markets can be kept competitive). In the egalitarian system equal opportunity of access for those in equal need would be the determining rule and, because this requires the establishment of a social hierarchy of need which is independent of who is paying for the care, it would be best accomplished in a publicly provided system (provided that the system can be kept responsive to social values and changing economic circumstances) .... Note that the success criterion to be applied to the egalitarian system is the level and distribution of *health* in the community.

Needless to say, in practice, neither system lives up to its ideals and most of the problems stem from (i) the peculiar role of doctors in health care systems, (ii) the problems associated with market deficiencies on the supply side and (iii) information problems on the demand side ....

In most countries health care is provided by a mixture of systems, with no common ideology. This may simply reflect the fact that we all live in pluralist societies which try to accommodate subgroups with incompatible ideologies. A hypothesis suggested by this analysis is that the structure of the health care system in each country is likely to be systematically related to the nature of the equity concerns that have been dominant in the (recent ?) past, and is also likely to reflect the ideology which generated those concerns. An obvious instance is the balance between public and private provision, which differs markedly between countries.

One interesting feature of this discussion is again the strong emotional reaction engendered by the notion of 'rationing' health care. If by rationing is meant deciding who shall and shall not get health care, what sort of health care they will get and when they will get it, then rationing is, and always has been, pervasive in all societies. Instead of asking '*should* we ration health care?' we should ask '*which method* of rationing health care is best?'. And in my preferred terminology the question would become 'how should we *determine priorities* in health care?'. One possibility is by willingness and ability to pay and another is by 'need', an elusive concept (Chapter 11) which I think can best be defined as 'a person's capacity to benefit from something, as judged by some other party'. Since needs outrun resources, some needs have consequently to remain unmet, so needs have to be prioritised according to some higher level criteria. So, in a 'needs-driven' system, when a doctor decides that I 'need' an operation and organises things so that I get it, he has judged that my 'need' is greater than that of those who are waiting or going without. This method of 'rationing' should then be compared with one that works on the willingness and ability to pay of individuals (or, more frequently these days, on the willingness and ability to pay of the insurance carrier, where the reimbursement rules are the expression of the priorities of the system), to determine which produces the socially preferred outcome.

As explained earlier, the EM was based on the assumption that for the individual the maximand is quality-adjusted-life-expectancy.[26] For the society as a whole the analogous maximand would then be the aggregate quality-adjusted-life-expectancy of the entire population.[27] But this aggregation process raises (at least) two further problems concerning distributive justice: what is to count equally between people and are some people to count for more than others?[28] The simple answer to the first question is that a quality-adjusted life-year is to be counted as of equal value no matter who gets it, an assumption that mimics the assumption behind the use of average life-expectancy as a measure of welfare, namely that an additional year of life (of whatever quality) is of equal value no matter who gets it. But this 'simple' answer seems much less acceptable when QALYs are the maximand than when (unadjusted) life expectancy (that is, life-years gained) is the maximand, though why this should be so I find puzzling. There are two common counter positions. The first is that the maximisation of QALYs (like the maximisation of life-years) discriminates against the elderly, because it is much more costly to provide them with an extra unit of 'health' (however you measure it) (for example, Avorn 1984). The second is that it discriminates against the poor (and those deprived in other ways) because their general health prospects are poor and also difficult to improve (for example, Harris 1987). So the assertion is that, in interpersonal allocation of health care resources, either there should be positive discrimination in favour of these groups or, more radically, access to health care should not be prioritised at all by 'need' as capacity to benefit, but according to 'entitlement' based on some notion of desert (Gillon 1985).

Such surveys as have been conducted (for example, Charny *et al.* 1989, Brakenheim 1990, Williams 1988c) eliciting the views of the general public, and of health care professionals, as to who should be given priority if not everyone can be treated, indicate a strong consensus on two points: the young should have priority over the old[29] and the parents of young children should have priority over their childless contemporaries. The interesting thing about the former finding is that it suggests that discrimination against the elderly · is not regarded as *unfair* discrimination.[30] The interesting thing about the second finding is that it mimics perfectly the priorities of the EM, as, to some extent, does the first finding.[31]

The case of the poor and deprived is a more difficult one, because the EM suggested that using the health care system as a compensatory mechanism to offset deprivations *which are not caused by deficiencies in the health care system itself* may be both inefficient and misguided. Suppose someone starts life in a poor family, with limited access to knowledge but in good health. Nutrition is likely to be poor, knowledge about health and hygiene is likely to be poor, educational opportunities are likely to be poor and getting money becomes the number one priority. So employment will be sought as early as possible (but may be harder to get than for the average person) and the individual will not be too fussy about occupational risks. As income rises the individual is likely to be in a culture in which smoking, heavy drinking and other hazardous pursuits will be the norm. Time off work on account of injury or sickness is costly in terms of forgone income (unless social security support is generous) and frequent visits to the health care system will be made to boost the overloaded repair system that evolutionary physiology has provided. The potentially favourable effects of the health care system upon the individual's health stock will be dissipated by continuing to transform health into wealth (by earning money in a hazardous way) and into current living standards (by pursuing hazardous consumption activities). So even if such an individual started life with average health (which is unlikely), the subsequent time path of current health is likely to be poor, as a direct result of the individual's attempts to optimise (within rather severe constraints) his or her 'portfolio' of health, wealth and wisdom. It therefore seems to me that it is likely to be both futile and inefficient to attempt to improve the health of the poor through special treatment within the health care system alone. The problem is much more pervasive and needs a more radical approach if inequalities in quality-adjusted-life-expectancy within a population are to be taken seriously as a public policy objective.

CONCLUSIONS

Anyone involved in public policy discussions about health and health care will be confronted almost daily with issues such as those outlined above and it is my belief that what we health economists are being called upon to do is to help our respective societies to come to terms with the consequences of the emasculation of natural selection. It is paradoxical that now that we are richer than we have ever been before and have more potentially beneficial health care activities available to us than we have ever had before, we seem to face more excruciatingly difficult decisions about resource use that we have ever had to face before. I think the reason for this is that our success as a species has given us considerable scope for deciding our own destiny and we are finding it too great a responsibility. We can now virtually choose our own demographic structure (within limits), either directly (and then seek to develop instrumental policies that will get us there) or indirectly (by choosing instrumental policies knowing the demographic structure they will lead to if successful).

Health economists are contributing to this discussion by stimulating a rather fierce debate about the objectives of health care (and especially about the efficiency/equity trade-off) and the implications of these objectives for 'traditional' medical decision-making. We are also helping the public and the health care professionals to compare alternative methods of 'rationing' health care, so that we can all come to a more informed choice about priorities and their consequences for the demographic structure and for the general welfare of the population.

But it is a mistake to confine the drama surrounding the question 'who shall live?' (and still more that surrounding the question 'who shall die?') to discussions about health and health care. As long ago as 1974 Victor Fuchs observed that:

> At the root of most of our major health problems are *value choices*: What kind of people are we? What kind of life do we want to lead? What kind of society do we want to build for our children and grandchildren? How much weight do we want to give to individual freedom? How much to equality? How much to material progress? How much to the realm of spirit? How important is our own health to us? How important is our neighbours' health to us? The answers we give to these questions, as well as the guidance we get from economics, will and should shape health care policy. (Fuchs 1974, p. 148)

He could have gone on to say that the answers we give to his questions will and should shape our destiny as humankind, for the emasculation of natural selection with respect to human populations [32] requires us to think much more fundamentally, and on a global level, about our relationship with our habitat and our relationships with each other, *in all realms of human activity*, not just in health care.

I think the first part of the ambitious programme on which we have to embark is clarification of what is involved in determining an optimum population size and structure. This should lead to a systematic comparison of the priorities or trade-offs that seem to be emerging, with a view to determining whether they are compatible with each other and what sort of population size and structure they are likely to lead to. It is too demanding an intellectual task for me personally but, perhaps, one of my readers will have the necessary talent and motivation to step into the breach.

NOTES

1. For a more comprehensive view of the scope of the subdiscipline see Blades et al. (1986), Parts 1 and 2 and the contents of the *Journal of Health Economics*.
2. I have done that in more popular vein in Chapter 4.
3. In this section I have drawn heavily on Sibley and Calow (1986). (Page references at the end of quotations in the text are to that work unless otherwise indicated.)
4. Further differentiation may also occur through mutation of the genotype.
5. Including holding back sufficient reserves to cope with future emergencies and routine maintenance.
6. It should be noted that by 'health care' is meant not only medical, nursing, remedial and rehabilitative interventions as typically provided by health services or nursing homes but also counselling and advisory services concerned with health that might be provided by other bodies. As will become evident shortly, the 'health care' provided within the household by lay people constitutes an awkward borderline case, which it is sometimes advantageous to separate out, and sometimes not. In general I shall here distinguish 'informal care' from 'health care', reserving the latter term for care provided by 'professionals'.
7. Here I am drawing on an earlier excursion I made into this territory (Williams 1988a).
8. Already we are in some difficulties, because health is an individual attribute, but economic behaviour is likely to be best explained using the household as the unit.
9. Which does not have to be a positive amount.
10. Though the effects of this irreversibility can in some cases be alleviated by organ transplantation, joint replacement and artificial support through intensive care.
11. This allows economic activities to have different implications for people's health according to their degree of physiological maturity and senescence, for which we usually (but dangerously?) take their age as a rough proxy.
12. Conversely health is a strong influence on an individual's economic situation.
13. In order not to clutter up the diagram too much, I have ignored formal on-the-job training provided by employers. I have also ignored occupational health care provided by employers.
14. It is of course now possible to separate these two sources of satisfaction and to pursue them independently.
15. There is one other comparative aspect of the rival maximands which is worth some attention, namely the representation of time preference. In the EM this

is picked up by using a discount rate which has the effect of making benefits more valuable the sooner you get them and costs less hurtful the longer you postpone them (other things being equal). In the BM similar effects are produced by the term in the fitness formula relating to the time interval between generations, which it is advantageous to shorten (other things being equal). Again the BM is more narrowly focused than the EM, in that it concentrates 'time preference' wholly on reproduction and adopts a unit of time that is very long (25 years?) by economist's standards, thereby excluding time preference *within* that period. As we have already noted, many people's subjective rate of time preference is so high that they do not pay much regard to things happening three months ahead!

16. See, for instance, the quotation in paragraph 8 above.

17. See, for instance, the observations about smoking behaviour in paragraph 23 above.

18. See Williams 1988e ('Economics and the Rational Use of Medical Technology', in Rutten, F.F.H. and Resider, S.J. (eds.), *The Economics of Medical Technology*, Springer Verlag, Berlin, 1988, p. 117.)

19. Peter Hogarth has observed (personal communication) that there are biological precedents for the care of aged parents, for example, in chimpanzees. The explanations given tend to be based on a total absence of supporting evidence but include (1) the idea that it is always worth while, if you are likely to be chased by a leopard, to have a few individuals around who can run more slowly than you can; (2) that an elderly and infirm chimp may still have something to contribute in the non-genetic transmission of experience; (3) that to transmit to your offspring the trait that ensures they look after you, you must carry the trait that you look after your own parents – which could work, but only if an elderly chimp still has *some* chance of reproducing, however small; or (4) that what has evolved is a tendency to invest to some extent in close relatives for the enhancement of enhanced inclusive fitness and the investment is not so great that it was worth evolving the ability to discriminate between one lot of close relatives (those still able to reproduce) and another (those no longer capable of reproducing). None of these arguments seems especially convincing, but it is at least possible that care of aged parents could enhance fitness. It would be hard to test any of these ideas rigorously.

20. I leave the family on one side partly because it is conceivable that effects of physiological evolution may still have a strong influence on family behaviour.

21. The key population group in this context being the nation-state.

22. The emphasis here will be upon health care rather than other contributors to health, because the problems of establishing priorities are particularly acute there. Similar arguments could be developed with respect to occupational hazards, road safety, consumer protection, health promotion, etc.

23. Or in the more complex world of modern health care, they are the facilitators of the supply of care, much of which is supplied by other health care professionals.

24. For instance, here is an extract from a particularly sharply worded protest along these lines:

> Of late an increasing number of papers in this and other journals have been concerned with the 'cost-effectiveness' of diagnostic and therapeutic procedures. Inherent in these

> articles is the view that choices will be predicated not only on the basis of strictly clinical considerations but also on economic considerations .... It is my contention that such considerations are not germane to ethical medical practice .... A physician who changes his or her way of practising medicine because of cost rather than purely medical considerations has indeed embarked on the 'slippery slope' of compromised ethics and waffled priorities. (Loewy 1980)

25. Though the contribution of GNP per head (that is, real income) to QALYs (that is, health) cannot be ignored, just as the contribution of health care to health cannot be ignored.

26. I am here passing over a complex literature on interdependent utility functions and altruistic behaviour which represents an intermediate stage between the altruistic individual and the collective agencies of the society.

27. And if one got really ambitious, of future generations too, thereby bringing us back closer to the biological maximand of reproductive capacity.

28. These issues are not avoided in a libertarian system but they are not problematical in that system because, provided that spending power was acquired in a legitimate manner, each unit of spending power is regarded as of equal (ethical) value provided it is spent in a legitimate manner. The resulting interpersonal distribution of health (or use of health care) is whatever it is and is of no further moral concern (unless it gives rise to external costs or benefits which need to be internalised in the pursuit of Pareto efficiency).

29. A view shared by the old.

30. A commonly cited justification for giving priority to the young is that it would be unfair to deprive them of the opportunities in life that the old have already had.

31. The reason for the note of reservation here is that it appears that the *very* young (for example, children under the age of 5) are given *lower* priority than older children, the reason usually given being that less has been invested in them and they are easier to replace.

32. It should be noted that there have been less than 100 generations of *homo sapiens* since the advent of settled agriculture, less than 15 since the industrial revolution and only about 5 since pre-maturity mortality rates dropped sharply in the richer nations of the world. This is but a fleeting moment in evolutionary terms and we must expect that our physiological inheritance lags way behind the current demands (and opportunities) of our economic and social situation.

7. Measuring the Effectiveness of Health Care Systems

THE PROBLEMS AND ITS SETTING

In my mind's eye I have an image of a health care system whose sole function is to ensure that the community it serves derives the maximum net benefit from its existence. The community it serves (its 'clients') comprise sick people and those who suffer pain, grief, anxiety, etc., because of sickness, both now and contingently in the future. It does not include, as such, those whose livelihood depends upon providing the inputs which the health care system requires; except in so far as they are 'clients' in the meaning given above. Moreover, this community wishes its health care system to be run in a manner which reflects the values of the community, despite the fact that the health care system is so large and complex that a great deal of decentralisation of resource allocation decisions is necessary. The basic problem I wish to examine is 'how can we measure the "efficiency" of such a system?' or, put in more everyday language, 'how can we tell whether that health care system is serving the community as well as it could?'

The simple-minded economist's formulation of this problem would be to establish production functions for the various health-affecting activities which the system might embrace and from these estimate marginal social costs, then elicit the preference function of the community for the outputs of the system and optimise (that is, maximise the difference between total social benefits and total social costs). At this level of abstraction, such a way of formulating the problem might be assigned to that class of economic propositions labelled 'true but unhelpful' but, because it is both true *and* fundamental, it is nevertheless a useful reference point for the discussion.[1]

In practice, we have difficulty implementing it as an operational research strategy for several reasons:

(a) Production functions are ill-defined, owing to our ignorance of the physiological, psychological and sociological influences affecting the efficacy of therapeutic or supportive activities.[2]

(b) In costing the inputs into these activities we are often unclear at a conceptual level as to the proper realm of discourse (that is, over what range of considerations we are suboptimising in an imperfect world) and, even when we are clear on that score, the data at our disposal are frequently inadequate (drawing heavily on financial data from large public agencies and from the well-organised markets which happen to exist, with rather poor data on true opportunity costs, especially as felt by the clients themselves).[3]

(c) The community's preference function is a will-o'-the-wisp, because we are not sure that the community actually has one, or even has a set of concepts and a language in which to discuss what is involved in developing one, not to mention much idea as to who should play what role in that process.

In this chapter I shall not discuss the difficulties arising under (a), though we shall have to take note of them. The agenda of business to discuss under (b) is long and fascinating and is the area of health studies where the economist's role is pre-eminent. My interests lie in area (c), because I think it is here that the most appalling weaknesses of current studies into the efficiency of health care systems (by economists and others) are manifest[4] and, although it is not strictly speaking 'economics', there appears to be some reluctance by other professional groups to plunge into the difficult territory of devising 'output' measure for health care systems which are going to serve the economists' purposes in answering the basic question posed earlier.[5]

I begin, therefore, by setting out certain desiderata which I think any such 'output' measures should fulfil, before going on to outline a strategy for devising such measures. I next report briefly on some work I am currently directing which forms part of this strategy and in the last section, I offer some observations on why I think economists need to devote more of their time, energy and skill to this kind of work.

DESIDERATA

Ideally, we seek an estimate of the benefits of health-affecting activities measured in monetary units commensurate with the relevant cost estimates. Before setting out the schemes by which such measures might be evolved, two cautionary disclaimers are in order: (a) using monetary units as a measure of value does not imply accepting any particular *mode of valuation* and, more specifically, it does not imply acceptance of market values and still less of a 'cash flow' approach to the evaluation of health care systems;

(b) in order to purge the subsequent discussion of any risk of misunderstanding on that score, I am going to use the convention that all the so-called 'economic' benefits (like getting people back to work more quickly, saving costs which would fall on other services, etc.) are treated as *negative costs* and offset against the items on the input side.[6] Thus, only 'humanitarian' benefits are considered here.

One other important preliminary matter needs to be made clear and that is that, because it is the client's state which is our central interest, the measures we are seeking must be to do with the client's state. I apologise if that seems too trite and banal to be worth mentioning but it is not as trivial as it seems at first sight. In this context, it is not only important to distinguish between indicators of state of health (like mortality and morbidity measures, days of restricted activity, etc.) and provision indicators (hospital beds per thousand population, doctors per thousand, etc.)[7], it is also important not to be led into believing that numbers of cases, episodes of treatment, hospital bed–days occupied, are true 'output' measures, even though they refer to clients, because they are concerned with throughput or workload, not 'output' in terms of amelioration of an individual's state of health compared with what it would otherwise have been. A still more tempting trap for the unwary is set by measures of 'need', such as nursing dependency scales, which indicate the quantity and quality of care required by a patient. Although these measures are closely associated with client state in terms of social functioning, they specify *inputs* needed, which is precisely what we must avoid prejudging.[8]

Moreover, it is the community's valuations of that state which are relevant. This may or may not coincide with the individual's own valuations of his own state, or indeed with any particular individual's valuations of any other client's state. I am going to duck the very large and difficult set of intellectual problems which surround the processes by which 'community' values emerge. I shall merely refer to a 'dialogue' between the parties, without investigating the power and influence relationships at work therein. What I shall do, instead, is indicate the proper *professional* roles of the various parties and try to provide them with a communication system which reduces the area of ignorance and risks of confusion in the conduct of that dialogue. As an 'efficiency' analyst I would be reluctant, in principle, to attempt more and enormously self-satisfied if I achieved anything significant even in that apparently limited task.

This measure of benefits should also be capable of wide application and not be specific to particular conditions or to the inputs of particular services, otherwise it will be of severely restricted value. This means that it has to refer to a context in which the client's state can be assessed in a fairly general manner and it leads, inexorably, to an approach based on the general *social* functioning of the client, that is, his or her ability to conduct the essentials of every-day living free of pain. This goes well beyond measurement of blood pressure, pulse rate, etc., and also beyond the identification of conditions (for example, pneumonia, pregnancy, arthritis) and even beyond the noting of physical disabilities (missing index finger, blind in one eye, loss of one leg below the knee), because it rests on the question: 'to what extent is this individual unable to carry out painlessly the activities which the community would expect individuals of that age and sex normally to be able to carry out?'

The final desideratum is that the measure should be capable of being made operational, which implies that it must be capable of routine application and use, which in turn implies that it must not make inordinate intellectual demands, nor intolerable demands in time and energy, upon those responsible for collecting and processing the information. Although high-powered research and development work may be required to test it out in the initial phase, it should be designed to run on low-powered inputs in the long run.

A SOLUTION IN PRINCIPLE

The basic elements in the proposed scheme are:

(a) a set of descriptive categories concerned with the client's state in terms of pain-free social functioning;
(b) a relative evaluation process that converts these states into index points;
(c) an absolute valuation of points in money terms.

In this section the general properties of each element will be described, before presenting a brief account of some empirical work designed to help us with element (a).

If we are to build up an index of health (or, in this case, of ill-health), we need to measure both intensity and duration. 'Intensity' is here interpreted as having two dimensions, 'pain' and 'restriction of activity'. The first step would be, therefore, to experiment with simple standardised descriptions of painfulness and of the extent to which activity is restricted, to see if there is any consensus among medical personnel as to how painful and how restricting particular conditions are, using these descriptive categories. The initial descriptive stage may be represented as in Figure 7.1 below.

Figure 7.1 Descriptive categories for a health index

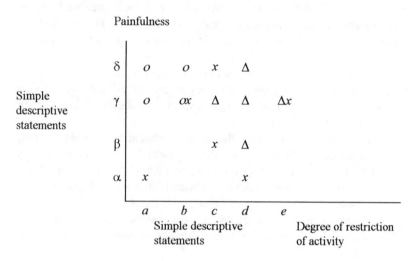

α, β, γ and δ are simple descriptive statements concerned with painfulness (such as 'mildly uncomfortable', 'very comfortable', 'extremely painful', etc.). *a, b, c, d* and *e* are simple descriptive statements concerned with restriction of activity (such as 'light work only', 'confined to house and immediate vicinity', 'confined to house', 'confined to bedroom', 'confined to bed', etc.). The symbols *o, x* and Δ each refer to different medical conditions. Each *o* plotted in Figure 7.1 represents a different expert's assessment of the most appropriate description of the effect of that condition (for example, one says, *a,* δ, another says *a,* γ, another says *b,* δ and yet another *b,* γ). Similarly, each *x* and each Δ represent corresponding judgements by various experts of the most appropriate descriptions of each of those conditions. If there is any consistency in these judgements, some 'norm' will be indicated as the standard description for that condition; where no consensus exists (as with *x* in the example), it is likely that the condition under study needs to be more closely specified.

However, supposing that we had each condition clearly ascribed to a pain category (α, β, γ, etc.) and a restricted-activity category (*a, b, c,* etc.), we should now need to establish the trade-off between them (for example, is the combination γ α better or worse than the combination β*c*?). This pairwise comparison is, essentially, a *social* judgement and should be recognised as such, but may have to be made in practice by medical people. This first evaluative step is set out diagrammatically in Figure 7.2, where the lines are embryonic indifference curves and it is desirable to be as 'close' to zero as possible.

Figure 7.2 Trading off the two dimensions

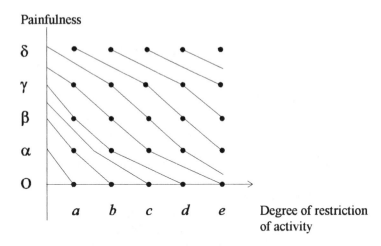

In the example shown, the combinations (β, O), (α, *a*) and (O, *b*) are equivalent to each other, but better than (γ, O) and (O, *c*) which are equivalent to each other. Those, in turn, are better than the next group of equivalents (β, *a*), (α, *b*) and (0, *d*), and so on.

Despite the fact that describing the intensity of pain is notoriously difficult and that interpersonal comparisons are bound to be rather arbitrary, medical personnel can and do make such comparisons between stages and classes of condition and such comparisons already have to be assimilated into judgements about 'acceptable' degrees of physical disability and pain at the diagnostic and therapeutic level when determining courses of treatment. It is, therefore, suggested that it should be possible to move to the second (partial evaluative) stage and construct, say, a 10-point scale of intensity of ill-health along the following lines:

  0 = normal
  1 = able to carry out normal activities, but with some pain or discomfort
  2 = restricted to light activities only, but with little pain or discomfort
3–7 = various intermediate categories reflecting various degrees of pain and/or restriction of activity
  8 = conscious, but in great pain and activity severely restricted
  9 = unconscious
 10 = dead

Since it is intended to use these numbers as *weights*, and not simply as *rankings*, it is important to stress that society's judgements concerning the relative importance of avoiding one state rather than another are represented by the actual numbers attached to each respectively, for example, state 2 is *twice as bad* as state 1, and state 10 is ten times as bad. This implication must not be shirked and it must be regarded as a statement about *health policy* and is to be made by whoever is entrusted with that responsibility, for example the politicians and not a technical statement about *medical condition*.[9]   In terms of Figure 7.2, this would be represented by attaching index numbers to each of the contour lines.

As to duration, this will be based on the outcome of scientific investigations, cast in statistical terms.   For instance, recovery from a particular disease will follow one time path (incorporating both intensity and duration) in 90 per cent of the cases, another in a further 9 per cent, and yet another in the remaining 1 per cent.   Chronic cases where no (or little) improvement in intensity is to be expected will have a duration equal to life expectancy of that class of individual and the duration of the 'gain' from postponing death where successful treatment is possible will be similarly measured.   A 'successful' treatment is *not only* one which reduces intensity and duration but could also be one that reduces intensity without affecting duration, or vice versa; or even that increases one at the expense of the other, provided the net outcome is to *reduce* the index number (a product of intensity *and* duration).   The important sources of information here are the medical statisticians, since it is purely empirical information that is required at this stage in the process.

Figure 7.3 illustrates how this would work for any particular condition. The diagram starts at a point of time, 0, when the condition in question is diagnosed.   In the illustrative example the first two weeks are spent in further observation, decision as to appropriate treatment and waiting for therapeutic facilities to become available.   The prognosis without treatment (or with the best treatment other than that under consideration) is represented by the broken line and may be described as a steady deterioration from approximately week 7 until death in week 12.   This would be the standard prediction for this class of case.   The average expectation of life for a person of that age/sex, etc. is represented as $n + m$, which may be rather large if necessary (for example, 50 years).   The prognosis with treatment is represented by the solid line and may be described as two weeks of severe restriction of activity (in the pre-operative, operative and immediate post-operative phases), plus possibly considerable pain, with a steady improvement in condition during the ensuing three weeks, a convalescent phase from weeks 7 to 9, and a further two weeks taking it easy in the normal environment, after which the patient is completely normal (as far as this condition is concerned).

Figure 7.3 Net benefit of treatment over no treatment

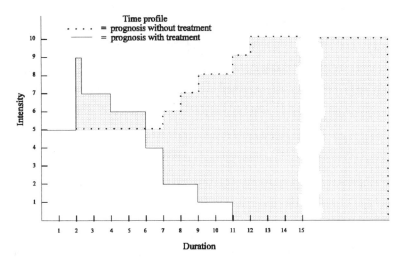

The index *score* (representing the 'effectiveness' of this treatment) would be the area unshaded *minus* the area shaded, obviously including in the former the interval omitted in the horizontal scale as drawn. This particular example would, obviously, be a highly effective treatment if applied to people with long life expectancy and less so the shorter life expectancy. Both the time profiles used would be derived from statistical analyses of clinical results, or on experimental data if the former were lacking. It is up to the medical statistician to provide these key data. A further sophistication which could be introduced, if necessary, would be to apply a discounting factor which would give less weight to future states of health compared with present states and, hence, reflect the greater weight which people seem to attach to the 'here and now' rather than to more distant prospects.

Certain features of this system are noteworthy:

(a) It enables preventive as well as therapeutic activities to be incorporated.
(b) Although much more difficult in practice, in principle it can embrace mental illness.

(c) It treats one week of suffering at any particular intensity level as being equally undesirable irrespective of the identity of the patient. Other distributional assumptions are possible in principle, but would make the analysis much more complicated.[10]

(d) It relates only to patients and does not include infectivity, or the pain and suffering caused to others by the patient's condition. Neither of these shortcomings are insuperable in principle but as a practical matter they will be difficult to overcome in the near future.

(e) The *satisfaction* felt by patients themselves (or their friends and relatives) is not regarded as an independent consideration in this formulation and to do so would raise such enormous difficulties for any health indicator that the matter is mentioned here only so that it is not lost sight of.

The primary purpose of such an indicator is to facilitate cost-effectiveness studies, by providing a quantification of the purely humanitarian benefits to be used in conjunction with economic costs (and benefits) in order to improve the effectiveness of health services in the face of severe resource limitations. But it could also generate, as a by-product, improved indicators of the state of (ill) health of the community if used as part of the basic information matrix in a national survey of the state of health of the community.[11] If successful, this would fill an important gap in our present knowledge, for it would include cases where people had not presented themselves for treatment, or where those giving treatment were unaware of the patient's condition between episodes of treatment.

This leaves us with the final step, which is attaching money values to index points. There are various ways of approaching this thorny issue. The one least likely to raise strong emotional objections is simply to calculate the *implied* values placed on marginal points by existing allocations of resources,[12] and to argue that if there are discrepancies and the relative points values are right, then resources should be shifted from activities yielding high cost points at the margin to those where marginal points are relatively cheap. A more direct method has been used in the field of crime, by getting respondents of various kinds to say how serious they thought one crime was relative to another, from which an index of crime seriousness was constructed.[13] Once people get used to this kind of discourse, one could then pose the question 'are these marginal points values of the right order of magnitude relative to other things?' It might be possible to help inform such judgements better by pointing out the kinds of compensation which the community regards as fair if someone is moved from health state to health state for a given duration through no fault of his own[14] and compare this with the sums the community seems willing to devote to preventing similar changes from happening by the provision of health care. Before long, we may even reach a stage when some bold politician[15] will have the nerve to put the issue to his constituents in these terms, for example, 'if elected, my policy will be based on a 5 per cent increase in the value of a health point' (that is, 'I think we ought to be prepared to devote an extra 5 per cent of the community's resources to improving health by one unit at the margins'). This 'price' thus becomes a 'cut-off' point for decentralised resource allocation decisions in health care systems and one which will force decision-makers to pay regard both to medical effectiveness (expected points score) and social valuations (as indicated by the 'price' of a point).

WHERE SHOULD WE START?

I am not so crazy as to imagine that by this time next year (or even in 10 years time) we could have the provision of health care rationalised in this way, even if all the parties whose co-operation is necessary were sold on the idea, so we need to face the problem of research priorities and testing during the developmental learning phase. I am convinced that the major stumbling block at present is the absence of any widely used standardised descriptive categories of social functioning,[16] for without these we cannot get off first base, so I have become a major advocate (and a minor organiser) of *longitudinal* surveys[17] of clients, these surveys needing to be broad in coverage both as regards the clients' conditions and the health care agencies treating them.

For tactical reasons, the care of the elderly has been selected as the test area, covering hospitals, general practitioners, welfare homes, domiciliary support, etc., in both urban and rural areas of England and Wales. In order not to try to clear too many hurdles at once, no index is being calculated, so we are operating at the level of Figure 7.1 above. This does not, however, preclude individual agencies from supplying their own relative valuations of client states if they wish and it will give them a set of categories across which they can sensibly do so. Nor does it preclude statistical analysis of the results designed to see what transitions from state to state are associated (*ceteris paribus*) with the various treatment modes, so that one can begin to build up the basis for prognoses along the lines of Figure 7.3, though in several (incommensurable) dimensions rather than in the common currency of 'health points'.

In our study, client state has three dimensions (physical mobility, capacity for self-care and mental state), each of which is ranked on a four-point scale, so that there are 4^3 (= 64) logically possible different states.

Clients are to be reassessed at three-monthly intervals, so that (if we add the state 'dead' to those already mentioned) it will be possible to compile a 64×65 matrix of 'transitions', which can be repeated (and combined) at three-month intervals to give three-month, six-month, nine-month, twelve-month, etc. transitions. The fundamental research task will be to try to identify (from a sizeable package of background information on clients, which is also being collected at each assessment) what statistically significant associations exist between the different transitions experienced by people starting in the same state and the different treatment modes provided by the health care system. A more immediately important operational payoff may well be the focusing of routine records on client states and a more general dissemination and testing of category systems of this kind (see Appendix 7.1 for a brief outline of the contents of schedule to be used).

This project will be at the pilot stage (with some 500 test observations) during 1973 and, if no fatal flaws emerge, will run on a fairly large scale (10,000 or so observations) during 1974 and 1976. But it will still be but a beginning in the task ahead, because it still will not include any agreed set of index weights across client states (still less their evaluation in money terms), nor will it provide detailed knowledge of production functions but only broad indicators of the relative effectiveness for one client group of whole packages of care and it is not concerned with the measurement of costs. However, it is focused on the appropriate primal element in the problem and is the first stage of a carefully mapped escape route from the intellectually imprisoning confines of the measures of workload and throughput (and even input) which too often are pressed into service as measures of the effectiveness of health care systems.

CONCLUDING OBSERVATIONS

No one step in this 'escape route' seems infeasible, though when viewed *in toto* the whole programme may appear a daunting task. Fortunately, economists need not (and, indeed, should not) play a central role in every phase of the work, though I think it would be extremely valuable if the work undertaken by others were well informed as to the data requirements and underlying philosophy of the economists' optimising models of health care systems and it may well be that this can only be achieved by having economists involved in an advisory way with work which they (and others) may not regard as their own particular speciality.

I am fortified in this view by the great dangers I see in the opposite course (which is that economists concentrate entirely on the estimation of costs and 'economic benefits'), because this, unfortunately, taints us with the stigma of a 'commercial' or 'GNP'-orientated approach to health care systems[18] and, while I should not wish anyone to think that I believed them to be predominant, it seems, therefore, to be extremely important that we demonstrate our practical concern for health service effectiveness in broader terms than these and the plan of work set out in this chapter is designed to do precisely that.

APPENDIX 7.1

Outline contents of assessment schedule and background information concerning care and the elderly

A. ASSESSMENT SCHEDULE

The assessment schedule has three divisions concerned with mobility, capacity for self-care and mental state. The factors considered in each of the divisions are as follows:

(1) *Mobility*
    (a)   Ability to get in and out of bed and/or chair
    (b)   Ability to negotiate a level surface
    (c)   Ability to climb stairs
    (d)   Ability to walk outdoors

(2) *Capacity for self-care*
    (a)   Ability to feed self
    (b)   Ability to dress self
    (c)   Ability to wash self
    (d)   Ability to make a hot drink
    (e)   Ability to cook a meal
    (f)   Ability to light a fire
    (g)   Ability to shop
    (h)   Whether or not continent

(3) *Mental state*
    (a)   Intellectual processes – memory, orientation of person and place
    (b)   Loneliness and desolation
    (c)   Depression
    (d)   Boredom

    (e)   Motivation towards independence
    (f)   Anxiety
    (g)   Antisocial or self-harming behaviour

Care has to be taken in this division to distinguish general feelings of mood from pathological mental disorders which are to be recorded in the background information.

B. BACKGROUND INFORMATION

(1)   *Personal and socio-economic status*
    (a)   Age
    (b)   Sex
    (c)   Marital Status
    (d)   Occupation or previous occupation of self or spouse
    (e)   Income

(2)   *Domestic environment*
    (a)   Type of house
    (b)   Facilities, for example, hot water, location of toilet, ⎤ for
          bath, etc.   people
    (c)   Shared or sole use of accommodation   living
    (d)   Warden or other people officially in attendance ⎦ community
    (e)   Presence of alarm system
    (f)   Residential or nursing home:
          public or private
          type of home
          facilities available
    (g)   Hospital ward:
          general, medical or surgical
          geriatric – assessment or
          rehabilitative – continuing care
          psychiatric

(3)   *Medical condition*
    Based on broad systematic classes, for example,
    injuries or diseases of:
          cardiovascular system
          locomotor system
          central nervous system

(4)   *Social contacts*
    (a)   Visits received from

> relatives, friends and neighbours,
> tradesmen, etc.

(b) Visits made to
>   relatives
>   friends and neighbours

(c) Casual visits to
>   social clubs
>   shops
>   church, etc.

(d) Other 'non-face-to-face' contacts, for example,
>   telephone calls
>   letters

(e) Interest in current affairs – newspapers, radio, TV

(f) Use of spare time

(5) *Use of services*
Use of local authority, hospital, general practitioners, etc. and voluntary services.

NOTES

1. This framework is developed further as a means of disentangling notions of 'need', 'trade-off', etc., in Appendix A of Culyer *et al.* (1972).

2. A particularly pungent exponent of this view is Cochrane (1972).

3. For instance, the simple notion that the client's time is also valuable is slowly gaining explicit recognition in 'scheduling' and 'location' studies in the health care system. It does not always spill over as fully as it should into comparisons of institutional versus domiciliary care, however, with some odd effects on policy choices. See, for instance, Wager (1972).

4. It even defeated the ingenious Feldstein (1967), who was forced to accept the convention that 'a hospital's outputs are measured as the numbers of cases treated in each of several categories' (p. 169) and among his list of ways of improving the model is included the tentative hope that one day 'it may be possible to relate selected measures of population health to the care being received' (p. 286).

5. Which is not to claim that no one else is active in this field. Besides the work cited later, note should be taken of Sullivan (1965) and Fanshel and Bush (1970).

6. This obviously leaves the calculation of 'net benefits' unaffected, but means that we might have to think of some activities as having negative *net* costs if, for instance, the 'economic' benefits outweighed the costs. Such cases I do not regard as very interesting from our standpoint, since all such activities should obviously be pursued in a rational society, as the 'humanitarian' benefits can be obtained only by the sacrifice of other good things (that is, where their costs are positive).

7. See, for instance, Culyer *et al.* (1972) for a fuller discussion of the shortcomings of some of the existing work based on these approaches.

8. This distinction is clearly in Carstairs and Morrison (1971).

9. Magdelaine *et al.* (1967), in their work on a similar indicator, fail to make this clear.

10. This is a particular manifestation of a more general problem in compiling social indicators, which is that we may not wish to count each individual on a one-for-one basis when certain

people register poor readings on many different indicators, for example, health, education, crime, poverty.

11. For a brief description and methodological critique of such surveys, see Linder (1965). It may be that the 'General Household Survey' will become a suitable vehicle for such an investigation in the UK; see Moser (1970).

12. For example, Hawood and Morley (1969), subsequently applied to health services by Porter, 'Planning Maternity Care', unpublished M.Phil. thesis (University of Durham).

13. See Sellin and Wolfgang (1964) and Chambers (1973). This has been carried a stage further to derive an actual tariff of money values by Shoup and Mehay (1971).

14. Some preliminary analysis of this kind has been done for the UK by Rosser and Watts (1971). It is likely that more will be undertaken as a result of an SSRC-financed project on 'The Cost of Human Impairment' under D. S. Lees at Nottingham University and the recent setting-up of a Royal Commission on Civil Liability and Compensation for Personal Injury.

15. So far we have at least reached the stage where the Minister responsible for road investment is willing to publicise the sums he uses in investment appraisals as the value of avoiding fatal accidents: in 1968 it was £8,800, of which £3,800 represented tangible costs (including loss of output) and £5,000 an allowance for 'warm-blooded' costs (pain, grief, suffering, etc.). See *Annual Reports of the Road Research Laboratory* (London: HMSO).

16. There are plenty of special-purpose ones, many of which do not even get into the literature. A few of the published ones which seem capable of wider application are: Garrard and Bennett (1971), Gruenberg and Brandon (1966), Hamilton (1960), Harris (1971), Isaacs and Walkley (1963, Katz *et al.* (1963), Tunstall (1966).

17. These are not unusual in the field of child care, but much rarer elsewhere. See Wall and Williams (1972).

18. See, for instance, Packer (1968).

PART TWO

The Measurement and Valuation of Health

8. Welfare Economics and Health Status Measurement

SETTING THE SCENE

Welfare economics can variously be interpreted as a search for:

(a) some 'objective' (that is, value-free) criterion for measuring improvements in social welfare;
(b) a criterion which satisfied pre-specified conditions of 'reasonableness';
(c) a criterion which commands 'general' assent;
(d) the value judgements implicit in any proposed criterion.

The first of these interpretations is now generally recognised to be futile and most of the action in recent years has been motivated by a search for the second (and to some extent the third) goal. The general conclusion appears to be that it is extremely difficult to devise criteria for social choice which simultaneously satisfy economists; scruples about the nature of the interpersonal comparisons of utility which they are willing to make; which survive careful tests for internal consistency from a logical viewpoint; are capable of handling all possible configurations of preferences and choice possibilities and satisfy individualistic/democratic notions of justice and fairness.

This chapter does not set out to construct, *de novo*, a criterion of social welfare (for the dimension of health) which satisfies all these desiderata but looks instead at some existing criteria in order to elicit any value judgements contained within them (or implied by them) which are of special relevance from a welfare economics viewpoint. The use of any measure of output implies valuation as well as description. To identify the value content of an index the welfare economist has two possible strategies: either to review all existing indexes and comment on their value implications or to specify a particular context and then explore the feasibility (and implications) of using some existing indexes for the purpose. This chapter adopts the latter strategy and the specific context chosen is the following: the benefit measure is to be used in cost-effectiveness studies in which all changes in resource use and in production are picked up elsewhere, while the benefit measure shall not depend on the wealth or 'economic' value of the affected individual.

Before proceeding, this context is worth reflecting on further, if the subsequent discussion is not to be misunderstood.

When we assume that all changes in real resources associated with the alternative under investigation have been measured, it is useful to classify them as:

(a) changes in resources used in service provision;
(b) changes in resources used by patients and their helpers;
(c) changes in gross domestic product.

Item (a) should include all affected services (for example, local authority welfare services, ambulance services, primary care services, etc.) and not just hospital services or whatever service is most closely involved in the particular option under investigation. Item (b) must include the value of any time which has opportunity cost but, to avoid double-counting with what is in the benefit measure itself, it will concentrate on the value of the patient's time to other people (as well as including the value of other people's time to themselves). The patient's own value of own time will not be included under this heading. Item (c) should include non-marketed outputs (such as housewives' services) which do not form part of GDP as actually measured but should do so in principle. Note that in each case it is the 'change' in the value of the relevant category of resource that is to be measured and this can be positive (hence a 'cost') or negative (hence a 'benefit'). Note also that associated tax and transfer charges are not included, though under a more ambitious rubric they would be separately picked up in a matrix of distributional consequences designed to establish the incidence of the economic costs and benefits and not simply their nature and totals which, for simplicity's sake, are all that will be considered here. Although these 'economic' costs and benefits go well beyond mere GDP calculations, they do deliberately exclude changes in the health of patients[1] *per se*. This is because these 'humanitarian' benefits (or costs) are to be the main focus of discussion here and occupy the 'other side' of the cost-effectiveness account. Thus the dichotomy will not be between 'costs' (= bad changes) and 'benefits' (= good changes) but between changes (good and bad) in resources and changes (good and bad) in health itself.

The benefit measure sought is one which embodies the ethical principle that any interpersonal comparison of the value of health shall not depend upon the wealth or economic value of the people concerned. This reflects not only the ostensible ethic of the medical profession itself but also the putative political principle on which health services are organised in many countries.   It is one of the major reasons why the provision of health services has not been left to the market but treated more as a 'citizen's entitlement' than as part of society's 'reward system' (Donabedian 1971). In that context, measures of the value of health which reflect people's ability to pay will be irrelevant.   The interest lies seeing how far we can get in establishing an 'uncontaminated' set of values which give as much rein as possible to differences in the *relative* values people attach to different attributes of health, while pursuing some kind of egalitarian principle between the weight given to one person's valuations compared with another's.   This is the central task of the chapter.   To be of general use, the measure has, in addition, to be a cardinal measure, since in the cost-effectiveness context the basic mathematical manipulations we may need to conduct may depend upon cardinality.   The properties of a wide range of extant indexes have been comprehensively reviewed by Rosser (1979) and a regularly updated cumulative annotated bibliography of work in this field (National Centre for Health Statistics, *Clearinghouse on Health Indexes*, has been maintained by the US Department of Health and Human Services since 1972).

The next section of the chapter outlines the manner in which one could undertake the measurement of the *relative* value attached by an individual to different health states, once one had postulated a *numéraire* health state to act as a unit of value and a zero-valued state so as to ensure full cardinality.   The third section outlines the ethical postulates required to justify various methods of aggregation of these individual valuations in order to derive a community valuation and explores their implications.   The fourth section shows what happens when the 'health status index', so derived, is put back into the cost-effectiveness framework, with the apparent result that the carefully specified egalitarian–humanitarian ethic built into the third section seems suddenly to be undermined by the resource side of the equation.   The final section contains some general observations and suggestions for further work.

THE CALCULUS FOR ONE INDIVIDUAL

There are myriad ways in which health can be characterised and, important and controversial though the topic is, here I shall merely assert that, for the purposes to which I see this health status measure being put, measures based simply on presence or absence of disease, or on changes in mortality, are inappropriate.   My general position is that the best measure of health for the purpose of economic evaluation must be a 'feeling-functional' one, in which the presumed ideal is a long life in which each individual is able to undertake the normal pattern of- activities free of pain and distress. Suppose we designate that ability as being 'healthy' and assign it (arbitrarily) the value 1 for a particular individual.   Since 'normal' functioning is a socially conditioned notion, this notion of healthiness may well fall short of 'perfect' health, in the sense of the maximum attainable by anyone, anywhere, ever. Rather it will have the more modest (and more useful) connotation of accepting that there is a threshold below which a society considers someone as 'to all intents and purposes' healthy (warts and all, although not 100 per cent fit, as judged by Olympic standards).

But what lies at the opposite extreme from healthy? The obvious answer seems to be 'dead'. Herein lies a hornet's nest of problems which are not at all easy to resolve. For instance, it could be argued that the opposite of 'painfree ability to conduct normal activities' is 'in very severe pain and totally unable to conduct normal activities'. It may further be asserted that this is worse than being unconscious (that is, in no pain but totally unable to conduct normal activities) and perhaps even worse than being dead (as witness: the proponents of voluntary euthanasia). There are two ways of resolving this dilemma in the present context.   One is to constrain individuals to conform to society's view (whatever that is) by postulating what the worst state is and assigning that the value 0. The other is to let each individual choose which is the worst (that is, zero-valued) state and let that be part of the realm of individual valuation. For the purposes of this exercise it will be postulated (because it is simpler) that 'dead' is the worst state, which still leaves individuals free to value other states also at zero if they wish. The other important characteristic of the state 'dead' is that it is, in the jargon of the trade, an 'absorbing' state, that is, once you are in it, the transition probabilities from it to any other state are all zero. This will not be true of any other states, even those which individuals also value at zero.

With these two fixed points to work from, individuals would be asked to value a large selection of intermediate states, each of which is a (different) combination of attributes in dimensions such as pain/distress, physical mobility, capacity for self-care and ability to play normal social roles. These valuations are usually made in a context in which it is assumed that:

(a) each health state is to be thought of as persisting for the same length of time;
(b) the actual length of time must be specified because it can affect the relative valuations of different states;

(c) the state is to be evaluated for its own intrinsic 'enjoyability' and not for any instrumental purpose it might have in enabling the individual to earn money, etc.;

(d) all states must be evaluated as if the respondent were in them now (and not in earlier or later life);

(e) no element of prognosis must seep in (for example, being in any particular state now should not be thought to imply anything about the relative probabilities of being in any particular state in the future);

(f) the valuation of each state is independent of the states which precede or follow it.

These are very stringent conditions, which make empirical work to elicit such valuations rather difficult. They are needed in order to avoid the logistically impossible alternative, which is to present subjects with all possible future time profiles of health states and then to get them to value each profile as a whole relatively to each other profile as a whole.

Even in the more limited exercise there are difficulties at both theoretical and technical levels, which begin with the selection of the dimensions of health. These must emerge from empirical investigation in which respondents are permitted to reveal their own constructs of health. In the foregoing discussion the results of this process were prejudged (on the basis of work already done) to lead to the dimensions mentioned. The choice of valuation technique is a second important area, where the strategic choice is between behavioural and psychometric techniques. The former are largely factorial and difficult to purge of 'welfare effects'. Psychometric methods therefore seem to predominate in this kind of work, the precise method being determined partly by the kind of measure required (for example, interval or ratio scale). It is important, however, to ensure that what you get is what you set out to get, whether people's valuations are internally consistent, or different methods (for example, category rating, magnitude estimation, or equivalence assessment) are consistent with each other when applied to the same subjects, or respondents seem equally able to respond to each. It is not sufficient simply to apply statistical tests of correlation to establish such concordance, because one also needs the more pragmatic test of whether, where they differ, these differences are crucial in their consequences for actions in the field, or whether they make no significant difference as to what people do. Current testing is deficient in this latter respect.

The kinds of measurements on which economists can draw for this purpose are epitomised in the work of investigators such as Patrick, Bush and Chen (1973), Card *et al.*, (1977), Rosser and Kind (1978) and Torrance (1976a) though none of them would probably accept my specific formulation. I have deliberately steered away from the technical and practical problems of conducting this kind of work. In this connection, Johnson and Huber (1977) includes an extensive bibliography. I have instead concentrated on eliciting the underlying value judgements, because it is important that these be recognised if the work is to be used in an ethically appropriate context.

THE PROCESS OF AGGREGATION AND ITS IMPLICATIONS

In an exercise of this kind an aggregation rule designed to elicit the community's relative valuation of different health states must of necessity imply an ethical postulate and cannot be regarded as a mere computational device. In the foregoing section, individual valuations were constrained in such a way that one unit period of time (say a year) in the state 'healthy' was to be valued at 1 and being dead was to be of zero value (from the viewpoint of our earthly existence at any rate). We now have to make a heroic leap and postulate some relationship between Individual A's one year of health (or death) and Individual B's one year of health (or death).

The simplest postulate consistent with the egalitarian–humanitarian ethic is that a year of healthy life is to be regarded as of equal intrinsic value to everyone, irrespective of age, sex, etc. This is equivalent to regarding one year of healthy life expectation as *numéraire* commodity, exchangeable on a one-for-one basis within the community but with individuals free to establish their own *relative* values of all other health states (except dead) which they might experience. This seems to be the commonest aggregation rule and is explicitly adopted in the work of Patrick *et al.* (1973) and Torrance (1976b). This rule implies that an extra 10 years of healthy life for one individual are of equal social value to one extra year of healthy life for each of ten different individuals. A second, more complicated postulate, would be to argue that an additional year of healthy life expectation is to be regarded as of differential *intrinsic* value according to the age and/or sex of the beneficiary. Thus, if we established as the *numéraire* 'one year of healthy life for a man aged 40', we would need to elicit the values placed on all the individuals in the community on the value of one year of healthy life at each other age for each sex.

It seems to me to be a weakness of the existing work on health indexes that a disproportionate amount of attention has been paid to the eliciting of relative values from individuals and to comparison of these relativities between individuals. Very little has been given to eliciting people's views about how the valuation of one person should be weighed with those of another in order to reach a set of 'social' values. The reason for this relative neglect is probably that most such indexes have been designed for purposes other than planning or resource–allocation. (See, for example, Thouez (1979).) Those that have had this latter orientation, have tended to add (or average) data in a rather unreflective way (as shown by Culyer (1978), and a common escape route in the more theoretical literature, as Whitmore (1973) has observed, is simply to postulate that everyone has the same utility index (an assumption I am here seeking to avoid).

On whichever basis it is decided to proceed, the derivation of a scale for the whole community requires only one further step, that is, defining the relevant community. This could, in principle, range from all humanity (including generations as yet unborn) to a small number of people living in one small geographical area at some specific time, for whom some health-related choice has to be made between alternative uses of a given amount of resources. This 'choice of community' is an issue that I intend to duck, since it is essentially a political decision which does not affect the essence of the argument, only the precise form in which it would be applied in a given situation. It might mean, for instance, that for purposes of health service planning at national level, it would be thought advantageous to elicit the valuations of a large random sample of the entire citizenry (or at least that sub-set eligible to vote). Adopting an entirely different approach, it might be argued that 'proxy' or 'representative' or 'expert' valuers might be chosen as the respondents, on some such grounds as the following: 'proxies' might be required for infants and very young children, for the mentally ill and handicapped and other non-competent subgroups; 'representatives' might be required through the political process because of the practical difficulties or organising the frequent redrawing of large samples of the population for direct consultation on these matters; while 'experts' might be required because the issues involved are so complex that only people with the right training, experience and intellectual skills could respond properly. (Although I have views on these matters, they are set aside here as separable from the main theme.)

Having come this far, the final step is aggregation itself. The simplest (and commonest) procedure is to take each health state in turn and add up the score assigned to it by each of the n individuals and take the average as the community's valuation of that health state. What emerges may not coincide with the actual valuations of any particular individual, of course, and it is a fascinating exercise in its own right to investigate what are the correlates of any significant deviation from the average, for example, whether they are associated in any systematic way with the respondent's age, sex, family circumstances, political or religious beliefs, educational level, occupation, or their recent or current experience of illness. (See, for example, Rosser and Kind (1978)). It has frequently been suggested in casual conversation (though I have not found an example in the literature) that distributional equity dictates that it is better for many people to get a little than for a few people to get a lot (that is, contradicting the notion that one person getting 10 years' additional life expenditure is the same as 10 people getting one each). To pick up that ethical postulate we would have to take (say) the logarithm or the square root of the 'quantity' of benefit gained by each individual and then work with that transformed value in the subsequent analysis.

There remain only three further matters for discussion before this 'tariff' of values is ready to be fed back into the framework of a cost-effectiveness study. The first of these concerns whether 'time preference' does or should exist and, if so, whether it is for each individual or for the community to choose the precise discount rate to be applied in order to give effect to it. The argument for individuals each applying their own discount rate is essentially that there may be great variation in private time preference rates for health states across the community and a valuation scheme, which allowed individuals to compare health states only at a point in time, but not between present and future periods, would be seriously defective. The counter-argument is based on the observation that because it is possible, at the margin, to transform health into wealth and vice versa, at any point in time and since the 'wealth' is (ideally) allocated through time with reference to the rate of social time preference, then it would be inconsistent to apply a different rate of discount to 'health' from that being applied to 'wealth'. I incline to the latter view.

This leads on to the second consequential matter, which is the respondents' attitudes to 'risk' and whether the 'certainty' approach employed here is not seriously defective because it fails to elicit risk-aversity. A similar line of argument to that in the preceding paragraph can be employed here. The 'risks' concern the likelihood that an individual will be in one state rather than another. This is a matter on which others are likely to be more knowledgeable than the individual is and one about which, for the community as a whole, there may be virtually *no* uncertainty whatever (for example, over age/sex specific death rates). Since the decisions on which this 'tariff' of community valuation of health states is to bear are ones which will typically involve choices for largish populations, the uncertainty is more relevant to distributional issues than to those addressed here. It therefore seems quite defensible to take advantage of the considerable simplification facilitated by *valuing* the states as if they were certain but generating the probability distribution of states separately.

The third and final consequential matter is the fear that, once respondents realise how their responses are going to be used, they will engage in some kind of 'false signalling' in order to manipulate the outcome to their own advantage.   For instance, if someone is already (chronically) in a particular health state, then it might be thought advantageous for that person to undervalue that state (that is, declare a value lower than the 'real' one) in the hope that this will help draw more resources into any activity which would move him into a better health state (and it certainly would help to do that in the circumstances postulated, although only to a very small extent if the total population of respondents is fairly large).   But the danger to the individual with this kind of false signalling is that it automatically entails the respondent undervaluing the *difference* between this current state and various *worse* states, so that he or she might find resources *drawn away* from activities which would *prevent a worsening* of his or her situation.   It does therefore seem to be a rather risky business.   Moreover, it is generally very difficult for individuals to be at all sure which health states they will be in, even in the immediate future, which adds yet another deterrent to false signalling.   All in all, it does not seem to be serious danger.

ON MIXING SUGAR AND SAND

Suppose we now have a set of relative values for different health states, on a scale 0 to 1, which can be used as the building blocks for constructing a time profile of expected future health states and the impact of a treatment upon that profile.   The difference between the two profiles is the value of the treatment in humanitarian terms, best described as expected years of life expectation adjusted for health quality.   These will then be aggregated according to whichever principles were deemed appropriate in the light of the considerations set out earlier.   When this is juxtaposed with the resource cost data mentioned earlier, it would be possible to say that treatment A 'costs' 10,000y and yields health benefits of 1000z, while B 'costs' 7500y and yields benefits of 1500z and C costs 500y and yields benefits of 25z (where y is a unit of money value and z is the *numéraire* health state).   Thus each unit improvement in health costs 10 with A, 5 with B and 20 with C, so if resources are limited and there are constant returns to scale in each case, then the priorities for new investment would be, first, treatment B, then A, then C.   (In reality it is likely that as the scale of provisions increases, each successive increase in capacity will result in less suitable patients being treated, so costs per achieved unit of health improvement will actually be increasing.)

But now we come to some further problems.   Suppose that the reason why treatment B 'costs' only 7500 is that (a) it typically gets highly paid patients back to work quickly, (b) the associated GDP benefits have been 'netted out' and (c) had this not been so the costs would have been much higher, say 30,000. Suppose the other two treatments were for the chronically mentally handicapped and the very elderly, respectively, and generated no GDP benefits whatever. It now seems that, despite our humanitarian–egalitarian principles, priorities are still being affected by the very things we were trying to shut out for, without the GDP benefits, treatment B would have cost 20 per unit of health benefit achieved and would have been given the lowest priority instead of the highest. Was it all worth while? Did we labour so mightily merely to bring forth a mouse?

The root of the problem is that we have two valuation systems operating side by side. In the nonhealth sector we have valuations 'contaminated' by unequal ability to pay, whereas in the health sector we wanted a set of 'decontaminated' values. But which of these should we use to measure the opportunity cost of the resources used in the health sector? So long as the society thinks it right that non-health benefits should be valued relative to each other by the 'contaminated' set of values, then there is no valid argument for excluding the GDP benefits once you include benefits in the form of savings of service costs. So the question at issue is whether any of the resource costs should influence priorities in the health sector. If it is held that they should not, then it will be impossible to guarantee the maximisation of health benefits for any given level of resource use. If it is held that they should, then all elements of resource use will exercise an influence and we shall have to face up to the consequence that it is in society's interest to give some precedence in matters of health care to those members of the society whose activities others value most highly. Not to do so seems perversely masochistic.

This brings us to the final critical issue, that is, just how much weight should be given to the 'economic' as opposed to the 'humanitarian' benefits? This is another way of posing the question 'how much precedence should be given to those whose activities others value most highly?' It will also determine, indirectly, what fraction of the community's resources are devoted to the health sector. So far I have shied away from the question of money valuation of health benefits but now it has inescapably to be faced and there is no easy answer to it. It would be self-defeating if, at this stage, each individual's money valuation of a unit change in health were elicited from his or her own behaviour when confronted with situations in which money had to be sacrificed to improve health, since such values would be 'contaminated' with the wealth and income inequalities whose effects we are trying to neutralise. It might be argued that this would be less of a problem if each individual's marginal valuations of 'units' of health, derived in this way, were somehow averaged and that average then applied to every unit improvement in health, no matter who got it. But although these individual marginal rates of substitution between health and all other goods may be a useful guide to the approximate magnitude of the corresponding social marginal rate of substitution, ultimately the latter has to be a community-wide decision, just like the choice of a social rate of time preference. It brings us back squarely to the issues raised in the literature on the value of life (recently surveyed by Mooney 1977).

This should not be interpreted as meaning that the earlier exercise was fruitless, however, because a different set of relative valuations of quality and quantity of life will almost certainly emerge from the calculus compared with the 'economic' calculus. Moreover, if units of health are valued very highly at the margin, the 'resource' element in the calculation will carry correspondingly little weight. Thus a very 'humanitarian' health system will select a different set of patients from one more concerned with resource savings (that is, non-health benefits). It might be speculated that the economic side of the equation will have more weight in communities where the overall standard of living is close to subsistence levels, whereas the more humanitarian side of the equation will have more weight where communities can afford to carry a significant proportion of 'unproductive' people. It should be noted, moreover, that this issue is distinct from the one about how differently the humanitarian benefits should themselves be valued (that is, whether a person's income and wealth should influence the weight given to that person's views in that part of the calculus). Since the GDP consequences and the level of income will be highly correlated, if humanitarian benefits are weighted by such influences, this will tend to augment the influence of economic considerations on priorities in health care.

CONCLUSIONS

I have tried to work through, and expose, all the important ethical considerations underlying the computation of a health index which seeks to estimate, by psychometric methods, an individual's relative valuations of different health states and then to fuse these into a set of social valuations over the same set of health states, in a manner suitable for use in cost-effectiveness studies. I have imposed on the 'fusion' process an additional requirement, namely that it must be ethically appropriate, by which I mean that in my selected context the social valuations which emerge must not be vulnerable to influence from the distribution of income and wealth across respondents.

There obviously remains plenty of work to be done in developing and refining the approach here outlined. Tight and widely applicable descriptions of key health states need to be established as the basic building blocks. Epidemiological and other clinical research needs to be so designed as to measure the course of each patient's condition in terms of these states, so that we slowly build up a clear picture of the relative effectiveness of various interventions in fairly standardised terms. All this is needed before valuation can be brought to bear but, in the meantime, the testing of the valuation process itself can continue and the rival methods carefully appraised for their sensitivity, reliability and acceptability. The applied welfare economists could, at the same time, be assisting in this task, while the theoretical welfare economists could be grappling with the dilemmas of principle posed by this approach, both internally and when it is put back into the context of an economic evaluation which values resources on a different basis.

It seems to me a potentially rich field for economists, especially in view of the fact that practitioners (and researchers) in the health care field are (and will be) using very crude and often quite inappropriate proxies to fill the vacuum caused by the absence of theoretically satisfactory ways of measuring levels of health.[2] There are doubtless some who would argue that, since our interests and expertise lie primarily in assessing money valuations as elicited from market or quasi-market data, then if that is not what people want, they should look elsewhere for help. I believe this latter view to be misguided and, if widely adopted, would exclude economists from exercising their analytical skills over large areas of social policy where the ethic of the market (and consequently the market as a source of information about valuations) is rejected. That would be a great waste of our talents. I sometimes fear that things are worse still and that the way in which the welfare economics of social choice has developed recently, with almost exclusive concentration on the second of the four tasks set out at the beginning of this chapter, may have already so inhibited and emasculated the subject that the third and fourth tasks are no longer considered even legitimate and to practise them (and especially to go so far as to set about estimating interpersonal comparisons of cardinal utility) puts one beyond the pale. If so, so be it, but those who thereby find themselves in exile will at least be able to take comfort in the thought that though their light may be dim, it shines where the world is darkest.

NOTES

1. 'Patients' here means whoever are the subjects of the study. Since some 'treatments' are given to one person in order to improve the health of another person, the latter may be 'patients' as well as the former.
2. The contents of the joint publication by the World Health Organisation and the International Epidemiological Association (1979), called *Measurement of Levels of Health*, bear adequate testimony to the diversity of approach and content that represents the current state of 'best practice' in this field.

9. Valuation of Quality of Life: Some Psychometric Evidence

INTRODUCTION AND SUMMARY

In this chapter the following problem is addressed:

'How does psychometric evidence from individuals on the *relative* values they attach to different states of (ill) health vary from person to person and to what extent is the psychometric evidence consistent with the *relative* values implied in court awards?'

The objective behind this enquiry is to outline a method by which a rough tariff for valuing distress and disability might be established in the form of a set of coefficients which, in conjunction with life expectancy and the current monetary value ascribed to life itself in that community, could be used to estimate the compensation due for various states intermediate between fit and dead.[1] These valuations do not include the instrumental value of health in earning money, nor the out-of-pocket costs of coping with ill-health. They are what might be called 'warmblooded' or joy-of-living valuations, in the value-of-life jargon.

This chapter will present results from one psychometric study[2] in order to test the potential and attractiveness of the idea.[3] The results of an analysis of legal awards will then be presented and similarities and dissimilarities between the two sets of relative valuations identified. Finally, it will be argued that there are good *a priori* arguments for psychometric evidence to play a more influential role in the compensation process.

THE VALUATION OF DISABILITY AND DISTRESS: THE PSYCHOMETRIC APPROACH

Twenty-nine states of illness have been derived by combining 8 states of objective disability and 4 states of subjective distress. The disability states are:

1. No disability.
2. Slight social disability.
3. Severe social disability.
4. Choice of work or performance at work very severely limited. Housewives and old people able to do light housework only but able to go out shopping.
5. Unable to undertake any paid employment. Unable to continue any education. Old people confined to home except for escorted outings and short walks and unable to do shopping. Housewives able only to perform a few simple tasks.
6. Confined to chair or to wheelchair or able to move around in the house only with support from an assistant.
7. Confined to bed.
8. Unconscious.

The distress states are:

1. No distress.
2. Mild.
3. Moderate.
4. Severe.

Since disability state 8 implies no distress, states 8,2; 8,3 and 8,4 are considered void. State 1,1 may be regarded as 'fit'.

There are a variety of psychometric techniques available for eliciting relative valuations across this range of states. In the particular study reported here a method of magnitude estimation was used as follows.

Interviews were undertaken with 70 subjects, each interview including a discussion of the difficulties in measuring the relief of illness and of the difficulties of defining and assessing the importance of degrees of relief. The interviewer then explained that the interview would consist of a numerical exercise using a simplified concept of illness. Descriptions of states of illness typed on to cards would be used and the subject and the interviewer would work together to explore how the subject perceived the severity of each state relative to every other state. Details are provided in Appendix 9.1.

Here attention is confined to the valuations elicited for 'permanent' states (that is, those where the prognosis is that the state will not change until death intervenes). The results from different subjects were highly variable and the data were not distributed normally. The scale shown in Figure 9.1 was obtained using the medians as a measure of central tendency and standardising these so that death = 0 and fit = 1. This transformation converts the ratios into an interval scale which is adequate for the application discussed in this chapter.

Figure 9.1 Median valuations of states

Disability /Distress	1	2	3	4
1	1.000	0.995	0.990	0.967
2	0.990	0.986	0.973	0.932
3	0.980	0.972	0.956	0.912
4	0.964	0.956	0.942	0.870
5	0.946	0.935	0.900	0.700
6	0.875	0.845	0.680	0.000
7	0.677	0.564	0.000	−1.486
8	−1.028			

The variability in valuations is associated with subjects' current experience of illness either as patients, as health service professionals or as healthy members of the community. Scales derived from six groups with different experiences are shown in Figures 9.2–9.7.

Figure 9.2 Valuations of medical patients

Disability /Distress	1	2	3	4
1	1.000	0.992	0.986	0.977
2	0.987	0.982	0.968	0.936
3	0.980	0.966	0.958	0.915
4	0.954	0.951	0.937	0.893
5	0.924	0.910	0.903	0.840
6	0.863	0.848	0.760	0.440
7	0.640	0.371	0.000	−1.480
8	−0.422			

Figure 9.3 Valuations of psychiatric patients

Disability /Distress	1	2	3	4
1	1.000	0.999	0.993	0.989
2	0.995	0.994	0.990	0.971
3	0.991	0.990	0.985	0.946
4	0.985	0.982	0.976	0.935
5	0.972	0.963	0.964	0.836
6	0.909	0.893	0.775	0.500
7	0.675	0.679	0.000	−1.443
8	−1.571			

Figure 9.4 Valuations of medical nurses

Disability /Distress	1	2	3	4
1	1.000	0.995	0.989	0.973
2	0.992	0.989	0.982	0.953
3	0.980	0.978	0.972	0.941
4	0.975	0.972	0.963	0.911
5	0.956	0.949	0.879	0.739
6	0.890	0.876	0.724	0.496
7	0.621	0.583	0.000	−1.048
8	−1.258			

Figure 9.5 Valuations of psychiatric nurses

Disability /Distress	1	2	3	4
1	1.000	0.997	0.994	0.977
2	0.994	0.992	0.981	0.932
3	0.986	0.982	0.977	0.865
4	0.977	0.974	0.970	0.855
5	0.965	0.956	0.953	0.835
6	0.816	0.828	0.775	0.340
7	0.577	0.450	0.000	−1.926
8	−0.217			

Figure 9.6 Valuations of healthy volunteers

Disability /Distress	1	2	3	4
1	1.000	0.994	0.989	0.944
2	0.989	0.986	0.973	0.937
3	0.983	0.979	0.953	0.913
4	0.975	0.957	0.939	0.882
5	0.961	0.945	0.873	0.390
6	0.851	0.817	0.657	−0.624
7	0.733	0.716	0.000	−2.291
8	−0.326			

Figure 9.7 Valuations of doctors

Disability /Distress	1	2	3	4
1	1.000	0.992	0.946	0.793
2	0.981	0.973	0.865	0.766
3	0.946	0.913	0.848	0.668
4	0.923	0.888	0.760	0.187
5	0.873	0.865	0.692	-0.394
6	0.800	0.773	0.298	-0.803
7	0.505	0.452	0.000	-2.288
8	-1.077			

Doctors place relatively less emphasis on the importance of death in comparison with other states; that is, they regard *more* states as worse than death. Doctors also place more emphasis on the importance of subjective suffering.

THE VALUATION OF DISABILITY AND DISTRESS: THE ANALYSIS OF LEGAL AWARDS

The same set of 29 states of disability and distress has been used (by Rosser) to analyse over 200 British court awards for damages in personal injury claims. All the awards described in Kemp, Kemp and Liavery (1967) and in the eighth cumulative supplement to this edition were examined with the help of a barrister and, where sufficient information was available, the disability and distress suffered by the plaintiff were categorised by the author. For each category, it was then plausible to assume that the plaintiffs had suffered injuries of comparable severity. Reports were excluded from the analysis if insufficient information was available to permit classification of disability and distress. The rationale of this approach is amplified in Appendix 9.2.

Cases were excluded from the analysis if the proportion awarded in compensation for financial consequences of the injury or illness could not be separated from the proportion intended to neutralise the effects of suffering *per se* or if the prognosis was uncertain. 202 awards were identified which satisfied all the conditions specified. For each of the awards the following data were used in the analyses:

- age;
- sex;
- life expectancy – if this was stated in the details of the case this figure was used. Otherwise life expectancy tables were used;
- amount of award made for non-pecuniary loss and adjusted to 1970 values using the retail prices index;
- disability and Distress State.

The statistical nature of this sample is difficult to define. It cannot be a random sample drawn from any definable population of court awards since it is the population of all awards which satisfy the defined constraints. Furthermore, a plaintiff who brings a case generally has a good reason for believing that his case differs in significant respects from previous ones. Thus the awards which reach the courts are a statistically biased sample of all settlements. The nature and amount of this bias is unknown. This uncertainty limits the scope of statistical analysis of the data.

The analysis of the awards was simplified by considering only those which had the following characteristics:

1. The plaintiff had been healthy prior to the accident. This justifies the assumption that the description of the state of illness of the plaintiff is a description of the damage caused by the accident.
2. The state of illness of the plaintiff was clearly described and the component due to the accident was expected to remain unchanged until the plaintiff's death. This justifies the assumption that the description of the present state of illness is also an estimate of the damage caused by the accident in future time periods until the plaintiff is expected to die.
3. The life expectancy of the plaintiff was either unchanged by the accident or stated in the case report. The courts make a separate award for loss of life expectancy and this is distinguished under a separate heading. However, in addition the courts adjust the award to take account of reduced life expectancies; generally shorter lives receive smaller awards.

The results are again highly variable. They have been standardised in the same manner as before. This approach does not provide a value for death, which is awarded a standard nominal sum by the courts. The value assigned to death has been assumed to be the same as the value assigned to the state of permanent confinement to bed (disability state 7) in moderate pain (distress state 3) since this was the state chosen as equivalent by all 6 groups of subjects in the psychometric experiment. The results are shown below in Figure 9.8:

Figure 9.8 Valuations implied by court awards

Disability /Distress	1	2	3	4
1	1.000	0.990	0.980	0.973
2	0.977	0.973	0.942	0.923
3	0.951	0.930	0.854	0.662
4	0.851	0.839	0.690	0.641
5	0.384	0.388	0.476	0.370
6	0.476	0.346	0.340	−0.144
7	0.340	0.276	0.000	−0.371
8	−0.082			

For some states, only one or two cases were available for analysis. Where more cases were studied the results were somewhat skewed in distribution and for comparison with the psychometric scale, the median was used as the principal measure of central tendency. However, the skew was not as marked as in the psychometric data, the means lying close to the medians.

A COMPARISON OF THE TWO SCALES

The legal scale, in comparison with the psychometric scale from all subjects, has the following features:

1. A greater emphasis on distress. The legal evidence introduces this bias by encouraging the court to assume that the disabled person inevitably suffers mental anguish.
2. A discontinuity in the valuation placed on disability state 5 (that is, inability to work) compared with state 4 (that is, limited effectiveness at work). This may be an artefact reflecting some arbitrariness in the partitioning of the award between pecuniary and non-pecuniary

components, especially in cases where the total award has been adjusted after the calculation of the separate components.

3.   A more balanced assignment of lower valuations between the severe states judged to be somewhat less or more serious than death and, in particular, a less exaggerated valuation of the two most severe states.

4.   Some illogical reversals of ranking within the scale.

The scale is significantly correlated with the psychometric scales from all subjects, but correlates most with the doctors' scales and least with those from nurses as shown below:

Subjects	r
Medical patients	0.792
Psychiatric patients	0.792
Medical nurses	0.767
Psychiatric nurses	0.768
Healthy volunteers	0.800
Doctors	0.827
All subjects	0.816

all significant (p. 0.01)

The similarity in the views of members of the medical and legal professions is noteworthy and raises questions about how far the emphasis on distress is a socio-economic bias rather than the consequence of close contact with suffering.

Despite the broad statistical agreement between the two scales, the legal one must be handled with scepticism since it is based on some sweeping assumptions. The differences between the two are largely accounted for by differences in the four most severe states. It seems likely that the legal scale is distorted at this point because the money equivalents become so large as to be unacceptable to the courts, although recently an occasional case has substantially exceeded the magnitude of the awards which were available for this analysis.

CONCLUSIONS

It appears that, compared with the psychometric evidence, legal awards tend to assign more moderate relative valuations to the most severe states but otherwise there seems no identifiable source of systematic bias between the two. This raises the question of which of the two approaches one should have more confidence in when they differ (assuming that the differences are not artefacts of the investigation itself).

In favour of the legal approach is the much larger volume of particular circumstantial detail about each case presented to the evaluator (the judge), the recourse to precedent and building up of personal experience in analysing these cases. In favour of the psychometric approach is the fact that the evaluator has to derive a consistent set of relative values across all states simultaneously rather than approaching each 'case' piecemeal over a long interval of time. This enforced review, driving at internal consistency, is a very severe discipline, as anyone who has participated in such psychometric experiments will testify and it seems a major *a priori* argument for this approach. A further consideration is that the psychometric valuation takes place in a quiet and thoughtful context free of partisan advocacy and any overtones of blame or shame.

There is clearly too sketchy a data base available at present for it to be credibly argued that psychometric evidence should replace existing precedent in informing future legal judgements concerning this particular element in the quantum of damages but, if compensation for personal injury is to reflect the detached judgements of the citizenry, it does seem justifiable to argue that rather more effort should be devoted to discovering what they are and that psychometric methods offer a promising way of filling the void.

APPENDIX 9.1

THE PSYCHOMETRIC EXPERIMENT

The following extracts are taken from Rosser and Kind (1978).

This structured exercise included the following stages:

1. Ranking 6 'marker states' in order of severity of illness. These states, chosen for wide dispersion between the categories of disability and distress, were the same for each interview. Tied ranks were permitted.

2. Placing the 6 states on a scale. The subject was first presented with his first 2 cards, starting with the least ill states. He was asked 'how many times more ill is a person described as being in state 2 as compared with state 1?'. The following assumptions were specified:

(i)   The descriptions are of people who are all of the same age – either young adults or middle aged.

(ii) All states have the same prognosis. All can be cured if the sufferer is treated but, if left untreated, will remain static until some other condition supervenes.

The second of these proved difficult to maintain and had to be frequently reiterated.

The subject was told that this was the most difficult stage in the entire interview and that he should consider the time spent on it as unlimited. There were a variety of implications in his decision and he might like to consider these before he chose a number. The implications described were:

(a) The ratio will define the proportion of resources such as time of trained personnel, money, equipment, etc. That you would consider it was justifiable to allocate for the relief of a person in the more severe state as compared with the less ill.

(b) The ratio will define your point of indifference between curing one of the iller people or a number (specified by the ratio) of the less ill people.

The subject was invited to discuss these implications and any discrepancies in the conclusions he reached on the basis of the original question compared with the two corollaries. When he had chosen a number, its implications in terms of resource allocation and indifferences were reiterated.

3.   Ranking 23 further combinations of pain disability.

4.   Expressing all 23 states in ratio terms including assigning zero to the state to which it would be reasonable to restore all patients. At this point, subjects were permitted to adjust the ratios of the marker states if they wished.

5.   The subject was then asked to change his assumptions about prognosis. None of the states would now be treated: they were thus permanent states. He was asked to adjust his scale accordingly.

6.   He was then asked to place the state of death on the scale of permanent states and assign a value to it.

This interview lasted from one hour and a half to four hours and a half. Ten subjects participated in a further experiment on the validity of the method. All numbers were removed from the description of the states. Subjects were then shown the description of a person confined to bed in severe distress and asked to identify the state which was nearest to half as severe, bearing in mind that for some of them the last step on the scale was so steep that no description was scored as high as 50 per cent of its severity. When they had identified the appropriate state they were asked to identify a third description of a state half as severe as the second one. This is the method of fractionation. They were then shown the description of a person without disability but in moderate distress and asked to identify the state which was nearest to twice as severe as this and then a third state which was nearest to twice as severe as the second. This is referred to as the method of multiplication. Consistency between the results of these methods and the original interview would provide evidence in support of the validity of the interview.

The subjects

There were six groups of subjects with different experience of illness. These included:

Group 1   10 patients from medical wards.

Group 2   10 psychiatric inpatients.

Group 3   10 experienced state registered general nurses.

Group 4   10 experienced state registered psychiatric nurses.

Group 5   20 healthy volunteers.

Group 6   10 doctors sufficiently experienced to have gained a Membership of Fellowship of at least one Royal College.

Data about characteristics and experience of illness were collected as specified below in the section on the analysis of data.

Reliability

Measuring test–retest reliability was difficult because of the fatigue due to the long interview and because the subjects experienced the interview as traumatic and felt that it changed their perception of illness. The test–retest and inter-observer reliability study was therefore performed separately using the marker states. One interviewer obtained a scale of the 6 marker states from 50 healthy volunteers on two occasions, the two interviews being interspersed by long memory distracting exercises. A second interviewer later administered a memory distracting exercise followed by a re-estimation of the scale for the 6 marker states. Stability over longer periods has not yet been studied.

Results

The experience of the interview:

The subjects found the experience of the interviews both painful and relevant. In general, doctors found it particularly difficult and patients found it rather disturbing. For example, a research psychiatrist said: 'This is really painful. You don't give it to patients do you?' A surgeon who rated distress very highly as compared to disability said: 'Clearly I expect everyone to put up with pain but me'. His colleague, an experienced physician who put high emphasis on disability, said 'I've been treating the wrong things all these years. It's what you can do, not what you feel that matters isn't it?' Another surgeon said 'It's all obvious, pain's the thing that matters. What? Everyone doesn't agree! That's positively immoral'. A patient recovering from a failed kidney transplant said 'Is this a con? I've been in every one of these states in the past month. It's frightening. Did you know my wife and I discussed euthanasia? I know exactly when I'd want death'.

Reliability

Test–retest reliability as measured by percentage agreement was 97.2 per cent. Inter-observer reliability was 88.0 per cent, this lower figure being partially attributable to the subjects' fatigue.

Validity

The methods by which the internal consistency of the scale is checked by asking the subject to perform a different type of judgement are referred to as 'internal' tests of validity in contrast to comparisons with scales obtained by totally different methods such as the analysis of legal awards which we call 'external' tests of validity.

Nine of the ten subjects who performed this test were internally consistent. The tenth had produced an exceptionally steep scale and consciously modified this during the validation test in full awareness of the invalidating effect of this adjustment on his original results.

Effect of changing prognosis

Changes in the scale when the prognosis was changed from 'treatable' or 'permanent' were few and minor. The scale of permanent states has the advantage of readily incorporating a score for death.

Marker cards

Description	State disability	Distress
Can work normally, do everything at home and have a normal social life. In moderate pain which is not relieved by aspirin.	1	3
Can work normally and do all household tasks. Illness interferes with some hobbies and leisure activities. In severe pain for which heroin is prescribed.	2	4
Too ill to work but can move around independently. At home can only do a few light jobs. In moderate pain which is not relieved by aspirin.	5	2
Can only move around in a wheelchair. Has slight pain which is relieved by aspirin.	6	2
Confined to bed. Has slight pain which is relieved by aspirin.	7	2
Confined to bed. In severe pain for which heroin is prescribed.	7	4

APPENDIX 9.2

RATIONALE FOR ANALYSIS OF LEGAL AWARDS

Purpose of the awards

Kemp, Kemp and Havery (1967) comment that an effective tribunal would achieve 'a reasonable measure of uniformity between awards in comparable cases and sufficient flexibility to produce a just result in individual cases'. If this were true, the application to these cases of a classification of disability and distress would be expected to reveal a consistent trend in the awards for injuries of different severities but with considerable variation between the awards for cases in any particular class of severity. A further target is 'a general level of awards which satisfied the social conscience of the nation'.

Implicit scale

The need to a scale, the process by which the award is reached and the difficulty of this task, have frequently been reiterated and clarified in the courts as exemplified by the following quotations:

You very often cannot even lay down any principle upon which you can give damages ... take the most familiar and ordinary cases; how is anybody to measure pain and suffering in monies counted? Nobody can suggest that you can by any arithmetic calculation establish what is the exact sum of money which would represent such a thing as the pain and suffering which a person has undergone by reason of an accident ... but nevertheless the law recognises that as a topic upon which damages may be given. Halsbury, L.C. The Mediana (1900) AC113 at 116.

When courts have to assess damages they have undoubtedly an extremely difficult task .... In one way it is an almost impossible task, for human suffering or physical deprivation cannot be translated into or related to terms of money and cash. But the best that can be done is to have regard to the developing and accumulating experience of courts in regard to these matters and (provided that the facts of other cases can be appreciated) to have in mind what has been done in other cases. Morris, L.J. Crawford v. Erection and Engineering Services (1953) CA No. 254 July 23, 53.

... Slowly and painfully, English law has evolved ways of assessing the incalculable and it is important that they should be followed and applied as far as possible so that the law may be coherent. Devlin, L.J. H West and Son Ltd v. Shepard (1964) AC326 at 329.

The court is seeking, however imperfect the attempt may be, a just proportion as between the damages awarded to one plaintiff and those awarded to another. This involves attempting to find a just proportion between the damages awarded for one kind of physical injury and another kind of physical injury such as physical pain, mental anguish and painless physical disability. Diplock, L.J. Wise v. Kaye (1962) IQB638.

Summary

The award made to compensate for a state of illness is a function of the symptoms of the injury or the disease, of the characteristics of the judge and of the characteristics of the victim. So the awards made for any particular state of fitness will be subject to variation. However, if sufficient awards can be identified it should be possible to use an average to obtain a scale which could be compared with psychometric one.

NOTES

1. Compensation for disability and distress *itself* – (I–C)VL where V = value of one year of healthy life, C = coefficient for this study ($-\infty < C$ & I) and L = current number of years of life expectation.
2. Details of which are reported elsewhere in Rosser and Kind (1978).
3. There are many others (such as Patrick *et al.* (1973), Card, Rusenkiewicz and Phillips (1977) and Torrance (1976a and b)), but they do not lend themselves well to the particular theme of this chapter, though with further work it should be possible to bring their results into systematic comparison with those presented here.

10. The Measurement and Valuation of Health: a Chronicle

THE TASK

No country can afford to do all the things that might improve somebody's health so some systematic method of establishing priorities is required. Commonsense suggests that resources should be concentrated where they will do the most good. In health care, 'doing the most good' means maximising improvements in people's health. Improvements in health have two broad dimensions: improvements in life expectancy and improvements in quality of life. Various classification schemes have been developed to provide descriptions (or simple scores) for particular dimensions of health-related quality–of–life that are of particular relevance for patients with particular conditions (for example, arthritis, cancer, heart disease, kidney disease). Because these 'specific' schemes have much content in common, more general health 'profiles' have been developed which encompass a broader range of dimensions than any one specific scheme and which enable comparisons to be made in a more standard way across different conditions.

However, because of the multi-dimensional nature of profiles they do not generate an overall 'index' number to indicate the extent of any health benefit. Indeed, in cases in which a patient is better on some dimensions but worse on others, they will even fail to indicate whether on balance there is any benefit or not. For this purpose a 'global' index is required, the descriptive content of which is neither condition–specific nor treatment–specific and the valuation content of which is such that it generates a single number which summarises the relative value attached to each of the multi-dimensional health states it encompasses. The valuation process must also include the relative value attached to improved life expectancy on the one hand and improved quality of life on the other.

From its inception, the basic objective of the Research Group on the Measurement and Valuation of Health (henceforth 'the MVH Group') at the University of York (see Appendix 10.2) has been to find practical ways of measuring health-related quality–of–life (HRQOL) that reflect the salient features of health as perceived by ordinary people. With this in mind, an important task for the MVH Group has been to elicit the valuations that ordinary people attach to different (multi-dimensional) health states.

This extremely ambitious task has had to be broken up into segments for it to constitute a workable research programme. The key strategic decision made by the MVH Group was to investigate separately the choice of the *descriptive* system for health states and the method of *valuing* them. The first task involved the elicitation of lay concepts of health. The second task (which turned out to be the much larger one) involved the testing of various valuation methods (and different practical means of administering each of them). The valuation task faced two further requirements: firstly, in order to fulfil the basic objective, the valuations had to be capable of being represented on a scale in which 0 = being dead, and 1 = being healthy and, secondly, in order to be useful for general policy purposes, such valuations were needed from a representative sample of the general public.

BACKGROUND

The foundations for the work of the Group were laid by Rosser and her collaborators some years ago (Rosser and Watts, 1972, Rosser and Watts, 1978, Rosser and Kind, 1978, Rosser, 1983). The central feature of this approach is a simple descriptive classification defining 28 states in terms of disability and distress, plus a twenty-ninth state 'unconscious', with 'dead' as the (implicit) thirtieth state (Table 10.1). A valuation matrix across these states was elicited from 70 respondents (Table 10.2), who were not a random sample of the population at large but a selection of fairly accessible doctors, nurses, patients and healthy volunteers. The original objective was to use this system to measure the 'sanitive output' of a hospital, that is, the extent to which patients benefited from hospital treatment.

Meanwhile a separate stream of methodological work from within health economics was developing the concept of the quality-adjusted-life-year (QALY) as an outcome measure for use in health care. (Culyer *et al.* 1971, 1972; Torrance *et al.* 1973; Patrick *et al.* 1973; Sackett and Torrance 1978). Williams noted that the Rosser valuation matrix (suitably transformed to a scale in which dead = 0 and healthy = 1) had the appropriate properties for it to become the 'quality adjustment' in the QALY concept and this idea was then published jointly by Kind *et al.* 1982 and applied for the first time by a small group of economists in the now classic study of the economics of coronary artery bypass grafting (Williams 1985a).

Table 10.1 Rosser's classification of illness states

Disability	Distress
1. No disability.	A. No distress
2. Slight social disability.	B. Mild
3. Severe social disability and/or slight impairment of performance of work. Able to do all housework except very heavy tasks.	C. Moderate D. Severe
4. Choice of work or performance at work very severely limited. Housewives and old people able to do light housework only but able to go shopping.	
5. Unable to undertake any paid employment. Unable to continue any education. Old people confined to home except for escorted outings and short walks and unable to do shopping. Housewives able only to perform a few simple tasks.	
6. Confined to chair or able to move around in house only with support from an assistant.	
7. Confined to bed.	
8. Unconscious.	

Table 10.2 Rosser revised matrix (original values in parentheses)

I	[1.00]	.89	.89	.67	
			(.995)	(.990)	(.967)
II	.89	.81	.89	.56	
		(.990)	(.986)	(.973)	(.932)
III	.70	.63	.57	.44	
		(.980)	(.972)	(.956)	(.912)
IV	.63	.56	.51	.40	
		(.964)	(.956)	(.942)	(.870)
V	.44	.43	.44	.22	
		(.946)	(.935)	(.900)	(.700)
VI	.44	.44	.34	.22	
		(.875)	(.845)	(.680)	(.000)
VII	.38	.40	.33	.20	
		(.677)	(.564)	(.000)	(–1.486)
VIII	.01				
	(–1.028)				
	[Dead = 0]				

The enormous interest generated in the potential of this approach led the York Centre for Health Economics to devote more resources to its development and, in particular, to pursue the following activities:

(a) conducting surveys on people's attitudes to health (see, for instance, Wright 1986);

(b) developing a simple self-assessment questionnaire to describe patients' current quality–of–life (see Kind and Gudex 1994, Gater *et. al.* 1995);

(c) helping the various clinicians and medical researchers who have approached us to incorporate quality–of–life measurement into their studies;

(d)   incorporating such measures, where appropriate, in studies with which we were associated (for example, studies of CT and MRI) (Kind and Sims 1987, Hutton and Williams 1988);

(e)   setting up such studies *de novo* wherever the opportunity arose (for example, on the general surgical waiting list at Guy's Hospital and a nation-wide study of the Qol of patients in end-stage renal failure, in collaboration with EDTA) (Gudex *et al.* 1990, Gudex 1995).

(f)   assisting in the use of the QALY concept in priority–setting at management level in the NHS (for example, for the North West Regional Health Authority) (Gudex, 1986);

(g)   liaising more closely with other British researchers working on global indexes (for example, Buxton, Rosser);

(h)   seeking out other European groups engaged on parallel exercises (Dutch, Finnish, Swedish) later to become the EuroQol Group;

(i)   establishing direct face-to-face contact with the US and Canadian groups working in this same territory, to ensure prompt exchange of ideas;

(j)   devising ways of strengthening and developing the measurement of health benefits generally (Williams 1985b, Williams 1987b, Williams 1988b);

(k)   conducting a pilot study of people's attitudes concerning the extent to which the NHS should discriminate between different sorts of people (for example, old versus young) in the distribution of the benefits of health care (Williams 1988c).

Out of these diverse activities grew the MVH Group at York.

PHASE I – 1987 TO 1990

Against the background described above, in 1987 we proposed a major programme of development work to the then Department of Health and Social Security (DHSS), which would extend over four and a half years and the core of which would be 'a major survey designed to elicit the relative valuations of approximately 2000 respondents, to replace the current Rosser classification and its associated valuation matrix'. However, in order to prepare the ground as thoroughly as possible for this, the first two and a half years of the project (Phase I) was to be devoted to three preliminary tasks:

(a)   to establish whether a Rosser valuation matrix based on the views of a sample of the general population would be similar to or different from the existing matrix based on a convenience sample of 70 respondents;

(b)   to establish whether the descriptors used in the Rosser Classification were the most suitable ones to use in future work and, if not, what descriptive system should be used in their place;

(c)   to establish which of the available valuation/scaling methods would be the best one to use in the major study.

This preliminary work was to proceed on as large a pilot sample as could be afforded and to do this we were supported with core staff financed by the Economic and Social Research Council (ESRC) and fieldwork financed by the Nuffield Provincial Hospitals Trust (NPHT), as well as the DHSS support. It was envisaged that there would be a review of progress about 18 to 24 months into the study which, if favourable, would lead to the release of the funds earmarked for Phase II.

Replicating Rosser

For this task we used Rosser's descriptive system (Table 10.1) and the 'magnitude estimation' valuation method. In its conventional form this involves asking how many times worse each state is than some reference state. Rosser had posed the question in a different way, first asking how many times worse State X was than the reference state and then how much worse state Y than state X was and then Z compared with Y and so on. We used the conventional method, in which the same state is the reference state throughout, so it was not an exact replication. Moreover, it was not possible for each subject to value all 28 states, so they were divided into two subsets, with six states common to both. We got very different valuations from our sample of 140 members of the general public compared with those that Rosser had elicited, as can be seen from Table 10.2.

Finding the best descriptive system

This was a much more elaborate exercise. In order to establish our own baseline data on what the general population regard as the salient features of health, so that we could appraise different approaches to the construction and content of generic (health-related) quality–of–life measures, we conducted a survey (financed by the NPHT) in the west midlands. The aim was to recruit the following:

(a)   a random sample of 200 members of the general public aged 18 and over;

(b)   100 physically disabled young people aged between 18 and 25 years and living at home; with 100 able-bodied young people aged 18 to 25 and living at home, to act as controls;

(c)   100 individuals who have been caring (at home) for physically disabled children who are now young adults; with 100 individuals who have brought up able-bodied children who are now young adults, to act as controls.

The interview schedule was carefully designed so that it was identical for all respondents. The main body of the interview had three phases:

(i)   an unprompted section in which we elicited what individuals thought were the distinguishing features of good or bad health in themselves or in others;

(ii)   a prompted section in which they were presented with 37 statements about health which they were asked to endorse using a series of categories from 'very important' to 'not at all important';

(iii)   a section in which six groups of statements, each representing a particular concept of health, were presented to subjects, who were asked to indicate which of successive pairs of such concepts better represented their own notions of health.

As well as these core data, information was collected from each subject on sociodemographic characteristics, Health Locus of Control, Eysenck Personality Questionnaire and the EuroQol health questionnaire.

The most important general finding from our survey data was that although *initially*, when encouraged to offer unprompted ideas about health or ill health in self or others, the presence or absence of diseases or symptoms played a significant role, later, when offered items to rate in order of importance and, later still, when asked to choose between broad conceptualisations of health, the notion of health as simply *not* being ill faded into insignificance and notions of functional capacity, feelings and general fitness came to predominate.

For our immediate purpose, the data were used to appraise existing (and proposed) instruments for measuring health-related quality–of–life. The unprompted health items were used to see what proportion of the items expressed by the public are actually covered by the different instruments.

With one exception, the instruments included in this comparison were all generic measures designed to yield a single index number, for example, the Rosser Index, EuroQol, Sickness Impack Profile (SIP) and Quality of Well-Being Scale (QWB). The exception is the Nottingham Health Profile (NHP), which was not designed to yield an index though some users have nevertheless converted it from a profile to an index simply by adding together the different items across dimensions. The SIP and QWB are based on US weights (though there is a UK adaptation of the SIP called the Functional Limitations Scale). Rosser has English weights and, at the time, the EuroQol had English, Dutch and Swedish weights. The greater complexity of the NHP, QWB and SIP measures is largely due to their aim to pick up relatively specific variations in health status which may be of particular significance in particular circumstances. The EuroQol measure on the other hand was *not* intended as a 'stand alone' instrument but rather as a comparative tool to be used alongside more specific instruments which were not directly comparable with each other. It therefore had a less complex descriptive system.

Concentrating only on the general population subsample (of 196 respondents) and pooling all their unprompted responses (whether relating to health or ill health, or to self or others), we arrive at the data set out in the left–hand columns of Table 10.3. In the subsequent columns are indicated those items which are included in each of the five descriptive systems that are being compared. From the bottom line it will be seen that *coverage* generally increases as the number of items of information collected increases. It seems that the simple systems provide about 26–36 per cent coverage of the items mentioned spontaneously by our respondents and, the more complex systems, around 50–60 per cent. But about a third of the items mentioned are not relevant for a general health state index designed to appraise clinical interventions or variations in health-related lifestyle. Of the remaining omitted items the most significant is that relating to energy and tiredness (10.5 per cent of all mentions). The omission of this item accounts for most of the difference in coverage between the Rosser and EuroQol instruments on the one hand and the NHP and SIP on the other. Judging by our data, it was a strong candidate for inclusion. An experimental version of the EuroQol Questionnaire, with energy/tiredness included as a sixth dimension, was subsequently tested in a pilot study. The additional dimension was, however, found to have such a small impact on the valuations of the health states concerned that, in the interests of parsimony, it was not included in the standard EuroQol Instrument.

We also tested the *importance* of items as rated by respondents in the prompted section of our interview material. This was summarised as the percentage of the general population rating each item as 'very important' averaged over the items covered by each instrument. There was little to choose between the various instruments on these grounds.

Table 10.3 Coverage of various descriptive instruments

Item	Code No.	All Unprompted Responses (Gen Pop) N	%	Rosser (in 2 Items) Covers	Euroqol (in 5 Items) Covers	NHP (in 5 Items) Covers	Qty of Wellbeing (in 43 Items) Covers	SIP (in 136 items) Covers
Usual Activities	01	118	8.32	*	*	*	*	*
Gen Well/Ill	02	65	4.58				*	
Dying	23	0	0				*	
Gen Not Well/Ill	24	16	1.12				*	
Pain	25	96	6.77		*	*	*	
Named diseases	26	45	3.17				*	
Appetite	27	43	3.03					*
Resp. Symptons	28	37	2.61				*	
Other symptoms	29	41	2.89				*	
(Illn/symptoms)	SUM	343	24.2					
Feelings	03	157	11.0	*	*	*	*	*
Finance	04	6	0.42			*		
Gen Hlth Behvr.	05	8	0.56					
Diet	51	22	1.55					*
Smoking	52	15	1.05					
Exercise	53	22	1.55			*		*
Alcohol	54	11	0.77					
(Behaviours)	SUM	78	5.50					
Energy	06	150	10.5			*		*
Appearance	07	115	8.11					
Stress/coping	08	44	3.10					
Lucky	09	6	0.42					
Med/Doctors	10	34	2.39			*	*	
Mobility	11	107	7.55	*	*	*	*	*
Strength/Rstance	12	54	3.81					
Fitness	13	81	5.71			*		
Sens Impairment	14	12	0.84				*	*
Sleep	15	15	1.05			*	*	*
Weight	16	44	3.10				*	
Enj usual acts	17	3	0.21			*		*
Dependence	18	33	2.32		*	*	*	*
Cogn Impairrmemt	19	13	0.91				*	*
Lonely/helpful	20	4	0.28				*	*
	SUM	1417	100	26.9	35.9	58.07	58.62	49.11

A further consideration borne in mind was the sheer size of the classification system offered by each instrument. If it were very small, then fine discrimination between health states would not be feasible. If it were very large, direct valuation of each health state would become impossible and short-cut methods would have to be adopted to fill the gaps. Of those considered here, Rosser's was the most parsimonious, with only 29 states (excluding dead). The EuroQol system had 244 states (excluding dead), which meant that valuations could be conducted on only a subset of them, the rest (apart from 'unconscious') being estimated by a formula working in the five–dimensional space. The NHP generated a very large number of possible states (more than 10,000!) and the valuation problem was 'solved' by simply attaching a score to each item and using it additively whenever it appears. A similar consideration applied to the rather more complex two-stage system adopted by the QWB scale, which also generated over 100,000 different possible states and dealt with the valuation problems by using simple additive weights for the 40 adjustment factors in the 'symptom–problem complexes'.

Against this background, the task before us was to balance five considerations against each other with respect to each instrument, namely information demands, coverage, importance of items, complexity and scope for full valuation of states. In summary, the situation was as follows:

	Rosser	EuroQol	NHP	QWB	SIP
Information required for classification	2 items	5 items	45 items	43 items	136 items
Coverage	37%	39%	58%	59%	49%
Average importance	54	57	54	55	57
Complexity	29 states	244 states	>10,000 states	>100,000 states	>100,000 states
Valuation strategy	Interactive and complete	Interactive but selective	No overall valuations	Additive and complete	Additive and complete

Both the SIP and the QWB seemed to be too complex for our purposes, the NHP was inappropriate (being a profile measure) and the EuroQol seemed slightly better than Rosser if interpolated valuations derived from a subsample of directly valued states were acceptable.

Choosing a valuation method

The following criteria were employed in deciding which valuation methods would be studied:

1. Use in other relevant studies
2. Methodological importance
3. Efficiency
4. Ease of use in large scale studies
5. Type of method (direct or indirect valuations)
6. Orientation (individual vs. aggregate scales)

The Equivalence Technique was excluded because it introduced an additional element, interpersonal comparisons, which were to be taken up explicitly at a later stage in the research programme. Standard Gamble (SG), Magnitude Estimation (MG), Time Trade-Off (TTO), Pairwise Comparison (PC) and Category Rating (CR) had all been used in a number of relevant studies and all were considered to be methodologically important. It was decided to test two direct methods and two indirect ones: TTO-and ME fell into the former category, PC and CR into the latter. SG was rejected because it was considered to be too time-consuming and because it was concurrently being studied by Rosser's group at the Middlesex Hospital. These four were therefore the methods chosen for use in the pilot study. Furthermore, it was decided that *three* variants of CR would be employed: visual analogue (thermometer)) (CRT), labelled boxes (Likert version) (CRL) and numbered boxes (CRN). The various methods are described in Appendix 10.1. We envisaged that in the main survey we would eventually use one direct method as our 'main' method (for possible use in economic evaluations) and one indirect method (for possible use in other types of evaluation).

Once more we proceeded by conducting a survey, this time of almost 300 subjects in the City of York. We did not manage to achieve a close match to the population of England and Wales but we did achieve a fairly wide spread of characteristics amongst our respondents (apart from ethnicity). The core of each interview was made up *either* of an ME task, *or*, of a TTO task, accompanied by one of the three variants of the CR method, which sometimes preceded the other task and sometimes succeeded it. In addition the same supplementary data was collected as with the lay concepts of health study. By a complex factorial block design the different combinations of tasks were randomised between subjects, between interviewers and between geographical areas. To avoid subjects being overloaded, each valuation task could cover only a subset of all 29 Rosser

Health States but the factorial block design also incorporated a careful distribution of these subsets so that each state was valued by at least 35 people.

We devoted a great deal of attention to identifying 'inconsistent' responses (based on the assumption that in Rosser's classification the disability states get successively worse, as do the distress states). Rather than rejecting inconsistent data as unreliable, we decided to analyse it carefully to see what information it yielded and to retain it if at all possible. We found that the inconsistency rates of individual subjects were particularly high for older people from manual occupations. We also found that they varied by interviewer. Order of presentation of task made no difference. Although it is clear that some subjects (the older people from manual occupations) experience greater difficulties than most people, no clear explanation has been found for the inconsistencies produced in all methods. Judging by the experiences of others who have reported such inconsistencies (but then discarded the inconsistent data) some 'random noise' is to be expected and the problem is how to minimise it rather than how to eliminate it. One limited explanation for some of our inconsistencies is that some of the Rosser descriptors seem difficult to digest or conceptualise (for instance not everyone may regard 'moderate' distress as being worse then 'mild' distress).

Of the two direct scaling methods considered here (ME and TTO), there seemed little to choose between them on grounds of ease of completion. TTO, however, generated fewer inconsistencies. Both were more difficult for respondents than CR methods. Within the three CR methods, CRT generated the most inconsistencies, yet is rated the easiest to complete.

In the middle of this valuation study we obtained copies of the Manual and Standard Gamble Board, devised by the McMaster Group in Canada, which was intended to facilitate the use of the valuation method, which we had initially excluded as being too complex (see Appendix 10.1 for a brief description of this method). We decided to extend our study by inserting the SG method into our design combined with a CR task. Since the resources available to finance extra interviews were severely constrained, we were limited to 72 interviews (that is, only half the number who did TTO and ME) so we decided to present CR before SG and forgo any test of the effect of the order of presentation. As recommended by the McMaster Group, we limited coverage of the Rosser States to eight, which were always presented in the same order. Unfortunately, when we came to process the data we found an error in the manual that had been supplied to us and which we had followed carefully which rendered this supplementary study abortive. We were thus no further forward in judging the relative merits of SG.

To convert the raw data from each of our methods into a comparable valuation matrix required transformation of the data to a scale in which dead = 0 and health ('no disability and no distress') = 1. For the ME, TTO and CRT methods it is possible to make this transformation for each individual separately and then construct a group matrix from the medians of these transformed valuations for each state. For the CRN and CRL methods this transformation can be done only on the medians (or means) of the raw scores. The data were also processed as if ordinal rather than cardinal. The resulting valuation matrices contained a fair number of 'reversals' of logical orderings, some of which were due to the partitioning of the states into small subsets which were valued by different subsamples of the population. Again TTO showed up quite well and CRT badly by this test. Reversals tended to concentrate around certain Rosser states, suggesting that the descriptive system itself may have been partly responsible. But the rank ordering of states was similar for all valuation methods.

Our conclusion at this stage was therefore that ME should be discarded in favour of TTO but, since we had been unable to test TTO against SG, this task remained. CRT proved easier to do than TTO or ME but was much less reliable in the data it generated. It is, however, the preferred method for postal questionnaires in the EuroQol Group and needed to be kept in play for that reason. We therefore proposed to test SG against TTO in the early stages of Phase II and to use the 'winner', plus CRT, in the main valuation study.

PHASE II – THE PILOT STUDIES – 1991 TO 1993

At this point a major shift occurred in the way the MVH Group worked in that, instead of organising and conducting our survey work ourselves, we joined forces with Social and Community Planning and Research (SCPR), a London-based research-orientated survey organisation with a strong interest in attitudinal research. All future survey work was designed jointly between SCPR and the MVH Group but was carried out by SCPR field staff. Amongst the joint responsibilities were the training (and de-briefing) of interviewers and the quality control of data from the fieldwork, which became a central feature of the next stage of the work as we sought to find the best way of getting good quality data from the various valuation methods that were still in play.

The 1992 Pilots

As planned, the early stages of Phase II were concerned with testing TTO against SG as the main valuation method to be used in the main survey. From the literature it was evident that in principle neither method could be regarded as a 'gold standard', each having its advocates and detractors. We therefore decided to base our selection upon more practical considerations, namely:

1. **Logical consistency**: the extent to which the health states used were given a logical ordering *within* each method.
2. **Validity** – *Concurrent*: the extent to which people's valuations corresponded between methods with each subject. *Discriminant*: the extent to which valuations differed (in accordance with prior expectations) by respondent characteristics.
3. **Test–retest reliability**: the extent to which respondents' responses were stable *within* each method over a relatively short time interval.
4. **Completeness**: the extent to which each method produced a complete data set.

There were some further criteria which, although not sufficiently important to be used in choosing between methods, nevertheless offered additional evidence on the performance of the methods. They mostly concern the burden placed upon subjects and interviewers and they needed to be taken into account when designing the main survey. They were:

(a) the time taken to complete each task;
(b) the difficulty of each task as reported by both respondents and interviewers;
(c) respondents' willingness to be re–interviewed.

The principal research objective was to achieve *within–subject* comparisons for the two main methods under review. Each method was tested in two variants, one of which used specially designed boards and cards as an aid to decision-making by respondents (Props), and the other used a self-completed booklet (No Props).   Since we felt that the average length of an interview should be about one hour, we abandoned any attempt to generate enough valuations in this pilot survey to make it possible to estimate the valuation space generally, so we used only six health states (apart from healthy and dead).

The target population were adults aged 18 and over in the general population, with no upper age limit. A random sample of 700 addresses was drawn from 11 regional areas in the UK using the postcode address

File. The fieldwork was carried out between March and May 1992. Of the 525 'in scope' addresses, 190 (36%) yielded refusals and 335 (64%) yielded an interview. A subsample of those who had said they would be willing to be re–interviewed were approached again four to twelve weeks after the original interview. Respondents were asked to do exactly the same tasks as before, with the additional question of whether anything important had happened to them since the last interview.

Table 10.4 The Euroqol classification system

Mobility
1.   No problems walking about
2.   Some problems walking about
3.   Confined to bed

Self-care
1.   No problems with self-care
2.   Some problems washing or dressing self
3.   Unasable to wash or dress self

Usual activities
1.   No problems with performing usual activities (for example, work, study, housework, family or leisure activities)
2.   Some problems with performing usual activities
3.   Unable to perform usual activities

Pain/discomfort
1.   No pain or discomfort
2.   Moderate pain or discomfort
3.   Extreme pain or discomfort

Anxiety/depression
1.   Not anxious or depressed
2.   Moderately anxious or depressed
3.   Extremely anxious or depressed

Note   For convenience each composite health state has a five digit code number relating to the relevant level of each dimension, with the dimensions always listed in the order given above. Thus 11223 means:

1   No problems walking about
1   No problems with self-care
2   Some problems with performing usual activities
2   Moderate pain or discomfort
3   Extremely anxious or depressed

Each interview followed the same pattern: description and rating of own health state (using the EuroQol Classification as in Table 10.4); ranking of health states; category rating (using the thermometer in Figure 10.1 as a Visual Analogue Scale or VAS); SG followed by TTO (or vice versa); sociodemographic background date. The questionnaire had an additional section requiring interviewer feedback. Compared to the general population, there were more people in the survey population with no children at home and more with a degree or professional qualification. There were also slightly fewer people aged under 20, slightly more aged over 60 and fewer in paid work (20 per cent were retired). This seems to indicate that there may be some response bias in favour of the more educated and that people with children at home are less willing to undertake a rather time-consuming interview. However, there seemed to be no difficulty in eliciting the co-operation of older people (though they had greater difficulty with some of the tasks). On average the time taken for an interview was just over an hour but the time taken at retest was, on average, shorter. Of the 14 respondents with incomplete interviews, 71 per cent were aged 61 or over and none was in paid work. These were important pointers to problems that we might face in the main survey.

Turning to the criteria for choice set out above, our findings were as follows:

Logical consistency: there were no significant differences between the methods but TTO Props performed slightly better than the others and seemed more robust to characteristics such as age and education level.

Validity: TTO Props avoided most of the more extreme valuations and it also identified some of the significant differences related to respondents' own health states, but these observations are rather tentative in view of the difficulty in establishing 'correct' values.

Test–retest reliability: the median values for all states were extremely reliable for all methods but at individual level valuations were most consistent with TTO Props.

Completeness: TTO Props was the best of the four main methods.

Figure 10.1 Visual analogue scale

Best
imaginable
health state

To help people say how good or bad a health state is, we have drawn a scale (rather like a thermometer) on which the best state you can imagine is marked by 100 and the worst state you can imagine is marked by 0.

We would like you to indicate on this scale how good or bad is your own health today, in your opinion. Please do this by drawing a line from the box below to whichever point on the scale indicate how good or bad your current health state is.

Your own health state today

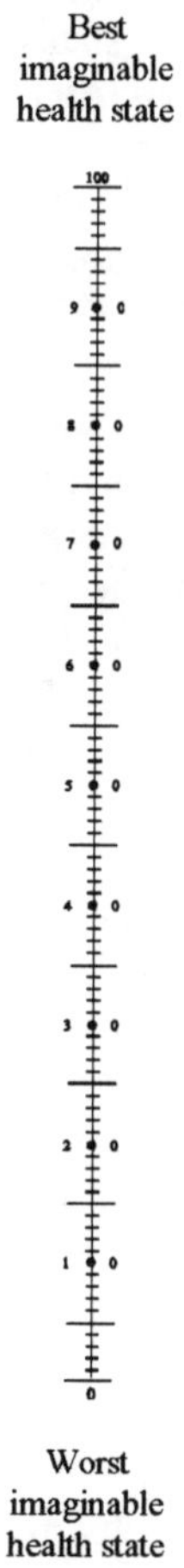

Worst
imaginable
health state

Thus, although there was not a lot in it, as regards the *quality* of the data, there was an accumulation of evidence in favour of TTO Props, so this was chosen as the best valuation method to use in the main survey.

Whichever method had been chosen, there would remain some consequential problems to be considered, mainly relating to the conduct of the interview itself. The 'Props' variants took four or five minutes longer on average than their corresponding 'No Props' variants. This may be because a larger percentage of states were considered to be worse than death in the 'Props' variants. In addition, more states were considered to be worse than death on TTO than SG. From this it will be noted that the method we chose is the one that took most time. This indicated a need to explore further ways of streamlining the presentation of the TTO Props method in the interview situation. TTO Props was not the easiest of tasks to understand from the respondents' point of view, although it was not the hardest. Interviewers considered TTO Props to be the most easily understood of the major tasks.

The 1993 Pilots

The number of states that could be valued within a single interview was a key variable in determining the required sample size for the main survey. So, although we had now chosen our preferred valuation method, there remained two outstanding problems that needed to be resolved before embarking upon it. The first of these was to work out with the interviewers ways of making the TTO Props method as user–friendly as possible, from both the interviewers' and the respondents' perspective. The second was to discover, in the light of this, the maximum number of states that could be valued by a respondent, within the context of a one–hour interview in which the TTO method would be preceded by a ranking and a rating exercise. These tasks occupied us for most of 1993.

To pursue the first task, a brainstorming session was held between the MVH Group, two senior members of the SCPR staff and six of the interviewers who had used the TTO Props version in the 1992 Pilot Study. This led to various suggestions for simplifying the choice process in the TTO method. These modifications were then tested in the field in the first pilot. This pilot also tested the bisection method for use with CRT scaling exercise (Appendix 10.1). The purpose of this 'bisection' process is to ensure that the resulting valuations have interval scale properties (Stevens 1971). Here, however, we were simply testing its feasibility. Finally, the number of states used by each respondent was raised to 15 to see whether this number could be handled under the new procedure.

The outcome was that the revised procedures worked well and appeared to be understood by the interviewers. The bisection approach also worked well in practical terms. Most respondents were able to evaluate the full set of states but interviewers felt that 15 states should be regarded as an absolute upper limit. Interviews took an hour on average, which was the target. One new problem arose with the TTO method, however, which was that at the mild end of the spectrum quite a few respondents were unwilling to give up any time to improve a health state, so it was decided to allow some extra 'fine tuning' at this end of the scale.

The second pilot in 1993 was designed as a 'full dress rehearsal' for the main survey. One of the purposes here was to test the procedures as stringently as possible by training a completely new batch of interviewers and exercising rigorous quality control over the data they generated. It transpired that the key aspect of the interviewer briefing was the conduct of practice interviews under the guidance of experienced staff who acted as dummy respondents. This was the system eventually adopted for the main survey.

Sample size

The final matter that had to be determined before embarking on the main survey was the sample size. In the TTO method the smallest difference that it would generally be possible to express would be .025 (3 months out of 10 years). To detect such a difference at the 5 per cent significance level with 80 per cent power would require 3235 valuations for each state. Although some states would be valued by all respondents, 36 states would be valued only by 25 per cent of respondents, so about 13,000 interviews would be required. In the end we settled for a sample size of 3235, which meant that we would have only about 800 valuations for most of the 45 states to be included in the survey. On that basis we expected to be able to detect a 0.1 difference in valuations between subgroups at the 5 per cent level of significance.

The Main Survey – Design and Execution – 1993

The objective of the main survey was to elicit the views of a representative sample of the non-institutionalised adult population of England, Scotland and Wales by interviewing them in their own homes. Broad geographical coverage was required in case it emerged that there were marked regional differences in valuations.

The EuroQol Classification generates 245 theoretically possible health states, some of which are unlikely to occur in practice. Respondents cannot handle more than 15 each and about 40 are required for modelling purposes (that is, to estimate valuations for the states that are not directly valued). Valuations for two of the states ('unconscious' and 'dead') cannot be estimated from the valuations given to any other state and so must be directly valued. The state 11111 ('healthy') is essential to the rescaling of the VAS (thermometer) data and so must also be directly valued by all respondents. For all other states we had discretion. In exercising that discretion we had several considerations in mind. Firstly, we wanted the states to be widely spread over the valuation space in terms of mildness or severity (as indicated from earlier valuation data). Secondly, we wanted the set of states to include all plausible combinations of 'levels' across each of the five dimensions, so as to be able to test for significant interaction effects (for example, to test whether the weight given to 'moderate pain or discomfort' is different if it is combined with 'some difficulty in walking' from what it is when combined with 'moderately anxious or depressed'). Thirdly, we wanted to stay as close as possible to the selection of states that had been used by Finnish EuroQol colleagues in a major postal survey which they had just conducted. Fourthly, we wanted to exclude states which seemed prima facie implausible to respondents, so as to sustain motivation and credibility. The result of applying these criteria was the selection of states shown in Table 10.5. The reason why the states in the table are stratified as they are is that, apart from unconscious and dead (for which we had to have valuations from everyone), we wanted the two 'reference' states (11111 and 33333) to act as a common frame of reference for all respondents. We also wanted each individual to have in their valuation set two of the five mildest states (11112, 11121, 11211, 12111 and 21111). Amongst the remaining 36 states we needed to ensure balance at individual level between the relatively 'mild', 'moderate' and 'severe' states. Thus 3 out of each group of 12 states were randomly selected within this stratification system for each individual respondent.

Table 10.5 Euroqol states valued in the main survey

A.   Each respondent valued all four of these key states:

    11111
    33333
    unconscious
    immediate death

B.   Each respondent also valued 2 of the following (very mild) states (selected at random):

    11112
    11121
    11211
    12111
    21111

C.   Each respondent valued 3 randomly selected states from Set 1, which are the mildest ones. Each respondent also valued 3 randomly selected states from Set 2 (the moderately valued ones) and 3 randomly selected states from Set 3 (the most severe ones).

SET 1	SET 2	SET 3
12211	13212	33232
11133	32331	23232
22121	13311	23321
12121	22122	13332
22112	12222	22233
11122	21323	22323
11312	32211	32223
21312	12223	32232
21222	22331	33321
21133	21232	33323
11113	32313	23313
11131	22222	33212

The core of each interview contained five elements:

Self-reported health
Ranking of states
VAS (Thermometer) rating of states
TTO rating of states
Personal background data

In order to test the reliability of the three valuation methods, a representative subsample of approximately 200 respondents was to be interviewed approximately three months after the original interview. Respondents were therefore also asked whether they would be willing to be reinterviewed at a later date.

A great deal of emphasis was placed on interviewer training. All interviewers attended personal briefings (held in Birmingham, London, Manchester and Newcastle) which involved intensive training in the three valuation methods. Any interviewers who appeared to be having problems with their first few interviews were asked to attend a half-day rebriefing session before they were permitted to carry on with their assignment (13 were so recalled).

The main fieldwork was conducted between August and November 1993 and the reinterviews during December 1993.

The Main Survey – Results – 1994

Of the 6080 addresses selected for sampling 706 (12 per cent) were found to be 'out of scope' of the survey, being non-residential, empty/derelict, untraceable, or even not yet built. Of the remaining 5324 addresses, 3395 interviews were achieved, giving a response rate of 64 per cent on in–scope addresses. After the survey data had been weighted to correct for the effect of varying household size on selection probabilities, the sample was found to have nearly identical characteristics to the general population.

There were few *missing data* from 3395 respondents and logical consistency within method was also surprisingly high, with an average of 97.5 per cent on the VAS and 93.8 per cent on the TTO. Four separate data sets were assembled: a ranking data set, a VAS data set, a TTO data set and a combined VAS and TTO data set. Some respondents have been excluded from each set on the grounds of missing data and logical inconsistency but, despite stringent criteria, the numbers were extraordinarily small: 107 (3.2 per cent) from the VAS data set, 58 (1.7 per cent) from the TTO data set and 398 (11.7 per cent) from the combined data set. Although the excluded respondents have tended to be those older than 60 years and with no educational qualifications, the respondents remaining in each data set are still representative samples of the general population. The entire data set has been deposited with the ESRC Survey Research Archive.

As an incidental by-product of our survey we assembled data on the self-reported health of a representative sample of the non–institutionalised adult population of the UK. Some illustrative examples of our findings are given in Figures 10.2 and 10.3. For example, we found 33 per cent of respondents reporting pain or discomfort and 21 per cent reporting anxiety or depression. Health problems generally increase with age and, within every age group,. By social class too. These data provide evidence for the validity of the EuroQol instrument as a measure of health-related quality–of–life.

Each respondent rated 15 health states on the VAS. In order to compare scores from different respondents, these 'raw' scores were adjusted relative to two states that all respondents rated: the state 11111 (full health) (set equal to 1), and death (set equal to 0). Both median and mean scores were logically consistent and median scores were all positive (that is, every state was rated as better than being dead by a majority of respondents). Significantly higher median scores were given by the lower social classes and by the less educated, meaning that they do not think poor health states are as bad as the others do.

The set of valuations emerging from the TTO task contained no logical inconsistencies but far more states were rated worse than being dead than in the VAS valuations. There were some significant differences in valuations between men and women and also according to marital status and employment status. The background factor which had the most

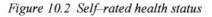

Figure 10.2 Self–rated health status

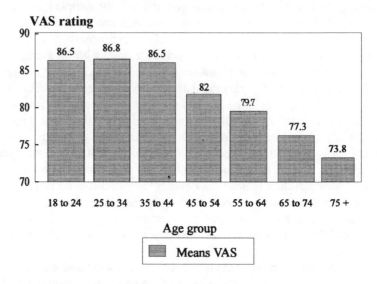

*Figure 10.3 Frequency of reported problem
distribution by age group*

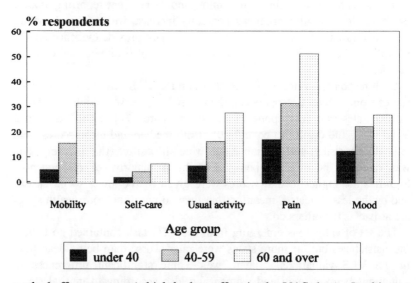

marked effect was age (which had no effect in the VAS data). Looking at
the age effect more closely it appears, however, that respondents over the
age of 60 gave significantly lower values to the more severe states than the

rest of the population. One possible explanation is that as people's life expectancy shortens, they see less reason to tolerate suffering during their remaining years. An alternative explanation might be that it is an artefact of the TTO method. If respondents do not believe that they have 10 years' life expectancy, they might willingly give up these 'excess' years, thereby depressing the apparent value attached to the more severe states. This puzzle we come back to later.

The relationship between the VAS valuations and the TTO valuations did not appear to be the power relationship found in earlier studies (see, for instance, Torrance 1976a and b), but a 'spreading' relationship, in which the TTO valuations are more extreme than the VAS ones at both ends of the valuation spectrum but especially with respect to the more severe states. This means that people are relatively unwilling to sacrifice life expectancy to improve mild states but relatively more willing to sacrifice it to avoid severe states, where 'mild' and 'severe' relate to their VAS ratings.

At retest all three methods proved very reliable at both group and individual levels and the already low inconsistency rates declined to still lower levels.

The Main Survey – Modelling the Tariff – 1994

At this stage in the proceedings we had valuation data on 45 states, from which we needed to interpolate values for the remaining 200 states in the EuroQol Classification. One of these states, 'unconscious', lies outside the five dimensional scheme and has been directly valued by all respondents. The states 11111 'healthy' and 'dead' act as calibration points in the valuation scale, so it is the valuations given to the other 42 that constitute the core data set. The data set used for all the modelling activity was that which contained the valuations of the 2997 respondents for whom we had complete data over both the VAS and TTO valuations (that is, the 'combined VAS and TTO data set' referred to earlier).

In estimating valuations for those EuroQol states on which we did not have direct valuations, we enlisted the collaboration of those members of the EuroQol Group with a special interest in modelling and also obtained the services of Mona Abdalla and Ian Russell to act as external statistical consultants during this phase of the work. Each of the four participating groups was given access to our data set and invited to enter a sort of competition to come up with the 'best' model. The following criteria were used to help us choose the 'best' model:

1. *goodness–of–fit* that is, how well the model explains the differences in the valuations given to those states on which there is direct data;

2. *parsimony* that is, the simplicity of the model;
3. *consistency* that is, states that are logically worse *must* have lower predicted values;
4. *transparency* that is, the ease with which non–experts can understand the manipulations made.

Another member of the EuroQol Group, who was not in the 'modelling competition', provided an independent critique of the various approaches when the results of all these different analyses were presented at the Plenary Meeting of the EuroQol Group, held in London in October 1994. At this meeting, it was decided that the model presented by Paul Dolan (a member of the MVH Group) satisfied the above criteria most fully. Before discussing this model in more detail, it is encouraging to note that the results presented by Abdalla and Russell using a different technique corresponded closely to the Dolan model.

Figure 10.4 Structure of the analytical work

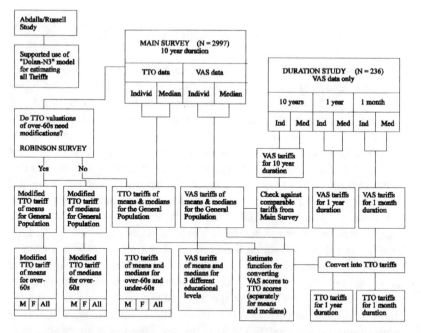

Essentially the preferred model (known as 'Dolan–N3') predicts the value of a health state from its components by attaching a value to each separate deviation from good health. In the EuroQol system there are 10 such 'decrements' in health, made up of a moderate and a severe level of dysfunction for each of the five dimensions (mobility, self-care, usual activities, pain/discomfort, anxiety/depression). The model contains two other terms, one of which (N3) is active whenever any of the five dimensions of health is 'severe' (that is, at level 3), while the other is simply a constant term (which might be interpreted as the loss of value involved with being in any kind of dysfunctional state whatever). This approach has been used by the MVH Group for the estimation of all its 'tariffs' of social values, so that they are all based on a common analytical approach. The data used for each tariff is different and there are some important matters of interpretation and use which need careful consideration. It is to these that we now turn.

The structure of the analytical work surrounding this modelling work is shown schematically in Figure 10.4. At top centre is the modelling data set, containing both VAS and TTO valuations from each respondent for the 15 states in that person's set. These data can be used in two different ways: *either* treating each individual as a separate observation, *or* taking the median values for each state, thus simplifying the situation but losing information. We have used the first approach when generating tariffs of mean values and the second approach when generating tariffs of median values. The main modelling activity is represented straight down the middle of the chart, in the generation of 'tariffs' of social values for all the EuroQol states, representing the mean or median values of the general population, as elicited by the VAS method or by the TTO method. The other items on the chart will be described later.

Our basis tariff, when a weighting system is required for use in an economic evaluation, is the one of *mean values based on the individual TTO scores*. The reason for taking this as the base case is that the use of individual scores retains the maximum amount of data and the use of TTO scores means using a valuation method which involves trade-offs which, in an economic context, is more appropriate than a simple rating scale. There remains the choice between means and medians. The median value for a state is the value given by the person in the middle of the distribution, so it is insensitive to the particular valuations provided by people at the extremes of the distribution. Many people prefer to use medians as the measure of central tendency when a distribution is strongly skewed (as these distributions are). By contrast, means give every respondent some weight, but are sensitive to 'outliers'. In this case the 'outliers' fell into two groups. There were those who rated the 'mild' states as exceptionally good, resulting in means which exceeded medians at the upper end of the scale. There were also those who rated the severe states as exceptionally bad, hence the mean values for the severe states tended to be much lower than the medians. The Dolan–N3 model is shown in Table 10.6. The basic tariff is given in Table 10.7.

Not everyone may wish to use this basic tariff, though we recommend for comparative purposes that, even if another is preferred for a specific purpose, the basic one is used too. The possible reasons for preferring a different tariff are many and varied and we have tried to cater for as many of them as our data permits. Thus, although not shown here, we have tariffs for medians as well as means, for VAS as well as TTO, VAS tariffs for different educational level and TTO tariffs for different age groups (and for each sex within age groups). All this is indicated at the bottom of Figure 10.7.

Table 10.6 A tariff of values from the general population
(for health states of 10 years duration)
Means based on time–trade–off valuations

Dimension		Coefficient
Constant		0.081
Mobility		
	level 2	0.069
	level 3	0.314
Self-care		
	level 2	0.104
	level 3	0.214
Usual activity		
	level 2	0.036
	level 3	0.386
Pain/discomfort		
	level 2	0.123
	level 3	0.386
Anxiety/depression		
	level 2	0.071
	level 3	0.236
N3		0.269

Note: Unconscious = – 0.402

The arithmetic needed to calculate the value for any health state from this table of decrements is given by the following example:

Taking health state 1 1 2 2 3

Full health = 1.0

Constant term (for *any* dysfunctional state)		(minus 0.081)
Mobility	level 1	(minus 0)
Self-care	level 1	(minus 0)
Usual activity	level 2	(minus 0.036)
Pain/discomfort	level 2	(minus 0.123)
Anxiety/depression	level 3	(minus 0.236)
Level 3 occurs within at least 1 dimension		(minus 0.269)

Hence **the estimated value for state 1 1 2 2 3 =**
1.0 – 0.081 – 0.036 – 0.123 – 0.236 – 0.269 = **.255**

Note   Some severe states will have negative values if they last 10 years, indicating that the general public regards such a prospect as worse than being dead.

Figure 10.7   TTO tariff of means: whole population – 10–year duration

	Level 2	Level 3
Mobility	0.069	0.314
Self-care	0.104	0.214
Usual activity	0.036	0.094
Pain/discomfort	0.123	0.386
Anxiety/depression	0.071	0.236
Constant = 0.081		N3 = 0.269

11111	1.000	11311	0.556	12211	0.779
11112	0.848	11312	0.485	12212	0.708
11113	0.414	11313	0.320	12213	0.274
11121	0.796	11321	0.433	12221	0.656
11122	0.725	11322	0.362	12222	0.585
11123	0.291	11323	0.197	12223	0.151
11131	0.264	11331	0.170	12231	0.124
11132	0.193	11332	0.099	12232	0.053
11133	0.028	11333	– 0.066	12233	– 0.112
11211	0.883	12111	0.815	12311	0.452
11212	0.812	12112	0.744	12312	0.381
11213	0.378	12113	0.310	12313	0.216
11221	0.760	12121	0.692	12321	0.329
11222	0.689	12122	0.621	12322	0.258
11223	0.255	12123	0.187	12323	0.093
11231	0.228	12131	0.160	12331	0.066
11232	0.157	12132	0.089	12332	– 0.005
11233	– 0.008	12133	0.076	12333	0.170

13111	0.436	21231	0.159	23121	0.244
13112	0.365	21232	0.088	23122	0.173
13113	0.200	21233	− 0.077	23123	0.008
13121	0.313	21311	0.487	23131	− 0.019
13122	0.242	21312	0.426	23132	− 0.090
13123	0.077	21313	0.251	23133	− 0.255
13131	0.050	21321	0.364	23211	0.331
13132	− 0.021	21322	0.293	23212	0.260
13133	− 0.186	21323	0.128	23213	0.095
13211	0.400	21331	0.101	23221	0.208
13212	0.329	21332	0.030	23222	0.137
13213	0.164	21333	− 0.135	23223	− 0.028
13221	0.277	22111	0.746	23231	− 0.055
13222	0.206	22112	0.675	23232	− 0.126
13223	0.041	22113	0.241	23233	− 0.291
13231	0.014	22121	0.623	23311	0.273
13232	− 0.057	22122	0.552	23312	0.202
13233	− 0.222	22123	0.118	23313	0.037
13311	0.342	22131	0.091	23321	0.150
13312	0.271	22132	0.020	23322	0.079
13313	0.106	22133	− 0.145	23323	− 0.086
13321	0.219	22211	0.710	23331	− 0.113
13322	0.148	22212	0.639	23332	− 0.184
13323	− 0.017	22213	0.205	23333	− 0.349
13331	− 0.044	22221	0.587	31111	0.336
13332	− 0.115	22222	0.516	31112	0.265
13333	− 0.280	22223	0.082	31113	0.100
21111	0.850	22231	0.055	31121	0.213
21112	0.779	22232	− 0.016	31122	0.142
21113	0.345	22233	− 0.181	31123	− 0.023
21121	0.727	22311	0.383	31131	− 0.050
21122	0.656	22312	0.312	31132	− 0.121
21123	0.222	22313	0.147	31133	− 0.286
21131	0.195	22321	0.260	31211	0.300
21132	0.124	22322	0.189	31212	0.229
21133	− 0.041	22323	0.024	31213	0.064
21211	0.814	22331	− 0.003	31221	0.177
21212	0.743	22332	− 0.074	31222	0.106
21213	0.309	22333	− 0.239	31223	− 0.059
21221	0.691	23111	− 0.367	31231	− 0.086
21222	0.620	23112	0.296	31232	− 0.157
21223	0.186	23113	0.131	31233	− 0.322

31311	0.242	32323	− 0.221
31312	0.171	32331	− 0.248
31313	0.006	32332	− 0.319
31321	0.119	32333	− 0.484
31322	0.048	33111	0.122
31323	− 0.117	33112	0.051
31331	− 0.144	33113	− 0.114
31332	− 0.214	33121	− 0.001
31333	− 0.380	33122	− 0.072
32111	0.232	33123	− 0.237
32112	0.161	33131	− 0.264
32113	− 0.004	33132	− 0.335
32121	0.109	33133	− 0.500
32122	0.038	33211	0.086
32123	− 0.127	33212	0.015
32131	− 0.154	33213	− 0.150
32132	− 0.225	33221	− 0.037
32133	− 0.390	33222	− 0.108
32211	0.196	33223	− 0.273
32212	0.125	33231	− 0.300
32213	− 0.040	33232	− 0.371
32221	0.073	33233	− 0.536
32222	0.002	33311	0.028
32223	− 0.163	33312	− 0.043
32231	− 0.190	33313	− 0.208
32232	− 0.261	33321	− 0.095
32233	− 0.426	33322	− 0.166
32311	0.138	33323	− 0.331
32312	0.067	33331	− 0.359
32313	− 0.098	33332	− 0.429
32321	0.015	33333	− 0.594
32322	− 0.056	Unconscious	[0.402]

OTHER MATTERS

The valuations of the elderly

Figure 10.4 also indicates some other matters that have required our attention. As mentioned earlier, when using the TTO method the older members of the population rated the more severe health states as very much worse than did younger people and we puzzled over this, wondering whether it was genuine or an artefact of the method (since we did not observe this from the same people when they were using the VAS method). Fortunately we had the services of Angela Robins (of Newcastle University) during the summer of 1994 to reinterview a sample of our respondents in the north east of England and get them to talk their way through the valuation process as they were doing it, to see whether this yielded any clues as to why their valuations were as they were. As far as possible her interviews followed the protocol originally used in the main study but, since the nature of her study was qualitative, rather than quantitative, respondents were asked to value only seven states (as opposed to 15 in the main study). Respondents were asked to 'think aloud' as they completed the interview and to explain *why* they made certain decisions during the TTO exercise. The findings with respect to the TTO valuations from the elderly can be summarised as follows:

(a)   no evidence was found to support the view that variation in values was *primarily* an artefact of the TTO method;

(b)   evidence was found of a 'threshold of tolerability', below which states would have to fall before some respondents were prepared to give up even a few days, let alone months or years of life, to get out of them;

(c)   no convincing evidence was found that the elderly were more concerned than younger respondents about becoming a burden to their families;

(d)   older respondents appeared genuinely more likely than younger respondents to consider severe states as worse than death.

It appears that the key artefactual element is that older people do not believe that after being in a very severe state for any length of time they will in fact recover full health, whereas younger people do believe this. So the time they are going to gain is thought of as being time in poor health not, as required by the method, time in good health.

Since it appears that the low valuations from the elderly are partly an artefact and partly genuine, we explored the possibility of modifying the TTO tariffs for the elderly (and the TTO tariffs for the general population) to eliminate the artefactual element. This was possible because there is nothing in the TTO method which would lead older people not to indicate accurately whether they considered a state to be better or worse than being dead. The problem arises when they come to attach a value to the states that are worse than dead. If at that stage we assumed that their valuations are the same as anyone else who rated that state worse than dead, we would have an estimate of the *maximum* effect that could be attributable to the artefactual element. This would still leave older people valuing the more severe states lower than younger people, because they are more likely to rate such states worse than dead. We have calculated such modified tariffs should anyone wish to use them.

The duration of health states

Another complicating factor is that the valuations were derived for states lasting 10 years, which means that the tariffs are most appropriate for chronic conditions, or for 'before' and 'after' comparisons when patients' health states have stabilised once more. It is reasonable to suppose that a severe condition lasting only a short time would be more tolerable than if it lasted a long time. To test this we conducted a supplementary survey of 312 subjects from the Main Survey, with a 76 per cent response rate. Only the VAS method was used and the supposed durations of each state were taken to be 10 years (to maintain comparability with the main survey), one year and one month. For 38 of the 43 states valued, the value attached to it when it lasted one month was significantly higher than when it lasted for 10 years but there is much less difference between durations of one year and one month. When modelled, the key element causing these changes seems to be the 'N3' element in the formula (see Table 10.6), indicating that it is the presence of an extreme level of *any* dimension which makes a state particularly intolerable if it persists for any length of time.

Converting VAS scores into TTO scores

Scaling methods which are utility–based, such as TTO and SG, tend to be resource–intensive in that they are often interviewer–based and may require special aids to present descriptions of health states and to enable respondents to record their valuations of them. Such methods, however, are favoured by many researchers who consider them to be well-grounded in theory, or who demand that expressions of preference must involve an element of choice. Simpler, less demanding methods have been utilised and foremost amongst these has been category rating, of which the VAS is one.

Given that different scaling methods applied to the same health states tend to yield different valuations, a question arises as to the form of any relationship between these values. Quite apart from this methodological interest, there is a strong practical reason for considering this question. If the results obtained using a 'simple', technically accessible, method could be systematically related to those obtained using a more 'complex' method, then the former could be deployed when resources precluded the use of interviewer–based methods.

For the purposes of this study it was considered to investigate only the form of any relationship that linked the *estimated* values for any health state. Hence two general sets of data exist – estimated scores for 243 health states produced using the standard models applied to the individual and aggregate (median) TTO data and an equivalent set of scores based on the VAS models. The general problem to be investigated amounted to seeking an arithmetic process by which TTO values (based on models of either individual responses or median values) for all health states, could be estimated, given knowledge of the corresponding VAS rating. The resulting equation is shown in Table 10.7, together with the coefficients for an estimate based on medians and for an estimate based on means.

Table 10.7 Converting VAS scores into TTO scores

Various functional forms were used to examine the nature of the relationship between the VAS and TTO valuations. A general equation of the following form resulted from this study

$$TTO_i = a_o + a_1 . VAS_i + a_2 . VAS_i^2$$

where
VAS_i is the 'observed' VAS score for health state I
TTO_i is the predicted TTO value for health state I
a_o a_1 a_2 are coefficients with different values assigned when individual–level or median VAS data are modelled.

The values of coefficients for the two forms of equation is as follows:

	Estimated values based on means	Estimated values based on medians
a_o	– 0.445	– 0.704
a_1	2.112	3.313
a_2	– 0.580	– 1.604
r^2	0.99	0.98

This conversion method could be used to convert our duration–specific VAS valuations into duration–specific TTO values, though the chain of reasoning to support such a process of recalculation is somewhat tenuous. It requires two key assumptions: firstly that the same relationship exists between TTO valuations for states of different duration as exists between VAS valuations and, secondly, that the relationship that exists between VAS 10 year valuations and TTO 10 year valuations is similar to that between the two methods for each of the other durations. Unfortunately we have no data shedding any light on the reasonableness of those assumptions because we were unable to devise a feasible method for using the TTO method for very short durations. We have, nevertheless, made the conversion in order to fill an important void until such time as someone can generate the extra data required to do the job more directly. Meanwhile we have estimated TTO tariffs for states of one year duration and one month duration.

UNFINISHED BUSINESS

There is plenty of work still to be done on the measurement and valuation of health, even in the particular domain of generic indexes of health-related quality–of–life, which is only one part of this burgeoning field. For instance, one difficult technical problem that needs investigation is the extent to which the value given to a health state is influenced by the health state that precedes it, or the health state that is expected to succeed it.

A very important policy issue concerns variations in values between different subgroups in the population. We have explored these with respect to the sociodemographic characteristics of our study population but a more important issue is whether doctors and nurses have significantly different valuations from the general public. If they do, it would raise serious doubts about the appropriateness of professionally defined measures of benefit from health care and about the basis of clinical priority–setting. In conjunction with SCPR, the MVH Group has worked up a detailed protocol which involves the replication of our main survey on 1000 doctors and nurses chosen to be representative of those currently working in the NHS.

One issue that was on our original research agenda (but which has been set aside to enable the main flow of work to proceed), is whether the value attached to a health benefit depends on who is to receive it. The general assumption underlying all measurement of health-related quality–of–life is that it is the nature of the change in the individual's situation that is the focus of interest, not who the individual is. This position has a strong ethical justification, as well as being a convenient simplifying assumption for research purposes. It enables simple aggregation of results to proceed untrammelled, a feature of all such measures in practice and, indeed, a feature of much cruder measures such as survival or mortality rates. But many people think that, in some circumstances at least, priority should be given to the young over the old, or the parents of young children over their childless contemporaries. And there are even more contentious issues often raised here, concerning those who have cared for their own health and those who have not (for example, by smoking, heavy drinking, or drug abuse). We need to know a lot more about the attitudes of the general public towards these matters, so that policy-making can be better informed. Methodological work to improve on existing survey work remains on our research agenda.

But the main future activity of the MVH Group is going to be the implementation of the benefit measures we have already generated which requires the production of user-friendly documentation and instruction manuals which are different from those designed for methodological work by the research community. It also requires a support facility to ensure that the instruments are used appropriately and their full potential exploited. This is our next challenge.

APPENDIX 10.1

A brief description of the valuation methods

Magnitude estimation

It was a variant of this technique that was used by Rosser in her original work as a means for obtaining direct valuations of health states. Subjects are asked to judge each health state (H_i) in terms of its perceived severity compared with the reference state of 'no disability, no distress', which was assigned a value of 1. Subjects were told that each health state would last for a period of 20 years after which time they would die. They were then asked whether each state was better or worse than the reference state (1A) and then how many times better or worse. The utility values for each state are calculated using the following formula:

(The value attached to the state H_i minus the value for 'dead')
(1 minus the value for 'dead')

A modified form of magnitude estimation was used in the scaling of the Rosser Index.

Time trade–off

When the Time Trade–Off method is applied to states considered by the subject to be better than death, subjects are asked to make a decision between two alternatives: either to remain in health state (H_i) for a period of time (t = 20 years) followed by death, or to be healthy for a shorter period of time (x) followed by death. The duration of x is varied until the subject is indifferent between the two alternatives and the utility value of the individual's preference for health state (H_i) is given by the rations x : 20. For the purposes of this study we simply observed the numbers of states considered by each subject to be worse than dead under the TTO method, but did not generate any actual ratings for such states. Time trade-offs has been used to examine valuations for health states in the general population of Canada and was being used in the UK by the Brunel Group.

Category rating: three variants

This is an indirect method of obtaining health state preferences in which ordinal data are generated. Three variants were used in the present study: (i) a graphical visual analogue scale in the form of a thermometer where 100 represented 'best imaginable health state' and 0 represented 'worse imaginable health state' (CRT); (ii) a column of numbered boxes in which the box numbered 1 represents the 'worst imaginable health state' and the box numbered 9 represented 'best imaginable health state', (CRN); and (iii) an aversion which used nine labelled boxes with the following descriptions opposite each respective box: best imaginable health state; very good; good; fairly good; neither good nor bad; fairly bad; bad; very bad; worst imaginable health state (CRL).

Respondents were asked to rate each health state according to the categories shown. In order to make comparisons with the direct methods the values generated *using the thermometer version* (CRT) were transformed to a 0–1 scale where 0 represented death and 1 represented 'no disability, no distress' using the following formula:

In the later stages of our survey work, when using the CRT (or 'VAS') approach we adopted the 'bisection' procedure. In this, after first rating the best and worst states on the 'thermometer', respondents are then asked to select the state which came closest to being half-way on the scale between where they had rated the best and where they had rated the worst state. After rating this state wherever they thought it should go, the process is repeated for the state which falls roughly halfway between the middle state and the best state and then for the state which falls roughly halfway between the middle state and the worst state.

Pairwise comparison

With this method individuals are asked to make judgements about pairs of health states by indicating which of the two states is worse, or whether the two states are considered equal in severity. The method enables measures of internal consistency to be calculated and it can be used to assess the quality of each respondent's performance as well as the extent of agreement between individuals. Paired comparisons methods have been used in deriving valuations for the Nottingham Health Profile.

Standard gamble

When the standard gamble method is applied to states considered by the subject to be better than being dead, individuals are asked to compare the certainty of remaining in that state with a lottery in which they risk immediate death in order to be restored to full health. The risk of immediate death is varied until the subject is indifferent between the lottery and remaining in the state in question. The greater the risk of death in this situation, the worse the state must be. For states considered worse than death by the subject, the comparison is between the certainty of immediate death and a lottery in which there is a chance of being restored to full health rather than remaining in the state in question. The chance of being restored to full health is varied until the subject is indifferent between the lottery and immediate death. The greater the chance of being restored to full health that is needed to get the subject to this situation, the worse the state in question must be.

APPENDIX 10.2

MVH Group Membership

Dalen, Harmanna van (1988 to 1991)
Dolan, Paul (1991 to 1994)
Durand, Mary-Alison (1988–1990)
Gudex, Claire (1987 to 11990 and 1991 to 1995)
Kind, Paul (1987 to date)
Lewis, David (1990 to 1991)
Morris, Jenny (1988 to 1991)
Williams, Alan (1987 to date)

SCPR Collaborators

Eren, Box
Thomas, Roger
Thomson, Katarina
and the key field workers and their managers

Official Reports

Jan 1991	Report on Phase I of Research
Dec 1992	Valuing Health States: A Comparison of Methods. Final Report – Part I
Dec 1992	Valuing Health States: A Comparison of Methods. Final Report – Part II

Dec 1992	Valuing Health States: A Comparison of Methods. Technical Appendix
May 1993	Supplementary report
Jun 1993	Second Developmental Pilot: Dress rehearsal
May 1994	First Report on the Main Survey
Oct 1994	Generating a UK EuroQol Tariff (Interim report on modelling)#
Jan 1995	Final Report on the Modelling of Valuation Tariffs

SCPR Report

Aug 1992	Health Related Quality of Life: Technical Report (Thomas, R. and Thomson, K.)
Aug 1992	Health Related Quality of Life: Comments on the Pre-Pilots and Pilot (Thomas, R. and Thomson, K.)
Sep 1993	Health Related Quality of Life: the 1993 Pilots (Thomson, K.)
Apr 1994	Health Related Quality of Life: General Population Survey Technical Report (Erens, R.)

Other publications arising from this work

Gudex, C. and Kind, P.
The QALY toolkit. York: Centre for Health Economics (Discussion paper 38), 1988

Kind, P.
The design and construction of quality of life measures. York: Centre for Health Economics (Discussion paper 43), 1988

Kind, P.
Issues in the design and construction of a quality of life measure. Baldwin, S., Godfrey, C. and Propper, C. (eds.), *The Quality of Life: perspectives and policies.* London, Routledge, 1990, pp 63–71

Kind, P.
Measuring valuations for health states: a survey of patients in general practice. York: Centre for Health Economics (Discussion paper 76), 1990

Williams, A.H., Durand, M.-A., Gudex, C., Kind, P., Morris, J. and van Dalen, H. *Lay concepts of health with special reference to severely physically handicapped young adults and their carers. Report to the Nuffield Provincial Hospitals' Trust.* York: Centre for Health Economics, 1990

Kind, P. and Gudex, C.
The HMO: measuring health status in the community. York: Centre for Health Economics (Discussion paper 93), 1991

Williams, A.H. and Kind, P.
The present state of play about QALYs. Hopkins, A. (ed.), *Measures of the quality of life and the uses to which such measures may be put*. London: Royal College of Physicians of London, 1992, pp 21–334

Kind, P.
Quality of life and the calculation of disability-free life years. Robine, J.-M., Blanchet, M. and Dowd, J.E. (eds.), *Health expectancy. First workshop of the International Healthy Life Expectancy Network (REVES)*. London: HMSO (OPCS Studies on Medical and Population Subjects, 54), 1992, pp 99–104

Gudex, C.
Are we lacking a dimension of energy in the EuroQol instrument? Bjork, S. (ed.), *EuroQol conference proceedings, Lund, October 1991*. Lund: Swedish Institute for Health Economics (Discussion paper 1, working paper 1992:2) 1992, p 97

Gudex, C., Drummond, M.F. and Williams, A.H.
Quality-Adjusted Life Years: a review, Committee on Core Health Services Report, Wellington, New Zealand. York: Centre for Health Economics, 1992

Brazier, J. Jones, N. and Kind, P.
Testing the validity of the EuroQol and comparing it with the SF–36 health survey questionnaire. *Quality of Life Research*, 1993, 2, pp 169–180

Kind, P. and Gudex, C.
The role of QALYs in assessing priorities between health care interventions. Drummond, M. And Maynard, A. (eds.). *Purchasing and providing cost-effective health care*. Edinburgh: Churchill Livingstone, 1993, pp 94–108

Gudex, C., Kind, P. and Dolan, P.
The valuation of death. Sintonen, H. (ed.), *EuroQol Conference Proceedings, Helsinki, Oct 1992*. Kuopio: Kuopio University Publications (Discussion paper 2, Kuopio University Publications E. Social Sciences 8), 1993, pp 23–39

Gudex, C., Kind, P., van Dalen, H., Durand, M.-A., Morris, J. and Williams, A.
Comparing scaling methods for health state valuations: Rosser revisited. York: Centre for Health Economics (Discussion paper 107), 1993

van Dalen, H., Williams, A. and Gudex, C.
Lay people's evaluations of health: are there variations between different sub-groups? *Journal of Epidemiology and Community Health*, 1994, 48(3), pp 248–253

Kind, P. and Gudex, C.
Measuring health status in the community: a comparison of methods *Journal of Epidemiology and Community Health*, 1994, 48(1), pp 86–91

Kind, P., Dolan, P., Gudex, C. and Williams, A.H.
Practical and methodological issues in the development of the EuroQol: the York experience. *Advances in Medical Sociology*, 1994, 5, pp 219–253

Gudex, C. (ed.)
Time trade-off user manual: props and self-completion methods. York: Centre for Health Economics, 1994

Gudex, C. (ed.)
Standard gamble user manual: props and self-completion methods. York: Centre for Health Economics, 1994

Williams, A.
The role of the EuroQol instrument in QALY calculations. York: Centre for Health Economics, 1995

PART THREE

Priority-Setting and Health Technology
Assessment

11. Need as a Demand Concept (with special reference to health)

INTRODUCTION

It is a well-known dictum in operations research and management consultancy in general that the problem the client thinks he has at the outset seldom turns out to be his real problem at the end. So it seems to me with this topic, for in composing my thoughts on the subject of 'Need as a demand concept' I have found it necessary to consider 'Need as a supply concept' and 'Demand as a need concept'! Although in the concluding section I offer a tentative resolution of the conflicting concepts and terminology employed in this fuzzy interface between analytical economics, social policy and political argument, I doubt very much whether I shall have satisfied all parties. Nevertheless, in the interests of efficient discourse, some attempt to bridge the gap between economics and 'needology' seems called for and I cheerfully offer myself for martyrdom in this worthy case.[1]

NEED AS A SUPPLY CONCEPT

A convenient starting point for an exegesis of the notion of 'need' is provided by the following quotation (Matthew 1971):[2]

> The 'need' for medical care must be distinguished from the 'demand' for care and from the use of services or 'utilization'. A need for medical care exists when an individual has an illness or disability for which there is an effective and acceptable treatment or cure. It can be defined either in terms of the type of illness or disability causing the need, or of the treatment or facilities for treatment required to meet it. A demand for care exists when an individual considers that he has a need and wishes to receive care. Utilization occurs when an individual actually receives care. Need is not necessarily expressed as demand, and demand is not necessarily followed by utilization, while, on the other hand, there can be demand and utilization without real underlying need for the particular service used. (p. 27)

179

In this context 'need' is not a demand concept at all, but a quasi-supply concept: it means that a 'need' exists so long as the marginal productivity of some treatment input is positive. Only when the efficacy of treatment has become zero at the margin does 'need' disappear. People may still be sick but, since there is nothing we can do for them, the implication is that they are not in need.[3]

This interpretation also means that the last sentence in the quotation above could be rewritten so as to read: 'people may demand and utilise particular services even though the latter are totally ineffective'.[4] There are, of course, some difficult problems here concerning the content of the notion of 'effectiveness' as applied to medical services, which relate partly to physiological versus psychological aspects of medicine and partly to the question 'who is the client?' A treatment, known to be ineffective in relation to the patient's physiological condition, may still be given as a demonstration (to him or his loved ones) that someone cares and this may give satisfaction to them and, indeed, to other members of the community unknown to the patient who sympathise with the plight of the sick generally, so that the treatment may be quite 'effective' in this broader sense. Strictly speaking, therefore, we would have to say that a treatment is ineffective only if it has none of the above good effects.[5]

Though seemingly odd in its implications, this usage of the term 'need' is common in political debate, which leads to a further important connotation, namely, that establishing 'need' is a factual, not an evaluative, matter. As one writer has said (Barry 1965):

When we see statements to the effect that human beings need so many calories per day (and that states should make every effort to see that everyone gets this number) or that University teachers need books (which should therefore be allowed by the Inland Revenue as a claim for expenses) we may at first suppose that here is a justification for policies which ... appears ... to an 'objective' or 'scientific' procedure by which 'needs' are established.

Whenever someone says 'X is needed' it always makes sense ... to ask what purpose it is needed for. Once an end is given it is indeed an 'objective' or 'scientific' matter to find out what conditions are necessary to bring it about .... The end in my first example might be mere survival, or good health, or the satisfaction of hunger; and differences in the 'needs' found by different studies might no doubt be attributed to differences in the end postulated.

When I say that 'need' is not by itself a justificatory principle, I mean that no statement to the effect that X is necessary in order to produce Y provides a reason for doing X. Before it can provide such a reason Y must be shown to be (or taken to be) a desirable end to pursue .... A *conclusive* reason would require showing that the cost of X (that is, other desirable things which could be done instead of X) does not make it less advantageous than some alternative course of action. (pp. 47–8)

Thus the evaluative stage in the argument is pushed back one stage, but not avoided and the statement 'X is needed' simply means 'X is conducive to the stated objective' or 'the marginal productivity of X is positive'. In order not to be pedantic in everyday discourse we may find that it is convenient to accept certain 'objectives' implicitly, and Barry sketches out the manner in which this is likely to manifest itself when we move from statements such as 'X is needed to produce Y' to statements such as 'A needs X', where A is a person:

At the core is physical health (for example, the diet example); this extends more weakly to mental well-being (for example, people need privacy, people need community). Then, spreading further out comes the performance of some function or the achievement of some object (the university lecturer example). Finally, we arrive at the fulfilment of some standard which can be independent of any function or purpose of the person to whom need is ascribed (old age pensioners need more money if their level of prosperity is to keep step with that of the rest of the community). The nearer to the core the use of 'need' is, the less linguistic propriety demands that the end be supplied in the sentence and, of course, the easier it is to suppose that a need can somehow be established independently of an end. (*ibid.*, p. 49)

But it also carried with it the danger that the evaluative process is perverted so that analysts consciously strain at gnats while unconsciously swallowing camels!

NEED AS A DEMAND CONCEPT

We have so far interpreted 'need' to mean 'A could benefit if he had X', that is, X would be productive of something that is good for A. This confronts us with an issue which can no longer be shirked: '*who* is to judge what is good for A?', or, in the terminology of need, '*how* are A's needs to be assessed?'

A useful starting point here is provided by the following schema,[6] in which three parties are distinguished, society (S), medical experts (M) and the individual (I), each party being asked two questions, namely:

(a)   Is the individual sick?
(b)   Is the individual in need of public care?
        A third question is also asked, namely:
(c)   Does the individual demand public care?

Question (c) is answered by observing 'those individuals who come in touch with the system of public care with a desire for consultation and treatment, and who are willing to wait if this cannot be provided at once' (Spek 1972, p. 265). If 'yes' answers to (a) and (b) are represented by S, M, I, and 'no' answers by $\bar{S}$, $\bar{M}$ and $\bar{I}$, for each respective party, the outcomes could be represented as in Figure 11.1 (Spek's Fig. 2, *ibid.* p. 266).

Figure 11.1 Sickness and need as seen by three parties

Is the individual sick?*

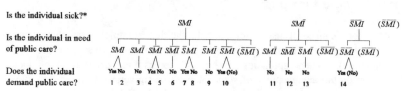

Is the individual in need
of public care?

Does the individual
demand public care?

*The four cases $\bar{S}\bar{M}I$, $S\bar{M}\bar{I}$, $\bar{S}M\bar{I}$ and $\bar{S}M\bar{I}$ are disregarded in spite of their great interest. They must not be forgotten in a more detailed analysis. In addition we do not discuss differences in agreement among representatives of society and amont medical experts.

Spek comments on this as follows:

> Case 1 represents *justified* demand, with society and medical experts in agreement. Cases 2, 3 and 11 represent *latent need* with society and medical experts in disagreement. Cases 10 and 14 represent unjustified demand with society and medical experts in agreement. Cases 4 and 7 represent demand with society and medical experts in disagreement.

> ... the answers to the first and second questions depend on knowledge and valuation, but they offer quite a different educational and informational problems: this in turn will affect the ease with which the latent need in the different cases can be converted to demand, as well as unjustified demand suppressed. Latent need may also be defined as need together with absence of demand for public care. It is partly known through population studies and from individuals who, having contacted the system of public care, refuse to wait. This is not the place to discuss the problems which arise when society and medical experts are in disagreement on the latent need. If Case 4 is regarded as justified active demand, then it represents the thorny problem of how to have the doctors furnish the right care. If Case 7 is regarded as unjustified demand, it represents the thorny problem of 'over-use'. (*ibid.*, pp. 265–7)

Bradshaw (1972) has a similar approach based on the presence or absence of the following four dichotomous discriminators:

(i)   Normative need (that is, 'that which the expert or professional, administrator or social scientist defines as need in any given situation. A (desirable) standard is laid down and is compared with the standard that actually exists').

(ii)   Felt need (that is, 'Here need is equated with want. When assessing need for a service, the population is asked whether they feel they need it').

(iii) Expressed need (that is, 'Expressed need or demand is felt need turned into action .... Expressed need is commonly used in health services where waiting-lists are taken as a measure of unmet need. Waiting-lists are generally accepted as a poor definition of 'real need' – especially for pre-symptomatic cases').

(iv)   Comparative need that is, 'obtained by studying the characteristics of the population in receipt of a service. If there are people with similar characteristics not in receipt of a service, then they are in need').

The first thing to note is that the judgements of 'society' play no role in Bradshaw's taxonomy, only those of the individual and the experts. It will also be seen that Bradshaw's 'normative need' is equivalent to Spek's medical experts answering 'yes' to both questions (a) and (b) (Spek's Cases 1, 2, 3, 7, 8, 9, 11 and 13.) Bradshaw's 'felt need' is equivalent to Spek's *individual* answering 'yes' to those two questions (Spek's Cases 1, 2, 4, 5, 7, 8, 10 and 14.) Bradshaw's 'expressed need' is equivalent to Spek's question (c) being answered affirmatively (Spek's Cases 1, 4, 7, 10 and 14.) Bradshaw's 'comparative need' seems to be approximately equivalent to Spek's 'latent need' (Cases 2, 3 and 11 but with 9 added since 'society's' adverse judgement is no longer relevant), where the individual does not demand the service even though the medical experts think he is sick and needs public care.

If we define 'need' as the situation in which the medical expert answers 'yes' to questions (a) and (b), we could rewrite the original quotation from Matthew, expunging the words 'need' and 'demand', as follows:

The medical experts' judgement as to whether or not a person is sick and capable of benefiting from medical care must be distinguished from that person's use of services or 'utilization'. A person is judged by medical experts to be capable of benefiting from medical care when the individual has an illness or disability for which they believe there is an effective and acceptable treatment or care. This phenomenon can be expressed either in terms of the type of illness or disability to which the treatment or care is directed, or in terms of the treatment or facilities for treatment themselves. An individual presents himself for treatment when he considers that he is sick and may be capable of benefiting from medical care and, and wishes to avail himself of that possibility. Utilization occurs when an individual actually receives care. An individual who would be judged by medical experts to be sick and capable of benefiting from medical care will not necessarily present himself and request treatment, and such requests are not necessarily complied with. On the other hand, there exist cases where individuals request treatment and get it even when medical experts (or society?) judge them not to be sick or not capable of benefiting from the particular service used.

So far so good, but we are still taking 'need' to be an on/off concept, that is, we are interpreting it in an absolute rather than in a relative sense and we need to move on to the language of 'priorities', already introduced tentatively and obliquely at the end of the first quotation from Barry above. Unfortunately this notion, which prima facie implies relative valuation, seems to be just as abused and hence meaningless as 'need' itself.

For instance, one eminent writer in this field (Butterfield 1968) managed to include all the following statements in his argument for better management and more research in the British National Health Service:

> Until there is some tangible prospect of our getting on top of the increasing demand by the public, often for conditions medical students despise, all imposed on limited medical personnel, there is scant possibility of securing the second prime priority, namely happy general practitioners and health workers (p. 77).

> Now the decision behind establishing priorities implies some objective .... Should one begin by asking, what are the urgent short term priorities in medical care of the health service? ... Or would it be wiser to take a longer view, and to consider the ultimate medical Utopia as the challenge, and then examine the present situation and see how far it falls short and ask what one can do to move in the desired direction? (pp. 78–9).

> The long term target must be perhaps an unattainable, but generally desirable, medical Utopia (p. 79).

> The inauguration of the National Health Service changed the relationship between the medical profession and the public. Before, the Hippocratic Oath notwithstanding, doctors could and did take up a position (in certain cases) of being unable to do anything. With the coming of the National Health Service, I think we must make the assumption that the profession as a whole took the Hippocratic Oath with society as a whole and such evasion is becoming progressively less easy. If the patient is really dissatisfied, he widens his contacts with the National Health Service, reduplicates his consultations in another hospital. He or she can demand and probably will get help somewhere in the end. It is not unreasonable therefore to rank the priorities in medicine in order of the distance from the symptoms, that is to say, first the patient, then his general practitioner, then the hospital and administrative organizations (p. 82).

This eventually (pp. 191–2) leads to 10 'selected priorities', characterised respectively by the labels (1) 'the first priority ...', (2) 'a very high priority',

(3) 'urgently needed', (4) 'will have to be undertaken', (5) 'causing serious losses', (8) 'there is a good case for', (9) 'funds should be available for', (10) 'there seems a strong case for'.

It is obvious to an economist that what is missing here is the notion of 'trade-off' *at the margin* between one good thing and another. Listing items in an order of priority implies that until the first is satisfied you will not allocate resources to the second and so on.[7] The oscillation, in the above quotation, between regarding the solution of short term problems as a *precondition* for tackling long term ones and fixing one's eyes on some unattainable Utopia and moving towards it willy nilly, is but a dramatic instance of this confused and unhelpful way of thinking. The question still remains, of course, as to *whose* judgements about these trade-offs we should accept and Butterfield recognises clearly that society, the medical experts and individual patients all play a role here and that their respective 'priorities' do not always coincide.

If we consider again the social context in which the term 'need' comes to be used by the protagonists in social debate, it is easy to see why someone like Marshall (1973) is led to conclude:

Needs, other than the basic needs of life and health, are subjective, that is, based on values. Economists can handle the concept of utility even though it is subjective because this concern is with conscious wants, or desires, which are expressed as demands ...

Social policy's concept of need is based on a collective value system, working with a norm of need satisfaction which is collectively subjective, that is, the norm reflects collectively acceptable views.

The aim of research and of social policy is to recognise this and to distinguish between unfelt needs, felt needs and conscious desires.

Considerations such as these led two of my colleagues and myself (Culyer, *et al.* 1972) to conclude that:

The word 'need' ought to be banished from discussion of public policy, partly because of its ambiguity but also because ... the word is frequently used in ... 'arbitrary' senses .... Indeed ... in many public discussions it is difficult to tell, when someone says that 'society needs ...', whether he means that *he* needs it, whether he means society ought to get it in *his* opinion, whether a *majority* of the members of society want it, or *all* of them want it. Nor is it clear whether it is 'needed' *regardless* of the cost to society.

But the literature of 'needology' does serve one useful purpose in alerting us to the complexity involved in expressing policy priorities in a key field of social policy such as health, where conflicting judgements are likely to be made by the various parties. Faced with this conflict, we should resist the temptation to allot any one party an overriding role by assigning to that party the sole right to use the prejudicial term 'need' with respect to its own judgements.

DEMAND AS A NEED CONCEPT

The time has at last come to permit the economists to occupy the centre of the stage and I propose to do this by analysing their notion of 'demand' in terms of the 'need' categories we have been using hitherto.

In its simplest form demand is one individual's ordering of his own priorities as he sees them, this ordering being constrained by the resources at his command. In principle it does not take account of the judgements of 'society' or 'medical experts', except in so far as these have affected the individual's 'tastes', which are regarded as exogenous.[8] This simple view of demand therefore places it as equivalent to Bradshaw's 'expressed need' (and Spek's Cases 1, 4, 7, 10 and 14). The 'resources at the individual's command' usually means income, though it occasionally means wealth, or exceptionally, information, skill, time, energy, etc. Thus in the ordinary discourse of economics, the demand for medical care would be taken to mean the amount individuals are willing and able to pay for at some going price. It will be noted that this is the first time the notion of price has entered the discussion (though we did mention opportunity cost earlier) and it warrants further exploration.

In the context of the British National Health Service medical care can be taken, as a rough approximation, to be available at zero price in the sense that patients pay virtually no fees. What the (potential) patient does have to be prepared to offer is time and energy (and he probably needs information about services and how to apply for them and skill in making his wishes known in some circumstances too). To keep the argument as simple as possible, let us assume that it is only time that is required. It is this notion which underlies Bradshaw's notion 'expressed need' and Spek's definition of demand as involving a willingness to wait. But this is itself ambiguous, since it fails to differentiate between two distinct situations, firstly, that where the individual is told to come back at some specified future time for treatment (for example, being placed on a waiting list for a non-urgent operation) but is able to pursue his normal activities perfectly well in the meantime and, secondly, that where the individual is kept waiting in circumstances (like a queue in a doctor's waiting room or where one is bedridden awaiting an operation) where normal activities are severely disrupted during the waiting period. Only the latter can really constitute a time–price for medical care, which fits in with the usual 'demand' notions in economics.

Although there is no reason in principle to deny the legitimacy of this interpretation of 'demand', it does have significantly different implications from the 'usual' case. Firstly, the time–price which is 'paid' by the demander is not 'received' by the supplier, as it would be with a money price, so that the informational content of the 'offer' is less accessible to suppliers. Secondly, the distribution of time resources is different from the distribution of money resources, so that the pattern of demand which emerges has a different equity interpretation.[9] Thirdly, since the money value of time is likely to differ significantly from one person to another, a constant time–price implies interpersonal variations in money costs depending on the value of time.[10]

In what we have said so far, both the money price and the time–price elements (and the respective distributions of 'purchasing power' associated with them) have been interpreted in a strictly individualistic manner, in which each person is assumed to be not only the best but the sole judge of his own welfare. If we now extend the realm of discourse somewhat, we are led into rather deeper waters.[11] One generally acceptable extension involves paying explicit regard to any external effects generated by an individual's consumption of medical care. Thus we could think of *other* people having a 'demand' for some individual having (or *not* having!) medical care! The operational problems of giving effect to this 'demand' are considerable, however, and may lead us directly into the more contentious second extension of the argument, where 'society' judges whether an individual's own assessment of his own 'need' is to be accepted as it stands, augmented, diminished, or even rejected. This gets even more tricky when 'society' decides that the best people to act as a social filter to approve or reject individual demands are the medical experts! Hence our initially tidy 'demand' concept leads step by step back into the morass of needology from which many economists fondly imagine that they have escaped!

But it is possible to go still further in undermining the relevance of the classical demand function, by denying that people are the best judges of their own welfare, *even when externalities are absent*. This seems to me to be the essence of the discussion of merit goods, which rests on the assertion that someone else (society or the medical expert) is better able to judge what is in an individual's own interests than he is himself.[12] Economists are so steeped in 'consumer sovereignty' notions that there is enormously strong professional resistance to 'interfering' with people's preferences, even though such 'interference' is a major avowed purpose of the educational system, commercial advertising and religious and cultural organisations. Moreover, even within these conditioning factors, the capacity for rational choice can legitimately be expected to diminish sharply when people are ill (and especially when they are psychologically disturbed), so that the marked resistance of economists to the substitution of societal or 'expert' assessments for the individual's own assessments in this field seems to be based on rather weak ground. Why then should we so strongly resist Spek's 'latent need' and 'unjustified demand' concepts?

I turn finally to issues connected with the prevailing distribution of 'purchasing power', or, more generally, of 'access rights' to medical care. The use of the simplistic demand function, which operates only with money prices, implies acceptance of the existing distribution of money income (and wealth) as the ethically appropriate basis for determining the right to formulate 'requests' for access to medical care. It is, of course, possible to filter these requests by requiring that some forms of treatment can be obtained only on the authority of a medical expert,[13] hence the expression of 'demand' in this case is a necessary but not sufficient condition for treatment. But this does not of itself improve access for 'the poor'.

A zero money price with queuing generates a different set of 'demands' and implies that the distribution of time is a better ethical base on which to formulate requests for access. Again, a filtering process may be interposed and British experience has led one well informed observer (Cooper 1974) to conclude that under such a system:

> In practice the scope for an individual to regard himself as in 'want' of healthcare is virtually unlimited .... The factors which determine whether an individual consults a doctor or not are highly complex and far from fully understood .... There are also real costs involved in converting wants into demands. Deterrents include the necessary expenditure of time and energy plus, for example, such factors as concern not to overwork the doctor .... The knowledge that there are queues will obviously deter many would-be patients from demanding care. Indeed, in the long run demand will tend to gravitate towards whatever level of provision there happens to be .... Need is, in any case, a medical opinion not a medical fact .... Collectively the profession appears to reassess its conception of need in line with actual levels of provision .... What emerges as certain ... is that there will always be an excess demand for resources and a problem of rationing. In the market place this would be achieved by a price rise .... In the Health Service the problem of rationing has fallen to the medical and allied professions. Rationing, however, has never been explicitly organised but has hidden behind each doctor's clinical freedom to act solely in the interests of his patient. Any conflict of interest between patients has been implicitly resolved by the doctor's judgements as to their relative need for care and attention. The clinical freedom to differ widely as to their conception of need has led to inconsistencies of treatment between patients and to the

allocation, without challenge, of scarce resources to medical practices of no proven value.

Thus changing the basis of formulating requests for access still leaves one facing the problem of assessing priorities ('relative needs').

There is a yet broader notion which could be adduced as a justification for 'need' as the central (though imperfect) concept in determining the pattern of medical care to be provided and its optimal distribution between individuals. This is that 'in communities and countries where egalitarian feeling is strong and a spirit of national sharing is general, a national health service may indeed be the most efficient means of satisfying these wants'.[14] Unfortunately this generalised appeal to notions of social justice and social solidarity plunges us into still deeper water, where we have to consider the political processes by which 'communities and countries' articulate, disseminate and discriminate between rival views on such matters and give them operational effect. It is not unlikely that this debate will be such that the *mode* of doing things becomes the central point, rather than what it is reasonable to expect will be achieved thereby, leading to that final reformulation in which someone asserts 'we *need* a National Health Service'. As I was saying at the beginning ....

A RESOLUTION?

The conclusion I draw from all this is that if economists insist on textbook notions of demand as the only proper way to go about assessing priorities in determining the amount and distribution of goods and services such as medical care, then they will miss important elements in the situation and (rightly) be discredited and disregarded by policy makers. If through appeal to complex notions of externalities and merit goods they attempt to go beyond this simplistic interpretation of demand, they will be forced to unravel the same tangled skein of conflicting roles and judgements that the 'needologists' have been grappling with and on which we economists have tended to pour scorn.

The heart of the matter, as I see it, is a societal judgement as to who shall play what role according to which rules. The parties in the 'who' bit are (a) patients and other potential beneficiaries, (b) 'experts', (c) politicians and (d) the electorate at large. The roles to be played are 'advertisers', 'applicants', 'diagnosers', 'priority–setters', 'treatment–assigners' and 'researchers'. The 'rules' consist of terms of reference and behavioural norms to guide choice within whatever area of discretion is so assigned.

An appropriate clarification and legitimisation of roles within a medical care system might, in skeletal form, be as follows:

(a)   The medical care system recognises that it has work to do *either* (i) because patients are brought to its notice 'spontaneously' *or* (ii) because it has been agreed, as a matter of policy, to go out looking for them.

(b)   Patients who present 'spontaneously' have to be assigned treatment modes (including 'no treatment') according to two distinct kinds of criteria: (i) whether they *could* benefit from treatment (a technical diagnostic judgement) and, if so, (ii) whether it is worth while offering that treatment (a *social–valuation* judgement).   Medical experts are clearly the appropriate people to play the former role but are not so clearly entitled to play the latter role, unless society says so and even then only within any 'guidelines' which society may lay down. This raises both (i) nice questions of principle about the limits of clinical freedom, the nature of professional ethics and the social, legal and moral sensibilities and responsibilities of medical experts and (ii) brutal practical problems concerning resource allocation, especially as regards investment (or non-investment) in facilities and skills which will enhance (or constrain) the capacity of medical experts to respond to the requests for treatment which are made to them.

(c)   The seeking of patients, via 'advertising' (that is, information dissemination, 'education' and persuasion) or via more coercive measures (requiring vaccinations, certifying patients insane, etc.), raises all these same issues in still more acute form and makes it still more important that one party does not usurp the legitimate role of another.   Thus surveys of differential patterns of use, or of expenditure, or of the incidence of sickness, do not of themselves constitute conclusive evidence that 'something should be done' and proposals to conduct 'screening' tests to identify presymptomatic illness, even when curable, are to be subject to wider social valuation than medical expertise to determine whether any consequential treatment is of higher or lower priority than other uses of resources or, indeed, whether the benefits of such campaigns are not outweighed by their 'costs' (in psychological as well as in material terms).[15]

(d)   There is clearly an important role for medical, psychological and sociological research in determining the 'technical' effectiveness of various modes of treatment (actual and potential) but again it needs to be stressed that to find a more 'effective' mode of treatment in this sense is not necessarily to recommend its adoption. It may be twice as effective as some existing mode but four times as costly!

(e) Even if 'demand priorities' emanating in a simplistic way from 'the market' are rejected (or modified out of all recognition) as a basis for allocation of treatment, there may still be an important role for economists as 'technical experts' in estimating costs and benefits of other aspects of the situation, provided that the market can be taken as a good first approximation to society's valuations of these elements (such as the labour and material inputs of the medical care system and loss of patients' and relatives' output from having people ill or under treatment). Again the well-recognised dangers exist of (i) failing to make clear the basis of valuation and illicitly supplying one's own values, (ii) becoming so obsessed with GNP oriented costs and benefits that others (pain, grief, suffering and loss of leisure) get neglected and (iii) overlooking distributional considerations.

(f) The heaviest burden of responsibility falls on 'policy makers', whose unenviable task it is to detect, clarify and give operational content to the 'wishes of society'. In this context the responsible policy makers may be full-time 'politicians' (at national or local level) or citizens on a governing board, an advisory panel or a finance committee. The proper discharge of their responsibilities is made extremely difficult by the fact that many of the judgements they should make are usurped by others and the information at their disposal is biased accordingly. They are the target of much of the confused and perjorative language of 'need' and are frequently poorly equipped to resist it in a constructively discriminating manner. They are thus encouraged (and often only too ready) to shuffle off their responsibilities to the 'experts', even though these 'experts' are *not* experts in the relevant matters.

My own conviction, therefore, is that we need to formalise these diverse elements in the medical care system in such a way that unambiguous communication between the respective parties becomes possible. This exegesis of need as a demand concept is one small contribution to that task.[16]

NOTES

1. This chapter was originally written for a conference organised by the International Seminar in Public Economics at Siena in September 1973. It reflects ideas generated in the course of work undertaken on the evaluation of public services financed by the Social Science Research Council (as part of the Public Sector Studies Programme at the Institute of Social and Economic Research in the University of York) and by the Department of Health and Social Security and the Institute of Municipal Treasurers and Accountants on output measurement in public services. I am grateful for comments and suggestions on an earlier draft received from Paul Burrows, Tony Culyer, Norman Glass, Alan Maynard, David Pole and Bernie Stafford.
2. Similar views are expressed in Office of Health Economics (1971).
3. There is a second stage interpretation of need which could be brought to bear here, which is epitomised in the statement 'we need to find some way of curing cancer'. Here we are saying 'if only we had an effective treatment, people could be better off'. Since this raises the same issues of principle as arise at the first stage, it is not pursued further here.
4. A pungent polemic concerning the ineffectiveness of much medical care is provided by Cochrane (1972).
5. It will be obvious that because this notion of 'need' relates only to the *productivity* of the treatment (requiring it to be positive), it does not say anything about its 'cost-effectiveness'. Positive marginal productivity for a treatment is thus a necessary but not a sufficient condition for recommending its utilisation.
6. See Spek (1972). See also the discussions of 'need and demand' on pp. 315–7 of Hauser (1972).
7. Tribe (1972) also appears to take this view, in a peculiarly modified manner, in his section on 'lexical ordering', where he writes 'both individual and societal preference orderings might well display significant discontinuities ... without any strictly lexical principles ever operating'. Marshall (1973) concludes that all such preference statements are ultimately relativistic because 'needs are relative to the social situation, not merely to a particular type of civilisation or national culture, but to the differing circumstances of groups within such a civilisation or culture. It is also suggested that the system of values in relation to which needs are defined becomes incorporated into the individual personality. This means that the social environment influences, or even up to a point determines, not only what is felt as lack or deprivation, but also what is not felt as such by one group although it is by another'.
8. This 'except ...' may, however, provide loopholes large enough to drive the proverbial coach-and-horses through. For instance, individual preferences may be strongly influenced by medical 'advice', and/or by the prevailing state of public opinion concerning the 'rightness' of particular courses of action (for example, abortion).
9. This is especially significant in systems which severely restrict the opportunities for substituting money prices for time–prices.

10. Smolensky *et al.* (1972) observe (p. 101) that 'the good provided publicly at a congested facility is not a "public good", but one whose implicit price is a function of the opportunity cost of time. A queue rations out users with a high opportunity cost of time, shifting them to quicker substitutes with a higher money cost. Instituting a money charge equal to the congestion costs of the marginal user will also ration customers out by the opportunity cost of time, but now it is those with high opportunity cost who will remain. In the absence of a social welfare function it is difficult to choose between these two devices'.

11. Charted more fully in Culyer (1971a).

12. An interesting recent examination of the implications of the Musgravian approach is to be found in Pazner (1972). Earlier explorations are to be found in Head (1966) and McLure (1968).

13. An interesting analysis of the implications of this sort of arrangement for the market for prescription-only drugs is contained in Liefmann-Keil (1974).

14. Lindsay (1973). But see also Culyer (1971b).

15. For instance, Thorner and Remein (1967) treat the exercise as a purely technical matter. A more circumspect view is expressed in Wilson and Jungner (1968), especially pp. 26–39.

16. Some of the other elements are set out more fully in my recent paper (Williams 1974a).

12. The Budget as a (Mis-) Information System

Budgets serve many purposes. The most widely known is that of cash control (who is spending how much on what and with whose authority?). The information deriving from that process is commonly used to forecast cash requirements, to set financial targets ('we really must try to do better than just break even next year') and, in a limited sense, to monitor the performance of those with budget responsibilities.

With increased emphasis on planning, there is immediately brought to light a serious deficiency both in the existing data base and in the information system that generates it. When it comes to telling us about the efficiency with which existing resources are deployed and on potentially better ways of deploying them, it is tempting to try to make good this deficiency by turning to the budgeting system for the missing information. It is on the dangers of uncritical use of this information that I shall concentrate in this paper.

In a paper on the problems of monitoring performance in the health service, Sir Richard Doll expressed scepticism about the extent to which it is possible to be vigorous and systematic in this difficult field (yet making it clear that he was sympathetic to such endeavours). He epitomised his position as being one in which he believed that this was a field in which 'gardening is real, but botany is bogus' (Doll, 1973). It is on that telling analogy that I have structured this essay, in the hope that by so doing I can demonstrate that, if we are ever to break out of the limitations of gardeners' folklore, we do need a strong dose of botanical knowledge.

CLEARING THE GROUND

Any rational planning process must provide for the following five elements:

(i)   the *recording* of the decisions that have been made (so that those who made the decisions shall be clear about what they decided and why and what they expect to happen as a result);

(ii)   the *transmission* of key information about these decisions to those who have to carry them out (since it is rarely the case that implementation requires actions only on the part of those who are parties to the decisions in the first place);

(iii)   the *monitoring* of performance (which means organising the collection and feedback of selected information about what actually happened in terms that relate to the intentions of decision-makers;

(iv)   the *evaluation* of outcomes (that is, requiring a judgement as to whether what happened is better or worse than what was intended and, if different, why, and what, if anything, should be done differently next time);

(v)   a fresh round of *decision* (which may simply be to repeat the decision made last time but may, at the other extreme, involve a radical re-thinking of objectives and policies in a particular area, entailing cessation of the activity).

At this point the cycle would be repeated and, in some complex situations, (iii) and (iv) might be going on continuously alongside (i) and (ii), with (v) feeding in only sporadically.   In some routine–hidden situations, (v) does not really exist because the system has so much momentum it can tick over anyway so long as the external environment remains stable and accommodating, hence no need is seen for (iii) and (iv); 'planning', in my understanding of the term, is absent (and, some would argue, unnecessary).

Before entering the main arena of discussion, there are two other basic notions that need to be introduced. The first is the distinction between an organisation and an activity and the second is the differences between input, throughput and output.

An 'organisation' is here taken to mean any legal–administrative entity that has a recognisable corporate identity (for example, a hospital, or a GP's practice, or the X-ray department of a hospital or a ministry of health). An activity, on the other hand, is some task-orientated grouping of resources, defined by what they are doing rather than what they are or to whom they belong (for example, diagnosing illness, or alleviating poverty or providing means for the elderly). Sometimes one particular organisation will carry out many (narrowly defined) activities; sometimes one particular (broadly defined) activity will require the collaboration of many organisations; and sometimes (but all too rarely) an organisation and an activity will exactly coincide.   The general non-coincidence of organisations and activities will prove to be a major source of difficulty for us, because it is broadly the case that budgets relate to organisations, but plans relate to activities.

By 'input' is meant the resources used in an activity or organisation. By 'output' is meant what has been achieved, in terms related to the objectives of the activity or organisation, whether those objectives (or outputs) are tangible or intangible. For instance, the 'input' of teachers disseminating information about personal hygiene into the activity of teaching children in school might yield an 'output' of reduced gastro-intestinal infection. Output, like objectives, can be articulated at different levels of generality. To take the latter example a bit further, it might be said the 'reduced gastro-intestinal infection' is only an 'intermediate' output, the 'final' output being an increase in life expectation free of pain and disability. Going back in the other direction, the extent of teachers' knowledge about personal hygiene will itself be an 'output' from some training programme. The general point is that, whether a particular element is input or output is relative to a *specified* focus of interest. Once that is defined, matters should fall into place. This still leaves us with one other definitional matter, the measure of work done (for example, numbers of children taught), which is neither input nor output. It is that which I shall call 'throughput'.

RECOGNISING WEEDS

Some recent correspondence in *The Times* on the alleged efficacy of talking lovingly to plants as a means of stimulating healthy growth led one disgruntled contributor to express disbelief in the whole notion on the grounds that his intemperate abuse of weeks over many years did not seem to have stunted their growth in the slightest! A later correspondent was then moved to suggest that perhaps a weed was, in essence, a plant impervious to insults. So, it seems, is the case with traditional budgets, which I shall now caricature in a slightly unfair but, I am sure, readily recognisable manner.

The traditional budget has three typical characteristics: it is annual, it is concerned solely with inputs and it relates essentially to those inputs for which cash is required. Its 'annualness' is probably a cultural hangover from the time when we were primarily an agrarian society and when the annual cycle of seasons was the dominant feature of all activities (this fits well, of course, with gardeners' folklore). Its survival has been reinforced by legal and administrative routines in parliamentary and statutory provisions but is of dubious relevance in a planning context. I shall not consider it in depth here, because it is peripheral to my main theme, but it does introduce some distortions in information flows which we cannot ignore.

The input orientation, especially its cash connotations, can best be epitomised by rehearsing the processes that this kind of budgeting generates when we go through the five elements set out above.   The *recording* of decisions tends to be a list of sums of money allocated to particular agencies (organisations) to be spent on particular inputs (for example, so much for wages and salaries, so much for supplies, so much for fuel).   The information *transmitted* tends to be a schedule of spending authorisations, in which a person is told what he has to spend on what items over what space of time (and this is frequently all he is told – he may now know what others have to spend, or what he is supposed to be achieving with his spendings).   *Monitoring* consists of keeping a running tally of how much has been spent under each head (plus, in the more sophisticated systems, how much has been committed and how this compares with previous years' patterns and with the current allocations) and this may or may not be fed back to the budget-holder on a continuous basis with appropriate words of encouragement or admonition.   *Evaluation* essentially comprises the comparison of estimates with 'outturn' and, if the latter exceeds the former by a sizeable margin, all sorts of 'tut-tutting' will occur, accompanied by varying degrees of inquisitorial beastliness.   If the figures are exactly equal, the books obviously need auditing to see where the spare cash is being stashed away or where the overspendings have been concealed.   If the outturn falls far short of the estimate, you suspect that you were taken for a ride at estimate time and sack the budget-holder for unprofessional practice!   *Decision* then consists of trimming the margins here and there but allowing a general overall drift in line with the change in total available.   Perhaps this is why it has recently become the vogue to refer to 'forward' planning, to distinguish it presumably from old-fashioned 'backward' planning, the basic ethic of which was 'to he that hath much, shall more be given: but from him that hath something left over, shall be taken away even that which he hath!'   Hence the proud claim of the astute financial gamesman, 'Never knowingly underspent'.

My prime purpose is not, however, to poke gentle fun at (or, as some might see it, to make cheap gibes at the expense of) traditional financial control systems.   My more serious purpose is to test it out as an information source for planning purposes.   It seems reasonable, for instance, to expect it to answer the question, 'What does X cost?', where X is a unit of activity, measured preferably in output terms but, failing that, at least in throughput terms.   For instance, we might reasonably ask how much it costs to get a man with a hernia working normally again, or what it costs to check whether a pale listless patient has anaemia. In fact, it is currently impossible to answer those questions, except by painstaking research. The reasons for this are well worth exploring.

In the first place, no one has had any great incentive to find out, since the organisations responsible for (different parts of) the activities mentioned do not think like that, do not plan like that, are not controlled like that and do not run like that. So 'managers' have had no reason to devote scarce resources to collecting that kind of information. Secondly, even if we asked them more limited questions, about what it costs to carry out their bit of the overall activity (say the inpatient treatment of the man with a hernia), they will have a job disentangling the hernia cases from the others, they will know only about the costs that show up in their budgets and, even then, they will probably be able to tell you only what the average costs are over all cases treated and not the incremental costs of a typical additional case. These are serious shortcomings from a planning viewpoint, because we shall usually not be considering an increase in all types of case equally and it will rarely be true that all relevant resources will show up in the 'manager's' budget (for example, the use of space, or of patients' time, or of various associated services provided 'free' by other organisations), or that there are no fixed costs (or economies of scale) so that the distinction between 'average' and 'incremental' costs would cease to be important.

A classic case of this sort has arisen over the pressure to reduce length of stay in hospitals. Some estimates of cost savings have concentrated on the average cost per occupied bed–day (that is, total hospital costs divided by total number of occupied bed–days). On this basis, reducing length of stay by 10 per cent reduces costs by 10 per cent. However, since the costliest parts of treatment are typically concentrated in the earlier parts of an episode of inpatient treatment, the days 'saved' are much 'cheaper' than the 'average' day, hence average costs represent gross overestimates of potential savings. Moreover, if bed occupancy rates are kept up by admitting more patients (that is, increasing 'throughput') although cost per 'case' will fall, cost per occupied bed–day will rise, since a larger proportion of the 'days' are 'expensive' days. More serious still, however, is the way 'cost per case' has slipped into the act, for it should really be *'hospital* cost per *episode* of inpatient treatment'*, since there is no guarantee that it is not the same 'case' coming in and out several times and, in any case, 'inpatient' episodes are usually linked with outpatient consultations before and/or after the inpatient episode itself and the costs of these are not included. Moreover, the patients' costs are being ignored in all this, as are the costs falling on non-hospital health services, local authority social services, voluntary bodies and friends and relatives who provide support. Thus, if we rely on our cash-orientated input budget for hospitals as the major source of information on costs for planning purposes (even with respect to hospitals), we are likely to be grossly misled by the information it generates, unless we are aware of its inherent defects and take pains to make them good.

Thus my summary judgement on traditional budgeting is that its prime defect is its *input* orientation and its secondary defect that it does not even count *all* the inputs. But is it possible to conceive anything better?

SOWING SEEDS

If we are to break out of this situation we need to have budgets related to objectives (output budgeting) or, as a step in that direction, to activities that have clear objectives (programme budgeting). Thus 'organisations' are not given pride of place, but are left to establish their role (if any!) in relation to the objectives/activities that are being promoted by the budget–holders. This apparently simple conceptual change has far-reaching implications.

In the first place, it forces people to think about objectives and where they fit into the order of things (a botanical rather than a gardening question). It also enables those controlling the system to ask whether there ought not to be brought into existence some organisations or activities that do not currently exist and, although one *could* rely on random mutation and natural selection to generate the required evolutionary adaptation to changing circumstances, a little botanical knowledge about hybridisation possibilities and a policy of deliberate interbreeding might speed the process up a bit.

The practical manifestations of this change in budgeting philosophy would be that we would have, as primary budget–holders, directors of agencies concerned, say, with the welfare of the elderly, or of children, or of some similar 'client group', who would then 'buy in' services from 'activity' agencies (for example, those providing home helps, domiciliary nursing, residential care, medical attention of various kinds). These 'activity' agencies will normally be providing services for more than one 'client group' agency (for example, home helps will be provided for maternity or acute and chronic medical cases for groups other than the elderly), so problems of cost apportionment and appropriate reimbursement mechanisms will immediately arise. These difficult problems are held out as reasons for sticking with a system that does not 'generate' such problems. Yet it is precisely the failure of the existing system to grapple effectively with these problems that underlies its weaknesses as an information system for planning purposes. It is not that the problems are not there; it is rather that they are not confronted.

That the gardener's life would be more uncomfortable under the watchful eye of the critical botanist cannot be denied but he should perhaps be made to pay that price if sufficient good comes out of it. Returning to the planning process set out above, let us see how it would now look. The *recording* of the decision would be more like a 'contract' between the buying agency and the selling agency, whereby the latter is not just told how much it will be 'paid' but also what services it is to deliver in return. The *transmission* of this information will similarly need to include both expected 'output' or 'activity' levels as well as expected remuneration for inputs used. *Monitoring* will, similarly, range over both output and input, so that *evaluation* can take the form of cost effectiveness studies and not just cost comparisons. In this way the higher-level agency would have a sensible basis for reviewing its decisions on the next round, by shifting resources towards those agencies that are proving relatively more cost-effective *in terms of the high level agency's objectives.*

There is still one weakness in this alternative view of the world that needs to be put right, namely the failure of agencies to report costs that do not fall on their budgets. This is a very serious and pervasive problem throughout the economic system and the solution obviously lies in finding some way of incorporating these notional (or 'shadow') costs into budgeting processes. One simple radical way of doing this is to make the agency actually pay for the resource in question (for example, patients are paid for the time they have to wait after the appointed time for a consultation). If that is considered too strong a dose of weedkiller for initial applications then, as a more moderate (but probably less effective) start, we could use sample 'audits' of such waiting times to estimate 'hidden' costs of various agencies and simply use these when it comes to the monitoring and evaluation phase of the planning process. Again, I do not under-estimate the conceptual and practical difficulties of establishing such 'shadow' costs; but it is no excuse for not grappling with them to say that the current budgeting system has the advantage that you do not have to bother with such niceties. I would rather say that the current budgeting system has the 'advantage' that it has systematically obscured such gross inefficiencies'.

STIMULATING HEALTHY GROWTH

Defenders of the present system will doubtless feel that it has been pilloried unjustly and that I am giving insufficient credit to the recent developments in hospital costing that have manifested a steady movement away from a sporadic, local, uncoordinated assembly of financial data based solely on broad input categories. By linking such data to hospital activity analysis, it becomes possible to cost out particular services over departments and, hence move the system in the direction of 'activity' budgets.

The difficulty here is the difference in focus between short-term 'managerial' considerations and long-term 'planning' considerations. For the former, it is confusing and irritating to the head of a department to have 'loaded' on to his budget items over which he feels he has no control (like a fixed percentage of the institution's total heating bill, or an imputed rental for the space he uses, or a share of central administrative overheads). What he wants is information only about the items that are susceptible to day-to-day or week-to-week 'management' and that is all that those monitoring his performance in context need to concentrate on.

When it comes to reviewing the overall scale of the activity in a planning context, however, we are going to need data on those variables that lie outside short-term managerial discretion but that could become possible objects for adjustment. Since hospital costing activity has (perhaps properly) been focused primarily on the former set of problems and not on the latter, it cannot be held up as an answer to my strictures, though it could be cited as evidence that the system is capable of healthy growth.

One fundamental limitation on such growth remains, however, and that is the notion that budgeting is 'treasurer's business'. In the old-fashioned sense, it was. In this broader sense it is not, because once budgeting is openly and inextricably intertwined with objective setting, priorities, performance targets set in non-financial terms and monitoring and evaluation in the broadest sense, it is essentially a multi-disciplinary process. Apart from economists and administrators, too little serious study and *informed* criticism of budgeting processes has been forthcoming from the other professional groups in the health service, which may partly explain their current frustrations. While the mystique of medicine has come under severe attack, the mystique of accountancy has hardly been touched. If this reappraisal is to be conducted in a constructive way, then it needs to be made clear what each contributes to the process and, in a system of output or programme budgeting, there is a very clear and central role for the non-financial information, whereas the traditional budget is simply finance, finance and still more finance. So in broadening our conceptual thinking, we also increase our capability to take on board new knowledge.

Here I return to the main theme. For me, evolution connotes accidental adaptation through unco-ordinated innovations. It has got us broadly where we are today (like it or not). With luck it will see us through tomorrow. Planning connotes deliberate adaptation through co-ordinated innovation. It has had some spectacular results (like them or not) in the last few centuries. With good judgement (and a little luck) it will ensure that tomorrow's world is a better place. Put starkly, then, the choice seems to be between gardening as best we can with what nature provides, or setting up a laboratory blueprint based on the best that the botanical sciences can offer. But even within this (by now rather strained) analogy, there is a middle way, that is, to adopt as our model the botanical garden. The objective then becomes to improve on nature in a scientific but realistically sustainable way (the plants will grow and survive, often outside their natural habitats).

Thus, what we seem to need in our approach to budgets is a planning tool in the health service. Let us try transplanting some promising 'stock' which has all the theoretical (that is, genetic) properties we need, so as to convert our present rather weedy growth into something we could be proud of. Why be content with a scruffy suburban allotment when you could make a play for Kew Gardens?

13. The Role of Economics in the Evaluation of Health Care Technologies

WHAT IS THE PROBLEM?

An electronic fetal monitor was developed in the mid- to late-1960s in the US. It was proposed for high-risk pregnancies. It undoubtedly had some dramatic successes in high-risk pregnancies and was used in major medical centres. (But) ... once something has had some success in one group of patients there is an inevitable tendency to use it with other patients. The definition of the patient group becomes fuzzy and may be eventually applied to all patients in that particular category. Fetal monitoring was expanded and expanded so that it is probably now used in almost every delivery in the US for the women who come in in time to have the machine hooked up. But controversy followed .... Women objected to being interfered with to this extent. So one of the advocates of the procedure decided to do a controlled clinical trial to prove that it worked. The results showed no benefit, much to his surprise. That opened up the question all over again, and questions that should have been asked 10 years earlier were asked for the first time. It did not appear to benefit every woman. Then, over a period of about a year, many of the advocates backed off from that position and went back to their earlier position that it is of benefit to high-risk women in terms of mortality. But again clinical trials and other analyses seem to show that even this is questionable. Then there is a great deal of talk about prevention of mental retardation and cerebral palsy via use of the fetal monitor. However, some very good epidemiological analyses of the causes of mental retardation and cerebral palsy indicate that this appears not to be the case. Finally the most recent assertion is that the electronic fetal monitor is for efficiency; that we can't provide the nursing care but that this is an adequate replacement for the nurse (Banta 1979).

This theme, with appropriate contextual variations, seems to reflect so common an experience that it has become encapsulated in a standard scheme to represent the "product life-cycle" curve for medical technology" (Neuhauser 1979). See also Stocking and Morrison (1978).

Neuhauser epitomises each phase as follows (see Figure 13.1):

Figure 13.1 Product life-cycle curve for medical technology

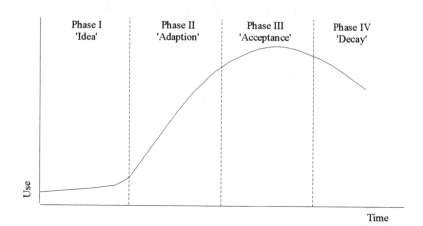

The idea period. Hundreds of new ideas occur all of the time. Only a fraction of these are actually pursued, and a fraction of these go into production or use. If the producer or innovator is required to pay for the careful, costly evaluation it is likely that fewer ideas will be turned into use .... A new procedure changes with time .... If evaluation is undertaken too soon these improvements will not be accounted for. At the end of Stage I the producing company has invested a lot of money and has yet to see the return on this investment. At this point articles are appearing in the medical literature. These articles are likely to be wildly enthusiastic but often based on faulty research design and inadequate evidence.

Adaptation. At the end of Stage I or at the beginning of Stage II the careful evaluative trials may have been started. These may be randomised trials with several years of follow up .... In Stage II there is growing acceptance. Sometimes this is by trial and error ....

Acceptance. Use has levelled off. Perhaps at this point the results of randomised trials are reported ... and ... the results are far less favourable than the initial enthusiastic reports. However, now hospitals have invested in equipment and doctors have learned these new skills and the climate has changed. These reported studies may be ignored or attacked with hostility. Unless the study shows clearly harmful results they may have very little effect on changing medical behaviour .... However, to delay adaptation until the trial results are available may be a political impossibility or, in the case of beneficial procedure, result in some patients being deprived of benefit.

Decay. Eventually the technology falls out of use. This may be easy if it is being replaced by something new and very difficult if there is nothing to replace it.

Among the important general principles Neuhauser derives from this analysis is that: 'The 'burden of proof' ... is important in the assessment of medical technology. The burden of proof in Phase I is on the innovator. In Phase II it is on the investigator who criticises the accepted practice', thus 'evaluation can be done too soon and too late'.

I, therefore, take the problem facing *economic* evaluation of health care technology to be to determine the optimum height, shape and timing of Neuhauser's 'product life-cycle curve' (in Figure 13.1) *from the point of view of a health care system*. This last phrase is of vital importance, because this particular viewpoint is to be differentiated from that of (a) the scientists who invented the technology, (b) the commercial interests which seek to make a profit out of it, (c) the practitioners who seek to add it to their armamentarium and (d) the patients who may benefit from it. Not even the impact on the latter category is to be seen as the 'touchstone' by which the optimum is to be determined, especially in a health care system that is financed to any significant degree by taxation of one kind or another. A favourable effect on patients is a necessary but unfortunately not a *sufficient* condition for the optimum level of utilisation of a new technology within a health care system to be positive (I am excluding from consideration here the use of these technologies within a research context, which raises different, but equally intractable, problems of evaluation).

SOME GROUND CLEARING

In the complex field of interdisciplinary endeavour it behoves us all to be careful about our jargon and I hope that the reader will bear with me if I devote a few pages of this chapter to defining certain key concepts, thereby removing in advance (I hope) some potential misunderstandings which experience has taught me have a better than even chance of obscuring our vision. I intend to work my way remorselessly through the key words in my title ('economics', 'evaluation' and 'technology' – only 'health care' am I taking for granted!) and then I am going to distinguish different types of evaluation, stressing the differences between medical, financial and economic, finally coming to some subdivisions within the latter category. Only then will we be in a position to tackle the substantive problems before us.

It is convenient for the flow of the discussion to dispose of the meaning of 'technology' first. It is tempting to include any innovation in the practice of health care delivery, be it a different arrangement of beds (or patients) in a ward, a different division of labour among staff, a change in the location of treatment, a new surgical procedure, a new drug, a new piece of diagnostic or monitoring equipment, a new prosthetic device, or even improved heating or ventilating systems in a hospital. Indeed, one could strike out 'innovation', 'different', 'new', or 'improved' in the foregoing sentence and include all existing ones too. In principle there is no reason to exclude from our remit *any* 'technique' bearing on health care delivery but it seems more appropriate to this occasion to limit ourselves to those with two particular properties: firstly, that if or when widely used they are (or will be) 'very costly' (a phrase I do not propose to define further at this stage), *and* secondly, that there are strong professional or commercial pressures on those managing health care systems to adopt them at a rate that is faster than their 'natural' inclinations would have generated. Thus my presupposed 'client' for the evaluation studies I am going to discuss shortly, is the health care system's management, acting on the English Common Law doctrine of *caveat emptor* (let the buyer beware) and seeking to establish what rate of investment (including zero) is in the best interests of the community it serves.

To understand the role that economics might play, we need some common understanding about what economics is. Elsewhere, (Chapter 3), I have gone on at some length about the distinctions between economics as a 'topic' and economics as a 'discipline'. The 'topic' of economics has as its subject matter everything to do with 'the economy' in general (inflation, unemployment, the balance of payments, productivity, etc.) and, for some people, therefore, economic evaluation means extending evaluation to take into account 'economic' variables, such as effects on employment, or earnings, or prices, or public expenditure, etc. Economics is certainly about those things, but that is not *all* that it is about and, although this interpretation does have important implications concerning the proper scope of evaluation in the present context, it is nevertheless *not* my central theme in this chapter. My central theme concerns economics as a 'discipline', that is, as a systematic body of knowledge with its own distinctive concepts and modes of thinking. I shall argue that it is these concepts and modes of thinking that need to be harnessed in a more sustained manner to the problems before us and especially those to be found in that branch of economics concerned with allocative efficiency (of which more anon).

I take 'evaluation' to be synonymous with evaluative research and, on the distinction between evaluative and non-evaluative research, I take my text from Suchman (1967) who writes:

> Evaluative research is a specific form of applied research whose primary goal is not the discovery of knowledge but rather a testing of the application of knowledge .... In contrast, non-evaluative research, while it may have practical implications, is primarily aimed at increasing knowledge rather than

manipulation or action .... A corollary of this distinction between understanding versus manipulation relates to contrasting degrees of abstraction versus specificity. Basic research aims at the formulation of theoretical generalisations or abstract predictions, while applied research stresses action in a highly specified situation involving concrete forecasts .... As one moves from the theoretical study to the evaluative study, the number of variables over which one has control decreases appreciably, while the number of contingent factors increases .... This is a major reason why so many evaluation studies appear repetitive – one can never be sure that a programme which works in one situation will work in another. To the extent that evaluative research can focus upon the general variables underlying a specific programme and test the effect of these variables rather than the effectiveness of the programme as a whole, it may hope to produce findings of greater general significance.

Note that Suchman is writing about evaluation in general, not particularly about economics, though much later in the book (in a passage concerning the role of operations research!) he writes:

One component of evaluative research that is often neglected ... is the cost of a programme. Few programmes can be justified at any cost .... Competition among service programmes sets the stage for a public demand for evaluation of results in terms of required resources .... In using these cost criteria, however, one must keep in mind the social aspects of such financial evaluation. As Flagle (1963) cautions: 'The construction of a table or scale of utilities, although cast in terms of economics, is essentially a psychological experiment involving social values'. (pp. 145–6)

Thus *all* evaluation studies need to be interpreted carefully with respect to the time and place at which they were conducted, the options that were considered, the range of factors that were taken into account, the criteria of success/failure that were adopted and how these were measured and weighed one against another. A 'definitive' evaluation is likely to be as elusive as an 'ultimate' truth and this is intrinsically so for *all* types of evaluation, economic or non-economic, a thought which brings me naturally to my next topic.

An important kind of non-economic evaluation is variously referred to as medical, clinical, or epidemiological. For my purposes any differences between them are unimportant because they all seek to answer the question 'will technology X improve the health of the patients?'. To answer that question involves a careful specification of: (a) the alternatives (to answer the supplementary question 'compared with what?'), (b) the patients (that is, the condition to be treated and all other significant personal or social variables) and (c) the precise nature of all the concurrent activities associated with the use of technology X. Still more, it requires some definition of what is meant by the patient's health, how we know when it has improved and what measurements are used to establish this, over whatever time horizon is taken to be appropriate for the investigation. It is a tough assignment and it is no wonder that a great deal of such evaluation is inconclusive or downright misleading because of weaknesses in research design or data collection and analysis.

Two of these common weaknesses in medical evaluation are particularly important for the subsequent discussion of economic evaluation and, for that reason alone, warrant further discussion here. They concern the measurement of health and the time horizon. The measurement of health is a vast topic, worth a conference of its own (see, for instance, Holland *et al.* (1979), Culyer 1978, Culyer 1983), however, I wish here to make a few simple but important points. First, clinical trials frequently concentrate on very narrow technical indicators (such as blood pressure, or tumour size, or presence or absence of some other biochemical abnormality), or upon some variant of change in life expectancy (such as case fatality, survival rates at some arbitrary future time). More complex measures such as relief of pain, date of discharge (length of stay), date of return to work, or other effect upon normal activities will more rarely be included and effects on relationships with others are still more unusual. Thus a treatment may be declared 'successful' by some narrow test, yet it might cause a patient's marriage to break up, or even lead to a patient's suicide and would clearly be judged 'unsuccessful' by these broader criteria. I would describe these phenomena as cases where the *costs* of treatment (in health terms) outweigh their *benefits* (in health terms, that is, where the treatment is not in the best interests of the patient – or the patient's family). This calls for an extension of clinical trials beyond the scope called for by *safety* regulations, which are typically concerned with clinical side effects of an equally narrow nature to the dimensions in which the benefits were measured (for example, reductions in blood pressure are associated with, say, increased risk of pulmonary embolism). However, in the general case, whenever multiple criteria are used (narrow or otherwise), some valuation process is inevitably entailed in weighing one against another, that is, in weighing 'costs' (in health terms) against 'benefits' (in health terms) and this process itself is of key interest to the economist and important grist to his mill.

The other important element in medical evaluation in which economists should show a great interest is the time horizon of the study. Two potential sources of trouble lurk here. The first is the tendency of such trials to operate on a crude 'before' versus 'after' basis, which obscures what happened 'during'. Thus a trial might show the patient better off *one year later*, without specifying what the patient's health state was like *during* the year-long 'therapeutic process' (and especially how much worse it might have been compared with the no-treatment option – after experiencing which patients might actually declare that it was not worth it – the cure is worse than the disease). The second trouble spot is the endpoint itself, since there is no natural terminating point for sequelae (not even the death of the patient if direct or indirect effects on others are significant). Thus there is always residual doubt and a natural tendency to weigh proximate effects more heavily than remote ones (other things being equal). The lessons for economic evaluation are that it is essential to ensure a full identification of *all* the implications of a technology, before, during and after its use and to attempt to get the *timing* of each change accurately established (and monitored) and not just whether or not it occurs sometime or other. This applies both to the changes in health and to the changes in resource use.

For the rest of this chapter I shall assume that, within its intrinsic limits and the usual limits of time and resources, the medical evaluation of any particular technology has been well done – but that it will not have concerned itself with any costs or benefits other than those upon health. Thus the stage is set for some kind of economic evaluation. Here our first problem is the common confusion between finance and economics due to the ambiguity in ordinary speech of the term 'cost'. At a superficial level, the question 'what will it cost?' is readily answered by specifying the price of the product in question, that is, the amount of money that will have to be handed over to the supplier to induce him to part with it. This is rarely the end of the story, however. Shopping itself costs money (for example, in terms of transport) but it also costs *time*, both in the act of purchasing and fetching but also in gaining information, making decisions, organising the 'use' of the purchase, etc. Some products occupy space, incur running costs, need repair and maintenance, take time to learn how to use, etc. Anyone who thinks that the cost of motoring begins and ends with buying a car has a shock in store! It is therefore important to rephrase the question 'what will it cost?' much more carefully if unambiguous answers are (even in principle) to be attainable. One such rephrasing would be 'how much of my money will I have to part with over the relevant time period?' This might be the relevant question for a budget-holder whose receipts are unaffected by the purchase and who did not care about the non-pecuniary costs falling on him or about any costs falling on anybody else. If a purchase affects receipts as well as outgoings, then the net effect on the budget-holder needs to be estimated, not just the expenditure effects. But even then we would still only have a *financial* appraisal, not an *economic* appraisal. To move to this higher plane of awareness the initial question would have to be rephrased 'what valuable things will I have to sacrifice over the relevant time period?'. The notion 'valuable things sacrificed' is much broader than money, including time and any other resource which has value, whether having to be bought or not (for example, something which you already have, but are willing to redeploy to this new use, like space). This notion of 'opportunity–cost' (the value of a resource in its best alternative use) as the relevant notion is what distinguishes an economic appraisal from a financial one (and there is no necessary presumption that a resource will be correctly valued in opportunity–cost terms by the price that is paid for it, though fortunately the two do frequently coincide in broad terms).

Within the realm of economic appraisal for example, Drummond (1980, 1981a) it is common to distinguish between cost-effectiveness analysis (CEA) and cost–benefit analysis (CBA). Within the context of this chapter the difference is best seen in terms of the treatment of health in the analysis. If the effects on health are left in terms such as blood pressure, mortality risk, or even quality-adjusted-life-years, but the effects on resources are valued in money terms, then we are operating within the realm of CEA (the hallmark of which is that 'costs' and 'effects' are non-commensurable). Thus the necessity to value health in money terms (or money in health terms!) is avoided, but at the 'cost' of being unable to offer any guidance as to whether *any* of the evaluated options should be accepted or not (because CEA can only tell you which of them is *more* cost-effective, not whether the effect is worth the cost!). In principle, CBA enables one to go to this last stage, but it forces the analysis to be explicit about the valuation of benefits, a matter which is left to the (implicit) judgement of the decision-maker in CEA.

ECONOMIC EVALUATION IN GENERAL

In order not to have to face too many difficulties at once, I shall initially assume that a cost-effectiveness framework is adequate for our purposes and that we have at our disposal medical evaluations that have used measures of effectiveness that are appropriate for our purposes (the frequency with which this last condition is *not* met is a strong argument for economics playing a role in the setting up of medical evaluations, counterbalancing the advantages of clinicians playing a role in economic evaluation!). I shall also ignore here general problems of research design not peculiar to the economic dimension (for example, the specification of options, controls, study size). I therefore assume that we know the likely distribution of benefits amongst a defined class of patients and that we know how many such patients there are likely to be in the population served. (I am only too keenly aware how heroic these assumptions are and I will come back to them later).

The broad format for a CEA is now simple, namely: *effectiveness* is measured by a change in some index of health (for example, quality-adjusted-life-expectancy); *cost* is measured by the net value of the changes in resource availability occasioned by the technology under appraisal and, in the case of mutually exclusive options, the preferred one will be that with the lowest ratio of cost to effect. If more than one option could be adopted, then it will normally be useful for the decision-maker also to know the *incremental cost* of improving effectiveness, as one moves from the most cost-effective to the least (especially if the most cost-effective option is limited in scope for any reason). Further complications arise if there is a special constraint on one or other of the resource inputs, making for large discontinuities in resource availability (for example, the use of a particular site). These complications over 'rationing' and 'lumpiness' can be handled (with difficulty!) but would take us well beyond the scope of this chapter. For present purposes I shall assume that comparative cost-effectiveness ratios are a satisfactory indicator.

The next important decision of principle is – what costs are to count? The managers of the health care system may decide that it is only a narrow range of costs that is to be considered, that is, changes in those resources for which they are directly responsible, for example, the land, buildings, staff and supplies, assigned for use by the service itself. They may, however, be under an obligation to take into account effects on related services, for example, upon other social services, upon the social security system, upon other public bodies such as local governments, or even upon voluntary bodies and charities. They may also feel that it is their duty to take account of all resource effects falling upon the population they serve, so that they will be concerned with effects on patients' time, patients' earnings, or other resource consequences for patients or their relatives and friends, etc. There may even be an obligation placed upon them by the government to consider rather broader economic considerations, such as effects on employment (local or national), on the balance of payments, though in my view these issues are best handled at a level other than that of the managers of a health care system. Even so, the appraisal might need to be designed to include such variables, if only for subsequent negotiating purposes. Thus the range of resource consequences considered has to be appropriate to the particular decision framework and will differ from one to another (see Suchman's comments earlier on in this chapter).

But there are nevertheless some general rules that are important whatever the range of variables included. Firstly, true resource costs must be separated clearly from 'transfer payments'. A transfer payment is a cash payment *not* reflecting any offer of a real resource in the opposite direction. Thus a wage payment establishes a claim on labour supply (which is a real resource, whereas a pension payment establishes no such claim (there is no quid pro quo, so it is a transfer payment). Thus sickness benefits and other social security payments are 'transfers', which merely *redistribute* the consequences of *real resource* changes (so that the loss of output and income occasioned by illness is redistributed in order that part of it is borne by the sick person and part by the taxpayers via the social security system) and to count *both* the loss of output from sickness absence *and* the compensating/redistributing transfer payments would be to count the same phenomenon twice, as it manifests itself in different guises. Thus it is advantageous to keep a 'real resource account' separate from a 'transfer account', because both may be interesting to policy makers (the former for efficiency, the latter for equity reasons). The other general point about the real resource costs is that irrevocable *past* resource commitments can never be a relevant consideration in an economic evaluation, since they will be the same whatever future course of action is chosen. Thus the commonly heard assertion that 'having gone this far we must go on' is quite specious (good money after bad?), as is the argument that we must include the recouping of those 'sunk' costs amongst the costs of any option that they helped to facilitate.

We come now to the central issue of resource valuation, remembering that this should, in principle, reflect the value of any used resource in its (facilitated) alternative use. If the resource in question can be bought and sold freely in a perfectly functioning market, then the 'price' of that resource can be taken to represent *both* of these values, providing that the changes in the volume of resource availability occasioned by the options under investigation are so small in relation to the total availability of the resource that they will not themselves affect the price. This is a point to which I will return later. Meanwhile, let us consider what has to be done in those cases where the market is not functioning perfectly. Possible reasons for considering 'adjustment' of market prices on these grounds are: (a) the existence of subsidies or indirect taxes on the resource, which mean that its true costs are, respectively, higher or lower than its market price; (b) monopolistic elements in the market, which imply that even at the margin the price charged for the resource will be higher than its cost; (c) not all costs incurred in supplying the resource may have to be met by the vendor, hence the price charged understates its true costs (for example, if the production of the good generates uncompensated air or water pollution, or if in supplying it congestion is generated in the transport system, which falls on others). Any or all of these 'efficiency' adjustments may be necessary in a particular study, or it may be demonstrated by prima facie argument that they are of such small quantitative significance it is justifiable to accept market prices as a good approximation for the true value of the resource. It is a judgement that will frequently be valid but which still needs arguing in each case. In labour markets there are special problems when skills are so specific that alternative use values are really very low or when unemployment in general is so high that it can genuinely be argued that if not used in the project the resource would not be used at all, hence its alternative use value is zero. This can be a valid argument, especially in the short run, but it is less likely to be the case over a time horizon within which retraining becomes possible, or within which some turnover of labour might be expected as people withdraw from the labour force and are replaced by (more versatile) newcomers.

But even 'adjusted' market prices may be rejected on the grounds *either* that people are not the best judges of their own welfare (and hence market prices generated, to some extent at least, by consumer choice are not 'reliable') *or* that the distribution of purchasing power is ethically unacceptable (hence market prices, which are partly influenced by it, are also ethically unacceptable). If the economic evaluation is to encompass these views about the valuation of resources, then the way forward lies along one or more of the following routes: (a) to conduct experiments in which people who are held to be 'good judges' of other people's welfare establish relative valuations and to test these for consistency and consensus, (b) to cast some 'responsible' person (or persons) in the role of the Delphi oracle to declare what the relevant values are to be (an obvious role for paternalistic politicians and philosopher kings); (c) to set up market simulations in which purchasing power is equalised (or whatever other distribution is held to be ethically acceptable) and then see what relative valuations emerge from this. This is controversial territory in which one school of thought holds that to go beyond efficiency-based adjustments of market prices is to go outside the proper scope of welfare economics and hence both professionally foolhardy and politically dangerous, while the other school argues that since people do hold such views about valuation it is better for economists to help sort them out than leave such valuations unanalysed. All agree that whatever basis of valuation is adopted, it must be made explicit. My own views are set out in Sugden and Williams (1978). For an opposing view see Mishan (1981).

Having identified and valued all the relevant resource changes in one way or another, we must now turn to the problems of timing and time horizons. Since a given quantity of a particular resource may have a different value according to *when* it is used (or released), we must consider the problem of intertemporal valuation. Here there are two phenomena to be distinguished – inflation and 'pure' discounting. There are two ways of handling the effects of inflation, *either* make all the valuation estimates on a 'constant price' basis (for example, 1984 prices) adjusting only for expected *relative* price changes through time, *or* estimate a fresh set of valuations each year, incorporating any expectations about changes in the *general* price level as well as specific changes in *relative* prices. But if the *former* approach is adopted, the discount rate, to be used in the phase to be described next, must be an inflation-free rate, whilst if the latter approach is used the relevant discount rate is one that incorporates inflationary expectations. What must *not* happen (but sometimes does) is to work with 'constant prices' and then use as the discount rate a market rate of interest that has a lot of inflationary expectations in it! The 'pure' discount rate is there to reduce to a common (present) value the value of resources at different points in time, where the differences in value are due *not* to inflation but to either of the following phenomena: (1) the fact that resources used (or released) earlier in time could be reinvested to produce returns elsewhere, hence releasing resources early is more valuable than releasing them later and, conversely, using them earlier is more costly than using them later, (2) people generally prefer early to late gratification of their wishes, hence expect greater 'returns' if they are to be induced to set aside resources for their own later use (or, *a fortiori*, for the use of later generations!), an observation which also leads to the use of a discounting procedure. There is much dispute amongst economists as to whether either, both or neither of these observations justifies discounting and, if so, what the actual appropriate discount rate should be. However, the majority view seems to be that a positive (pure) discount rate of up to 5 per cent is generally justifiable (with which I agree) and that this is based on the view that this is the range within which long-term productivity growth lies in most economies (a basis I personally do not accept!). However, the issue can be avoided in most practical studies by discounting at a range of alternative rates from 0 per cent to 5 per cent just to see if it makes any significant difference to the preferred option and then ask the policy-makers to worry about it if it does!

This brings me to my final general point about cost–effectiveness studies, namely, the treatment of uncertainty. There are some common devices in use here that need to be treated with the utmost suspicion. The first of these is to adjust for uncertainty by taking a very short time horizon (the 'pay-back period'). This is acceptable *if* what we face is the expectation that everything will go as planned for a (short) period but then the project might totally collapse (for example, fall down, get blown up, or be confiscated by a hostile power). This does not seem to be a very realistic scenario in most cases, it being more likely that things will go awry more gradually, possibly at an increasing rate as time progresses. This is what lies behind my next *bête noire*, which deals with uncertainty by a 'risk premium' to the discount rate. This has the effect of giving relatively more weight to proximate elements and relatively less to more distant ones, but its weakness is that it will treat all items at a particular point in time as *equally* susceptible to such risks, which is again unlikely to be the case. A better approach is to ask the 'experts' what is the range of alternative assumptions that prudence dictates should be considered as possible for the elements that enter the appraisal, so that we can see how sensitive the outcome is to such variations. This is even better, of course, if such a sensitivity analysis can be buttressed by explicit judgements of the probability distribution associated with each of the uncertain variables. The *very* ambitious analyst might go to the ultimate stage and attempt to check the degree of risk aversion of the decision-maker and then estimate certainty equivalent valuations but I personally would rest content with sensitivity analyses, with some gentle probing on possible probability distributions.

My last point in this section concerns the move from CEA and CBA and the problem of valuing health care *benefits*. This merely raises once more all the valuation problems discussed above but in a rather more highly charged atmosphere. First of all, the 'market price' valuation method has frequently been applied to benefit measurement in a context in which the value of health is taken to be solely its value as a means of earning money (the 'lifetime earnings', or 'productivity', or 'human capital' approach). This is unfortunate, because whilst effects on productivity are undoubtedly relevant, they are certainly not the *only* relevant 'value' outcome and may not even be the most important. However, finding a satisfactory way of valuing health *per se* (that is, other than as an instrument for earning money) raises sharply issues such as the proper role of the distribution of income and wealth in influencing such valuations, whether people's own valuations can be reliably elicited and, even if they could, whether they are a proper basis for *social* policy in this field, or finally, whether this is not essentially a political judgement, which is the responsibility of the legitimate political authority. I have views of my own on these issues (see Part two) and they cannot really be escaped even by resort to CEA, because at the end of the day, even if the analyst has shied off and left the health effects unvalued, the decision to accept (or not accept) one of the options *implies* such a valuation, however unwitting it may be.

SOME SPECIAL PROBLEMS WITH NEW TECHNOLOGIES

The special problems posed by new technology for the conduct of economic evaluation are caused by the conflicting demands of pervasive uncertainty and political urgency, which together generate the dilemma mentioned at the outset, namely, that economic evaluation can be done too soon or too late, so we need to think carefully about the 'right' time to do it. Uncertainty diminishes with the passage of time, hence the advantages of waiting. But in the meantime some decisions may have to be taken anyway and these may prove to be pre-emptive so that the subsequent analysis, however thorough and polished, becomes a mere footnote for history. So what useful role is there for economic or indeed any other kind of evaluation in this treacherous territory? Let us seek appropriate responses to three particular sources of difficulty in turn: (a) the absence of conclusive medical evaluation, (b) feedback from the decision itself to the assumptions on which the analysis informing the decision is to be based, (c) technology in state of flux.

Hitherto I have explicitly been assuming that we had at our disposal an appropriate body of medical evidence concerning effectiveness. This is unlikely to be the case, especially if we are considering *early* economic evaluation. It might be thought that in the absence of medical evidence on effectiveness, no economic evaluation is possible. This is not so. Just as it may be useful (but not conclusive) to do a medical evaluation without any economics, so it may be useful (but not conclusive) to do an economic evaluation without any effectiveness measures. For instance, if it is clear that technology X is going to cost 100 times what technology Y costs, this may be sufficient for immediate purposes because it may be quite evident to everyone that X cannot possibly be 100 times as effective as Y and so cannot be a cost-effective alternative. Similarly, it may be evident that X is probably somewhat more effective than Y, though how much more so is not clear. If an economic evaluation showed X to be less costly than Y, then we would not need a more accurate medical evaluation to decide to prefer X to Y, whereas we would need one if it turned out that X were 'somewhat more' expensive than Y. Since fairly crude economic evaluation, aimed at establishing broad orders of magnitude from desk-based guesstimates, is a relatively cheap and speedy activity, I am surprised that it is not more frequently used in these initial stages to keep a sense of proportion about the wider claims for cost-effectiveness, often based on poor identification of relevant costs and rather dubious valuation practices which could readily be spotted by a competent economic analyst.

The feedback problem is not so readily disposed of but I suspect that it could be handled adequately by appropriate sensitivity analysis, buttressed by some astute political negotiation! The archetypal assertion that we are now considering is that if the new technology is adopted quickly and on a large scale its costs will come down rapidly, whereas if it remains for years at a prototype stage or at restricted experimental levels of utilisation, then costs will remain high. It should not be difficult to work out how low costs would have to fall to generate a change in the preferred option (noting, however, that *other* associated costs may *rise* if a lot more resources need to be drawn in), but the difficult judgement is how likely this is to happen. There may be some advantage in testing out the likelihood that fixed price 'bulk purchase' agreements can be negotiated, as part of the evaluative process itself and not to accept such estimates of cost reduction as a basis for decision-making unless firm commitments can be obtained from suppliers. This is especially important where such a decision would leave the system dependent on one supplier and not able easily to switch to an alternative if the estimates proved wrong. 'Vulnerability' (defined as the cost of putting things right if things do not go according to plan) is an important subject for sensitivity analysis in a highly uncertain world populated by risk–averse people and organisations.

This is even more true of the continuously evolving nature of some technologies, and the consequent tendency for any substantial evaluation to be outdated by the time it is completed. Here I will assume that by careful technical assessment we have distinguished genuine evolutionary change from specious product differentiation. My response then would be that updating the economic aspects is likely to be rather easier than updating the medical evaluation, since so much of the work could be done on the 'quick and dirty' basis outlined above. It also seems to be a case where *crude* economic evaluation should be associated with strict limits on the rate of adoption of the technology, the pattern of introduction being designed to generate knowledge for the next round as well as to generate immediate health benefits. In this context it may be justifiable to adopt a deliberately varied pattern of introduction as part of the field evaluation process itself and *not* to aim at 'definitive' outcomes but 'provisional' ones with relatively short time horizons, which are designed to inform successive negotiations or policy reviews. In the flux of uncertainty about the rate of technological advance, pre-commitment to any one technology is to be minimised, even though a 'premium' has usually to be paid for such 'insurance' in the form of 'wasteful' spreading of one's portfolio of assets. Thus the analysis of risks in the sensitivity analysis should be used to guide the system away from concentration on a narrow range of options, *all* of which are vulnerable to an adverse movement in the *same* uncertain variable.

Against this background my final point will come as no surprise. It is that in this context the objective of economic evaluation should not be to decide what is the best technology to adopt, for in the presence of pervasive uncertainty that oversimplifies the problem. Rather, the objective is to explore the likely implications of variations in the scale and timing of the replacement of one technology by another. Moreover, since this sort of decision is best made on a provisional basis that commits the system as little as possible, it is essentially a search for a *robust next step*, that is, guidance on what we should do *now* so as to be in the best position to exploit whatever are the most likely (but still uncertain) developments that will constitute the options before us when we next have to make a decision. Thus economic evaluation, presented in summary form as a decision tree (Raiffa 1968) seems to me to be the most promising way to clarify thought in this confusing territory.

To structure a decision as a 'tree' involves being careful and explicit about the way in which possibilities are specified, probabilities are attached to contingencies and choices are made in a sequence that works back from multifarious end-points (in the future), each of which has a 'payoff' (positive or negative) and a probability of achievement (partly the result of chance, partly of choice). This focuses our attention on the choice we have to make now, which inevitably turns out to be a choice between rival 'lotteries' (that is, between one 'package' of risky outcomes and another). In simple cases this is easy to see but in the more complex real world, it is often incredibly difficult just to *structure* the problem correctly, let alone collect the economic data on costs and benefits and form a judgement about probabilities.

A problem that has characteristics similar to those we are interested in is described and analysed in detail in the Appendix to this chapter. It has a moderate degree of complexity and I defy anyone to 'spot' the solution simply by looking at the description of the problem. Needless to say, the 'answer' is sensitive to the payoffs, the probabilities, the problem structure *and* the decision rule employed. The important point it is designed to illustrate is that to know what to do *now*, one needs to think ahead as far as one can, though in a contingent manner. This does not imply that one has to have a 'plan' in the sense of a firm intention to do this now, that next year and then the year after. Facing a world that is in a state of flux, one avoids commitment as far as possible and makes those decisions one has to make in the light of all the information one can bring to bear (with the possibilities for recontracting built into one's analysis). This will not prevent us from making 'mistakes' (in the sense that with *hindsight* we may later see that there was something better we could have done) but it will enable us to *adapt* more readily to changing circumstances, so that our 'mistakes' should not prove fatal!

APPENDIX 13.1

AN ECONOMIC DECISION MODEL APPLIED TO INVESTMENT IN MEDICAL TECHNOLOGY

The problem

There is pressure to invest in a new medical technology, to which the following data relates:

1. The machine being offered now (period (1) costs £1 million to buy and install, and its running costs are £0.5 million in the first year, £0.6 million in the second year, £0.8 million in the third year and £1.0 million on the fourth year. The machine has an expected physical life of four years and running at optimum capacity it generates benefits of 1000 quality-adjusted-life-years (QALYs) per annum.
2. There is a slight chance that medical practice will change in such a way to make this machine (and its rivals) totally redundant within this four-year period. The (incremental) probabilities of this happening by the beginning of each year are as follows: 0.01, 0.03, 0.09 and 0.27.
3. It is much more likely that the present technology will be improved, with much more compact and economical machines becoming available shortly, reducing both the setting-up and running costs but generating the same benefits. There are known to be two rival developments being pursued by other firms, one of which (A) would generate a 20 per cent reduction in both setting-up and running costs compared with the machine currently on offer and the other (B) a 50 per cent reduction. The (incremental) probabilities of each of these being available in each of the first three years are as given in Table 13.1.
4. The firm supplying the current machine claims to be able to meet the rivals' claims by upgrading existing machines (these machines, unlike A and B, are subject to import licences and, when you have bought one, only upgrading is possible). Independent estimates suggest that there is a 50:50 chance that for an additional expenditure of £0.4 million it will be possible to obtain running cost reductions of the following order (benefits unchanged): year 1 10 per cent, year 3 30 per cent. For technical reasons, a machine can only be upgraded once, but if the

original machine is purchased later it incorporates any upgradings available to date (but no subsequent upgrading is then possible). These machines incorporating upgrading still cost £1.0 million.

Table 13.1 Incremental probabilities of machines A and B being available in the first three years

	Year 1	Year 2	Year 3*
Model A	0.1	0.2	0.3
Model B	0.02	0.05	0.13

* No capital expenditure will be permitted after year 3.

Table 13.2 The values at the end of year 4

	After 1 years' use	After 2 years' use	After 3 years' use
Model A	£0.45 million	£0.3 million	£0.15 million
Model B	£0.4 million	£0.3 million	£0.2 million

5. Neither upgrading nor replacement of machines interferes with the running of existing machines, which come on stream in the year *after* the one in which they are purchased. All costs are in year 0 present values at the relevant discount rate, and machines cannot be sold even if replaced. However, at the end of year 4 (the planning horizon) the values in Table 13.2 are to be credited.

6. Your organisation has firmly decided not to put a money value on a QALY, its policy being to go for that course of action that minimises the cost per QALY gained. It also takes the view that once it starts providing a service like this, it must keep going, even if it turns out to be more costly than expected.

7. You are asked to advise whether to buy the machine on offer now, or whether to wait until next year in the hope that things become cheaper.

Structuring the problem

We can set this up as a simple binary choice – 'buy now' *or* 'wait'. We can then systematically explore all the possible subsequent courses of action associated with each of these primary choices, work out the payoffs associated with each and the way chance impinges on them. Where we have scope for subsequent choices, we must think our way into that situation and decide what we would do *if the situation then is what we expect it to be now.* This is a conditional (and hypothetical) choice and is not a firm commitment. What we will actually do when the time comes will depend on what information (and objectives) we then have.

Table 13.3 sets out the structure of the 'buy now' branch of the decision tree. It contains 55 possible endpoints and is constructed as follows. The first column labels the alternatives 1 to 55, with the time path of decisions, year by year, coded as follows: 'red' = redundant (because of changed medical practice); 'run' = run the (last) previously purchased machine (and do nothing else); 'upg' = upgrade the previously purchased machine; 'A' = purchase the rival machine A (whilst running the previously purchased machine); 'B' = purchase the rival machine B (whilst running the previously purchased machine). In brackets, after each 'decision' (except 'run') is the probability of that 'decision' being available. The machine becoming redundant is, of course, an overriding 'decision', which, if it happens, takes the matter out of 'our' hands. The decision 'run' is always available, so will actually take on the residual probability if some other decision is preferred at that point or where 'red' intervenes. (This will become clearer when we look at Table 13.5.) Against each alternative the benefits and costs associated with it are shown, first in incremental terms ('ΔB' and 'ΔC') and then in cumulative terms thus far ('ΣB' and 'ΣC'). The benefits are measured in thousands of QALYs, the costs in millions of pounds, but to save space the units are suppressed in both cases. In the cost build-up, under 'upg', 'A' and 'B' the setting-up costs are added to the costs of running the previous machine a year in which that decision is made and only one inclusive cost figure is shown. By contrast, in year 4 the end–period values (if any) of existing machines are shown separately in the 'ΔC' column for the sake of clarity since the overall effect is sometimes to make 'ΔC' negative in that year.

Table 13.4 does the same thing for the 'Wait' alternative, which generates no less than 90 more alternatives (numbered 56–145 inclusive). The only new pieces of terminology here are pretty self-explanatory, namely: 'Wait' = do nothing: 'Buy Orig' = buy the original machine; 'Buy Upg' = buy the original machine with the latest upgrading built in (if available). Option 145 is the waiting game throughout, yielding no costs and no benefits whatever happens! In all other respects, Table 13.4 follows the pattern of Table 13.3.

Table 13.3 Decision tree for 'buy now' option

Yr	1					2					3					4				
no. alt	alt	Δ_B	Δ_C	Σ_B	Σ_C	alt	Δ_B	Δ_C	Σ_B	Σ_C	alt	Δ_B	Δ_C	Σ_B	Σ_C	alt	Δ_B	Δ_C	Σ_B	Σ_C
1 red	(0.01)			0																
2 run		1	0.5	1	1															
3						red (0.03)	1	0.6	1	1.5										
4						run			2	2.1	red (0.09)	1	0.8	2	2.1	red (0.27)	1	1	3	2.9
5											run			3	2.9	run			4	3.9
6											upg (0.05)	1	1.2	3	3.3	red (0.27)	1	0.7	3	3.6
7																run			4	4.0
8											A (0.3)	1	1.6	3	3.7	red (0.27)	1	0.4	3	3.7
9																run			3	3.65
10											B (0.13)	1	1.3	3	3.4	red (0.27)	1	0.25	3	3.4
11																run		(−0.4)	4	3.25
12						upg (0.5)	1	1	2	2.5										
13											red (0.09)	1	0.64	2	2.5	red (0.27)	1	0.8	3	3.14
14											run			3	3.14	run			4	3.94
15											A (0.3)	1	1.44	3	3.94	red (0.27)	1	0.4	3	3.94
16																run		(−0.45)	4	3.89
17											B (0.13)	1	1.44	3	3.64	red (0.27)	1	0.25	3	3.64
18																run		(−0.4)	4	3.49
19						A (0.2)	1	1.4	2	2.9										
20											red (0.09)	1	0.4	2	2.9	red (0.27)	1	0.48	3	3.3
21											run			3	3.3	run		(−0.3)	4	3.48
22											B (0.13)	1	0.9	3	3.8	red (0.27)	1	0.25	3	3.8
23																run		(−0.4)	4	3.65
24						B (0.05)	1	1.1	2	2.6										
25											red (0.09)	1	0.25	2	2.6	red (0.27)	1	0.3	3	2.85
26											run			3	2.85	run		(−0.3)	4	2.85

222

27	upg	(0.5)	1	0.9	1	1.9	red	(0.03)			1	1.9												
28							run		1	0.54	2	2.44	red	(0.09)			2	2.44						
29													run		1	0.72	3	3.16	red	(0.27)		0.9	3	3.16
30																			run		1		4	4.06
31													A	(0.3)	1	1.52	3	3.96	red	(0.27)		0.4	3	3.96
32																			run		1		4	3.91
33													B	(0.13)	1	1.22	3	3.66	red	(0.27)		0.25	3	3.66
34																			run		1	(– 0.4)	4	3.51
35							A	(0.2)	1	1.34	2	3.24	red	(0.09)			2	3.24						
36													run		1	0.4	3	3.64	red	(0.27)		0.48	3	3.64
37																			run		1	(– 0.03)	4	3.82
38													B	(0.13)	1	0.9	3	4.14	red	(0.27)		0.25	3	4.14
39																			run		1	(– 0.4)	4	3.99
40							B	(0.05)	1	1.04	2	2.94	red	(0.09)			2	2.94						
41													run		1	0.25	3	3.19	red	(0.27)		0.3	3	3.19
42																			run		1	(– 0.3)	4	3.19
43	A	(0.1)	1	1.3	1	2.3	red	(0.03)			1	2.3												
44							run		1	0.4	2	2.7	red	(0.09)			2	2.7						
45													run		1	0.48	3	3.18	red	(0.27)		0.64	3	3.18
46																			run		1		4	3.67
47													B	(0.13)	1	0.98	3	3.68	red	(0.27)		0.25	3	3.68
48																			run		1		4	3.53
49							B	(0.05)	1	0.9	2	3.2	red	(0.09)			2	3.2						
50													run		1	0.25	3	3.45	red	(0.27)		0.3	3	3.45
51																			run		1	(– 0.3)	4	3.45
52	B	(0.02)	1	1	1	2	red	(0.03)			1	2												
53							run		1	0.25	2	2.25	red	(0.09)			2	2.25						
54													run		1	0.3	3	2.55	red	(0.27)		0.4	3	2.55
55																			run		1	(– 0.2)	4	2.75

Table 13.4. Payoff tree: 'wait'

no.	1 alt	Δ_B	Δ_C	Σ_B	Σ_C	2 alt	Δ_B	Δ_C	Σ_B	Σ_C	3 alt	Δ_B	Δ_C	Σ_B	Σ_C	4 alt	Δ_B	Δ_C	Σ_B	Σ_C
56	red (0.01)	0	0	0	0															
57	buy orig	0	1	0	1															
58						red (0.03)	1		0	1	red (0.09)	1		1	1.5					
59						run		0.5	1	1.5	run		0.6	2	2.1	red (0.27)	1		2	2.1
60																run		0.8	3	2.9
61											upg (0.5)	1	1	2	2.5	red (0.27)	1		2	2.5
62																run		0.56	3	3.06
63											A (0.3)	1	1.4	2	2.9	red (0.27)	1		2	2.9
64																run		0.4 (−0.45)	3	2.85
65											B (0.13)	1	1.1	2	2.6	red (0.27)	1		2	2.6
66																run		0.25 (−0.4)	3	2.45
67						upg (0.5)	1	0.9	1	1.9	red (0.09)	1		1	1.9					
68											run		0.48	2	2.38	red (0.27)	1		2	2.38
69																run		0.64	3	3.02
70											A (0.3)	1	1.28	2	3.18	red (0.27)	1		2	3.18
71																run		0.4 (−0.45)	3	3.13
72											B (0.13)	1	0.98	2	2.88	red (0.27)	1		2	2.88
73																run		0.25 (−0.4)	3	2.73
74						A (0.2)	1	1.3	1	2.3	red (0.09)	1		1	2.3					
75											run		0.4	2	2.7	red (0.27)	1		2	2.7
76																run		0.48 (−0.3)	3	2.88
77											B (0.13)	1	0.9	2	3.2	red (0.27)	1		2	3.2
78																run		0.25 (−0.4)	3	3.05
79						B (0.05)	1	1	1	2	red (0.09)	1		1	2					

80													run		1	0.25	2	2.25	red	(0.27)			2	2.25
81																			run		1	0.3	3	2.25
																						(– 0.3)		
82	buy	(0.5)	0	1	0	1	red	(0.03)			0	1												
	upg																							
83							run		1	0.45	1	1.45	red	(0.09)			1	1.45						
84													run		1	0.54	2	1.99	red	(0.27)			2	1.99
85																			run		1	0.72	3	2.71
86													A	(0.3)		1.34	2	2.79	red	(0.27)			2	2.79
87															1				run		1	0.4	3	2.74
																						(– 0.45)		
88													B	(0.13)		1.04	2	2.49	red	(0.27)			2	2.49
89															1				run		1	0.25	3	2.34
																						(– 0.4)		
90							A	(0.2)	1	1.25	1	2.25	red	(0.09)			1	2.25						
91													run			0.4	2	2.65	red	(0.27)			2	2.65
92															1				run		1	0.48	3	2.83
																						(– 0.3)		
93													B	(0.13)		0.9	2	3.15	red	(0.27)			2	3.15
94															1				run		1	0.25	3	3
																						(– 0.4)		
95							B	(0.05)	1	0.95	1	1.95	red	(0.09)			1	1.95						
96													run			0.25	2	2.20	red	(0.27)			2	2.2
97															1				run		1	0.3	3	2.2
																						(– 0.3)		
98	A	(0.1)	0	0.8	0	0.8	red	(0.03)			0	0.8												
99							run		1	0.4	1	1.2	red	(0.09)			1	1.2						
100													run			0.48	2	1.68	red	(0.27)			2	1.68
101															1				run		1	0.64	3	2.17
																						(– 0.15)		
102													B			0.98	2	2.18	red	(0.27)			2	2.18
103															1				run		1	0.25	3	2.03
																						(– 0.4)		
104							B	(0.05)	1	0.9	1	1.7	red	(0.09)			1	1..7						
105													run			0.25	2	1.95	red	(0.27)			2	1.95
106															1				run		1	0.3	3	1.95
																						(– 0.3)		

Figure 13.4 Payoff tree: 'wait' (continued)

no.	Yr 1 alt	Δ_B	Δ_C	Σ_B	Σ_C	2 alt	Δ_B	Δ_C	Σ_B	Σ_C	3 alt	Δ_B	Δ_C	Σ_B	Σ_C	4 alt	Δ_B	Δ_C	Σ_B	Σ_C
107	B (0.02)	0	0.5	0	0.5	red (0.03)	1	0.25	0	0.5	red (0.09)	1	0.3	1	0.75	red (0.27)	1	0.4	2	1.05
108						run			1	0.75	run			2	1.05	run		(-0.2)	3	1.25
109																				
110																				
111	wait	0	0	0	0	red (0.03)	0	1	0	1										
112						buy / orig (0.03)														
113											red (0.09)	1	0.5	1	1.5	red (0.27)	1		1	1.5
114											run					run		0.6	2	2.1
115											upg (0.5)	1	0.9	1	1.9	red (0.27)	1		1	1.9
116																run		0.42	2	2.32
117											A (0.3)	1	1.3	1	2.3	red (0.27)	1	0.4	1	2.3
118																run		(-0.45)	2	2.25
119											B (0.13)	1	1	1	2	red (0.27)	1	0.25	1	2
120																run		(-0.4)	2	1.85
121						buy (0.5)	0	1	0	1	red (0.09)	1	1	0	1	run (0.09)				1
122											run					red (0.27)	1		1	1.4
123																run		0.48	2	1.88
124											A (0.3)	1	1.2	1	2.2	red (0.27)	1	0.4	1	2.2
125																run		(-0.45)	2	2.15
126											B (0.13)	1	0.9	1	1.9	red (0.27)	1	0.25	1	1.9
127																run		(-0.4)	2	1.75
128						A (0.2)	0	0.8	0	0.8	red (0.09)	1	0.4	0	0.8	red (0.27)	1		1	1.9
129											run			1	1.2	run		0.48	1	1.2
130																run		(-0.3)	2	1.38

Row															
131				B	(0.13)	1	0.9	1	1.7	red	(0.27)	1		1	1.7
132										run			0.25 (-0.4)	1	1.55
133	B	(0.05)	0	red	(0.09)	1	0.25	0	0.5	red	(0.27)	1		1	0.7:
134			0.5	run				1	0.75	run			0.3 (-0.3)	1	0.75
135			0											2	
136	wait		0	red	(0.09) orig	0	1	0	0					0	1
137				buy					1					1	1.5
138				buy	(0.5)	0	1	0	1	red	(0.27)	1	0.5	0	1
139				upg						red	(0.27)			0	
140				A	(0.3)	0	0.8	0	0.8	run		1	0.35	1	1.35
141										red	(0.27)	1	0.4 (-0.45)	0	0.8
142										run				1	0.75
143				B	(0.13)	0	0.5	0	0.5	red	(0.27)	1	0.25 (-0.4)	0	0.5
144										run				1	0.35
145				wait		0	0	0	0			0	0	0	0

Doing the Arithmetic: The Expected Value Calculations

Tables 13.5 and 13.6 set out the calculations concerning the options whose payoffs were calculated in Tables 13.3 and 13.4 respectively. The general principle underlying the technique is to start at the end (year 4) and work back to the beginning (year 1) so as to decide which 'lottery ticket' ('buy now' versus 'wait') has the highest 'expected value' (that is, the value of each payoff multiplied by the probability of obtaining it). Each payoff has two 'values', the benefits (in QALYs) and the costs (in pounds) and each pair of numbers is processed as if incommensurable, but in Tables 13.5 and 13.6 is merely the payoff data picked up from the final ('ΣB' and 'ΣC') columns of Tables 13.3 and 13.4.

The method then proceeds as follows (using options 1–11 initially as an illustrative example). Options 4, 6, 8 and 10 represent what will happen if the machine become redundant in year 4 and there is a 0.27 probability of this being the case, so the 'expected value' of this outcome is shown after the '=' sign. For instance, in option 4, the expected value of (3–2.9) 0.27 is (0.81–0.783), obtained by multiplying each term in the first set of brackets by 0.27. This leaves us with the residual probability of running the existing machine at 0.73, so against options 5, 7, 9 and 11 are shown the payoffs, probability and expected values of that eventuality. Since options 4 and 5 *jointly* constitute one option at an *earlier* stage in the decision-making process, we need to add together their expected values to see what their joint value is, and it turns out to be (3.73–3.63). If we do the same for the joint options 6 and 7, 8 and 9, and 10 and 11, we obtain their joint values as (3.73–3.892), (3.73–3.664) and (3.73–3.291). If we look at all four of these joint options, since they all generate the same expected benefit (3.73), we can choose between them purely in terms of their costs. The least cost option is the last one (10 and 11) at 3.291, but we have only a 0.13 chance of having that one available to us, so there is a 0.87 chance that we may have to finish up with something else. The next most favourable option is (4 and 5) and this will certainly be available if we have reached that section of the decision tree, so it attracts the residual 0.87 probability. The other two joint options, (6 and 7) and (8 and 9), are therefore discarded and play no further role in the calculations, hence the double vertical lines blocking them off at this stage. What does continue to the phase of decision is 0.13 of (3.73–3.291), which comes to (0.485–0.428) and 0.87 of (3.73–3.63), which comes to (3.245–3.158), the sum of these amounting to (3.73–3.586). This represents the expected value of options 4–11 inclusive. Similar calculations are made for options 13–18, 21–23 and 25–26, which constitute other basic choices at the beginning of year 4, within the earlier hypothetical) decision (spanning all the options from 2 to 26 inclusive) of buying the machine in year 0 and running it in year 1.

The next phase of the calculation incorporates the possibility that whatever we have done up to then, there is a 0.09 possibility that it will be found to be redundant at the beginning of year 3. This is what options 3, 12, 19 and 24 represent, so each of these has to be weighed against their respective complementary 0.91 probabilities that we shall be able to proceed along the options analysed above. Thus each of the earlier expected values for the joint options (4–11), (13–18), (21–23) and (25–26) has to be multiplied by 0.91 and their value, when added to the value of the redundancy risk, gives us (3.574–3.452), (3,574–3.592), (3.574–3.383) and (3.574–2.828). We can now choose between them on a least–cost basis as before, from which it is obvious that the last one would be our first preference. But there is only a 0.05 chance of this being available (expected value (0.179–0.142) so we now turn to our second preference, which is the penultimate one. But this is only 'available' with a 0.2 probability, so with a 0.95 chance that we may turn to it, its expected value is (0.679–0.643). The fall-back position, if neither of these is available, is 0.8 of 0.95 if (3.574–3.452), the first combined option, which generates an expected value of (2.716–2.624). Adding together these new expected values gives us (3.572–3.409) as the expected value of all the options from 3 to 26 inclusive (although the options 12–18 have now dropped out of the picture).

At this stage the possibility of redundancy at the beginning of year 2 has to be brought in and, in this part of the decision tree, this is represented by option 2, which has a probability of 0.03. When the expected value of this is added to 0.97 of our previously calculated expected value of (3.574–3,409), we obtain an expected value of (3.497–3.352) for options 2–26 inclusive. Similar calculations yield corresponding expected values of (3.497–3.616) for options 52–55, each of which represents a set of choices to be made in year 1. Again our first preference is the least cost–combined options and that is 52–55, but this has only a 0.02 probability of being available, generating an expected value of (0.07–0.053). Our second preference will be options 2–26, which, with the residual probability of 0.98, yields an expected value of (3.427–3.285). Adding these two together gives us (3.497–03.338), which then has a weight of 0.99 against option 1, which is the outcome (probability 0.01) that redundancy will occur at the start of year 1. Thus the expected value of options 1–55 is (3.497–3.348), and this is the key summary statistic for the 'buy now' decision, if we go for it.

Table 13.6 does the same thing for all the options under the 'wait' alternative. It proceeds in like manner and the only complication is that

Table 13.5 Buy now: expected value calculation

Option no.	Year 4			Year 3	
1					
2					
3					(2-2.1) 0.09 = (0.18-0.189)
4	(3-2.9) 0.27 = (0.81-0.783)	(3.73-3.63) 1.0 [0.87]	= (3.245-3.158)		
5	(4-3.9) 0.73 = (2.92-2.847)	(2nd preference)			
6	(3-3.6) 0.27 = (0.81-0.972)	(3.73-3.892) 0.5	‖		
7	(4-4) 0.73 = (2.92-2.92)			(3.73-3.586) 0.91 = (3.394-3.26)	
8	(3-3.7) 0.27 = (0.81-0.999)	(3.73-3.664) 0.3	‖		
9	(4-3.65) 0.73 = (2.92-2.665)				
10	(3-3.4) 0.27 = (0.81-0.918)	(3.73-3.291) 0.13	= (0.485-0.428)		
11	(4-3.25) 0.73 = (2.92-2.373)	(1st preference)			
12				(2-2.5) 0.09 = (0.18-0.225)	
13	(3-3.14) 0.27 = (0.81-0.848)	(3.73-3.724) 1.0 [0.87]	= (3.245-3.24)		
14	(4-3.94) 0.73 = (2.92-2.876)	(2nd preference)			
15	(3-3.94) 0.27 = (0.81-1.064)	(3.73-3.904) 0.3	‖		
16	(4-3.89) 0.73 = (2.92-2.840)			(3.73-3.699) 0.91 = (3.394-3.366)	
17	(3-3.64) 0.27 = (0.81-0.983)	(3.73-3.531) 0.13	= (0.485-0.459)		
18	(4-3.49) 0.73 = (2.92-2.548)	(1st preference)			
19				(2-2.9) 0.09 = (0.18-0.261)	
20	(3-3.3) 0.27 = (0.81-0.891)	(3.73-3.431) 1.0			
21	(4-3.48) 0.73 = (2.92-2.54)	(1st preference)		(3.73-3.431) 0.91 = (3.394-3.122)	
22	(3-3.8) 0.27 = (0.81-1.026)	(3.73-3.691) 0.13	‖		
23	(4-3.65) 0.73 = (2.92-2.665)				
24				(2-2.6) 0.09 = (0.18-0.234)	
25	(3-2.85) 0.27 = (0.81-0.77)	(3.73-2.85) 1.0	=		
26	(4-2.85) 0.73 = (2.92-2.08)			(3.73-2.85) 0.91 = (3.394-2.594)	
27					
28				(2-2.44) 0.09 = (0.18-0.220)	
29	(3-3.16) 0.27 = (0.81-0.853)	(3.73-3.817) 1.0 [0.87]	= (3.245-3.321)		
30	(4-4.06) 0.73 = (2.92-2.964)	(2nd preference)			
31	(3-3.96) 0.27 = (0.81-1.069)	(3.73-3.923) 0.3	‖		
32	(4-3.91) 0.73 = (2.92-2.854)			(3.73-3.783) 0.91 = (3.394-3.443)	
33	(3.-3.66) 0.27 = (0.81-0.988)	(3.73-3.55) 0.13	= (0.485-0.462)		
34	(4-3.51) 0.73 = (2.92-2.562)	(1st preference)			
35				(2-3.24) 0.09 = (0.18-0.282)	
36	(3-3.64) 0.27 = (0.81-0.983)	(3.73-3.772) 1.0	= (3.73-3.772)		
37	(4-3.82) 0.73 = (2.92-2.789)	(1st preference)		(3.73-3.772) 0.91 = (3.394-3.433)	
38	(3-4.14) 0.27 = (0.81-1.118)	(3.73-4.031) 0.13	‖		
39	(4-3.99) 0.73 = (2.92-2.913)				
40				(2-2.94) 0.09 = (0.18-2.65)	
41	(3-3.19) 0.27 = (0.81-0.861)	(3.73-3.19) 1.0	=		
42	(4-3.19) 0.73 = (2.92-2.329)			(3.73-3.19) 0.91 = (3.394-2.903)	
43					
44				(2-2.07) 0.09 = (0.18-0.243)	
45	(3-3.18) 0.27 = (0.81-0.859)	(3.73-3.538) 1.0	= (3.73-3.538)		
46	(4-3.67) 0.73 = (2.92-2.679)			(3.73-3.538) 0.91 = (3.394-3.220)	
47	(3-3.68) 0.27 = (0.81-0.994)	(3.73-3.571) 0.13	‖		
48	(4-3.53) 0.73 = (2.92-2.577)				
49				(2-3.2) 0.09 = (0.18-0.288)	
50	(3-3.45) 0.27 = (0.81-0.932)	(3.73-3.278) 1.0	=		
51	(3-3.45) 0.73 = (2.92-2.346)			(3.73-3.538) 0.91 = (3.394-2.983)	
52					
53				(2-2.25) 0.09 = (0.18-0.203)	
54	(3-2.55) 0.27 = (0.81-0.689)	(3.73-2.697) 1.0	=		
55	(4-2.75) 0.73 = (2.92-2.008)			(3.73-2.697) 0.91 = (3.394-2.454)	

Year 2   Year 1

(0–1) 0.01 → (0–0.01)

(1–1.5) 0.03 (0.03–0.045)

(3.574–3.452) 1.0 [X0.95X0.8] → (2.716–2.624)
(3rd preference)

(3.574–3.591) 0.5

(3.574
–3.409)
X0.97
→ (3.467
–3.307)

(3.497–3.352) 1.0 [X0.98]
→ (3.427–3.285)
(2nd preference)

(3.574–3.383) 0.2 [X0.95] → (0.679–0.643)
(2nd preference)

(3.574–2.828) 0.05 → (0.179–0.142)
(1st preference)

(1–2.9) 0.03 → (0.03–0.057)

(3.574–3.663) 1.0 [X0.95] → (3.395–3.480)
(2nd preference)

(3.497
3.338)

(3.497
–3.348)

(3.574–
3.638) 0.97
→ (3.467
–3.529)

(3.497–3.586) 0.5

(3.574–3.725) 0.2

(3.574–3.168) 0.05 → (0.179–0.158)
(1st preference)

(1–2.3) 0.03 → (0.03–0.069)

(3.574–3.463) 1.0 [X0.95] → (3.395–3.290)
(2nd preference)

(3.574–
3.454) 0.97
→ (3.467
–3.350

(3.497–3.419) 0.1

(3.574–3.271) 0.05 → (0.179–0.164)
(1st preference)

(1–2) 0.03 → (0.03–0.06)

(3.574–2.657   0.97 → (3.467–2.577)

(3.497–2.637) 0.02
→ (0.07–0.053)

Table 13.6 Wait: expected value calculation

Option no.	Year 4			Year 3
56				
57				
58				(1–1.5) 0.09 = (0.09–0.135)
59	(2–2.1) 0.27 = (0.54–0.567)	(2.73–2.684) 1.0 x 0.87	= (2.375–2.335)	
60	(3–2.9) 0.73 = (2.19–2.117)	(2nd preference)		
61	(2–2.5) 0.27 = (0.54–0.675)	(2.73–2.909) 0.5	‖	
62	(3–3.06) 0.73 = (2.19–2.234)			
63	(2–2.9) 0.27 = (0.54–783)	(2.73–2.864) 0.3	‖	(2.73–2.658) 0.91 = (2.484–2.419)
64	(3–2.85) 0.73 = (2.19–2.081)			
65	(2–2.6) 0.27 = (0.54–0.702)	(2.73–2.491) 0.13	= (0.355–0.323)	
66	(3–2.45) 0.73 = (2.19–1.789)	(1st preference)		
67				(1–1.9) 0.09 = (0.09–0.171)
68	(2–2.38) 0.27 = (0.54–0.43)	(2.73–2.848) 1.0 0.87	= (2.375–2.478)	
69	(3–3.02) 0.73 = (2.19–2.205)	(2nd preference)		
70	(2–3.18) 0.27 = (0.54–0.859)	(2.73–3.144) 0.3	‖	(2.73–2.838) 0.91 = (2.484–2.583)
71	(3–3.13) 0.73 = (2.19–2.285)			
72	(2–2.88) 0.27 = (0.54–0.778)	(2.73–2.771) 0.13	= (0.355–0.360)	
73	(3–2.73) 0.73 = (2.19–1.993)	(1st preference)		
74				(1–2.3) 0.09 = (0.09–0.207)
75	(2–2.7) 0.27 = (0.54–0.729)	(2.73–2.831) 1.0	= (2.73–2.831)	
76	(3–2.88) 0.73 = (2.19–2.102)	(1st preference)		(2.73–2.831) 0.91 = (2.484–2.576)
77	(2–3.2) 0.27 = (0.54–0.864)	(2.73–3.091) 0.13	‖	
78	(3–3.05) 0.73 = (2.19–2.227)			
79				(1–2.0) 0.09 = (0.09–0.18)
80	(2–2.25) 0.27 = (0.54–0.608)	(2.73–2.251) 1.0	= (2.73–2.251)	= (2.73–2.251) 0.91 = (2.484–2.048)
81	(3–2.25) 0.73 = (2.19–1.643)			
82				
83				(1–1.45) 0.09 = (0.09–0.131)
84	(2–1.99) 0.27 = (0.54–0.537)	(2.73–2.515) 1.0 0.87	= (2.375–2.188)	
85	(3–2.71) 0.73 = (2.19–1.978)	(2nd preference)		
86	(2–2.79) 0.27 (0.54–0.753)	(2.73–2.753) 0.3	‖	(2.73–2.497) 0.91 = (2.484–2.772)
87	(3–2.74) 0.73 = (2.19–2.000)			
88	(2–2.49) 0.27 = (0.54–0.672)	(2.73–2.300) 0.13	= (0.355–0.309)	
89	(3–2.34) 0.73 = (2.19–1.708)	(1st preference)		
90				(1–2.25) 0.09 = (0.09–0.203)
91	(2–2.65) 0.27 = (0.54–0.716)	(2.73–2.782) 1.0	= (2.73–2.782)	
92	(3–2.83) 0.73 = (2.19–2.066)	(1st preference)		(2.73–2.782) 0.91 = (2.484–2.532)
93	(2–3.15) 0.27 (0.54–0.851)	(2.73–3.041) 0.13	‖	
94	(3–3) 0.73 = (2.19–2.19)			
95				(1–1.95) 0.09 = (0.09–0.176)
96	(2–2.2) 0.27 = (0.54–0.594)	(2.73–2.2) 1.0	= (2.73–2.2)	= (2.73–2.2) 0.91 = (2.484–2.002)
97	(3–2.2) 0.73 = (2.19–1.606)			
98				
99				(1–1.2) 0.09 = (0.09–0.108)
100	(2–1.68) 0.27 = (0.54–0.454)	(2.73–2.038) 1.0	= (2.73–2.038)	
101	(3–2.17) 0.73 = (2.19–1.584)	(1st preference)		(2.73–2.038) 0.91 = (2.484–1.855)
102	(2–2.18) 0.27 = (0.54–0.589)	(2.73–2.064) 0.13	‖	
103	(3–2.02) 0.73 = (2.19–1.475)			
104				(1–1.7) 0.09 = (0.09–0.153)
105	(2–1.95) 0.27 = (0.54–0.527)	(2.73–1.95) 1.0	= (2.73–2.038)	= (2.73–2.038) 0.91 = (2.484–1.089)
106	(3–1.95) 0.73 = (2.19–1.424)			

Year 2   Year 1

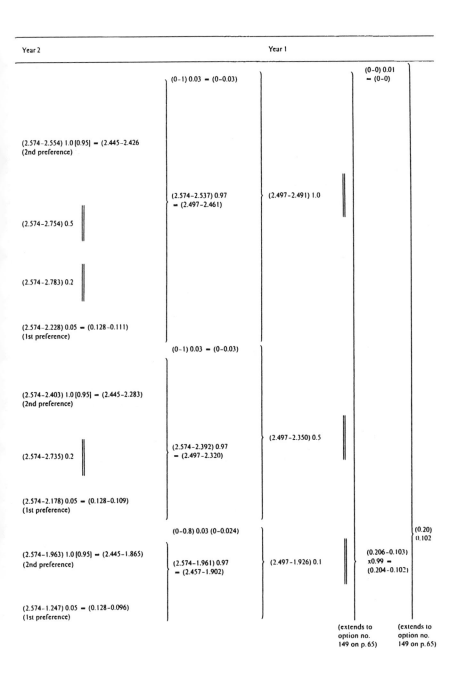

(0-1) 0.03 ━ (0-0.03)

(0-0) 0.01
━ (0-0)

(2.574-2.554) 1.0 [0.95] ━ (2.445-2.426)
(2nd preference)

(2.574-2.537) 0.97
━ (2.497-2.461)

(2.497-2.491) 1.0

(2.574-2.754) 0.5

(2.574-2.783) 0.2

(2.574-2.228) 0.05 ━ (0.128-0.111)
(1st preference)

(0-1) 0.03 ━ (0-0.03)

(2.574-2.403) 1.0 [0.95] ━ (2.445-2.283)
(2nd preference)

(2.574-2.392) 0.97
━ (2.497-2.320)

(2.497-2.350) 0.5

(2.574-2.735) 0.2

(2.574-2.178) 0.05 ━ (0.128-0.109)
(1st preference)

(0-0.8) 0.03 (0-0.024)

(0.20)
0.102

(2.574-1.963) 1.0 [0.95] ━ (2.445-1.865)
(2nd preference)

(2.574-1.961) 0.97
━ (2.457-1.902)

(2.497-1.926) 0.1

(0.206-0.103)
x0.99 ━
(0.204-0.102)

(2.574-1.247) 0.05 ━ (0.128-0.096)
(1st preference)

(extends to
option no.
149 on p.65)

(extends to
option no.
149 on p.65)

Table 13.6 Wait: expected value calculation (continued)

Option no.	Year 4		Year 3	
107				
108				(1–1.75) 0.09 = (0.09–0.158) }
109	(2–1.05) 0.27 = (0.54–0.284) }			
110	(3–1.25) 0.73 = (2.19–0.913) }	(2.73–1.197) 1.0	= (2.73–1.197)	= (2.73–1.197) 0.91 = (2.484–1.089) }
111				
112				(0–1) 0.09 = (0–0.09)
113	(1–1.5) 0.27 = (0.27–0.405) }	(1.73–1.938) 1.0 0.87	= (1.505–1.686)	
114	(2–2.1) 0.73 = (1.46–1.533)	(2nd preference)		
115	(1–1.9) 0.27 = (0.27–0.513)			
116	(2–2.32) 0.73 = (1.46–1.694) }	(1.73–2.207) 0.5 ‖		(1.73–1.932) 0.91 = (1.574–1.758) }
117	(1–2.3) 0.27 = (0.27–0.621) }			
118	(2–2.25) 0.73 = (1.46–1.643)	(1.73–2.264) 0.3 ‖		
119	(1–2) 0.27 = (0.27–0.54)	(1.73–1.891) 0.13	= (0.225–0.246)	
120	(2–1.85) 0.73 = (1.46–1.351) }	(1st preference)		
121				(0–1) 0.09 = (0–0.09)
122	(1–1.4) 0.27 = (0.27–0.378) }	(1.73–1.75) 1.0	= (1.73–1.75)	
123	(2–1.88) 0.73 = (1.46–1.372)	(1st preference)		
124	(1–2,2) 0.27 = (0.27–0.594)			
125	(2–2.15) 0.73 = (1.46–1.570) }	(1.73–2.164) 0.3 ‖		(1.73–1.75) 0.91 = (1.574–1.593
126	(1–1.9) 0.27 = (0.27–0.513) }			
127	(2–1.75) 0.73 = (1.46–1.278)	(1.73–1.791) 0.13 ‖		
128				(0–0.8) 0.09 = (0–0.072)
129	(1–1.2) 0.27 = (0.27–0.324) }	(1.73–1.331) 1.0	= (1.73–1.331)	
130	(2–1.38) 0.73 = (1.46–1.007)	(1st preference)		(1.73–1.331) 0.91 = (1.574–1.211)
131	(1–1.7) 0.27 = (0.27–0.459) }			
132	(2–1.55) 0.73 (1.46–1.132)	(1.73–1.591) 0.13 ‖		
133				(0–0.5) 0.09 = (0–0.045)
134	(1–0.75) 0.27 = (0.27–0.203) }			
135	(2–0.75) 0.73 = (1.46–0.548)	(1.73–0.75) 1.0	= (1.73–0.75)	= (1.73–0.75) 0.91 = (1.574–0.683) }
136				(0–0) 0.09 = (0–0)
137	(0–1) 0.27 = (0–0.27)			
138	(1–1.5) 0.73 (0.73–1.095) }	(0.73–1.365) 1.0 ‖		
139	(0–1) 0.27 = (0–0.27)			
140	(1–1.35) 0.73 = (0.73–0.986)	(0.73–1.256) 0.5 ‖		
141	(0–0.8) 0.27 = (0–0.216)			
142	(1–0.75) 0.73 = (0.73–0.548)	(0.73–0.764) 0.3 ‖		(0.095–0.051) 0.91 = (0.086–0.046) }
143	(0–0.5) 0.27 = (0–0.135) }	(0.73–0.391) 0.13	= (0.095–0.051)	
144	(1–0.35) 0.73 = (0.73–0.256) }	(1st preference)		
145	(0–0)	= (0–0) 1.0 (2nd preference)	= (0–0)	}

Year 2   Year 1

(0-0.5) 0.03 → (0.-0.015)

(2.574-1.247) 1.0 → (2.574-1.247) 0.97   → (2.497-1.210)

→ (2.497-1.225) 0.02
→ (0.050-0.025)
(1st preference)

(0-0) 0.03 → (0-0)

(1.574-1.848) 1.0

(1.574-1.688) 0.5

(0.161-0.08) 0.97
→ (0.156-0.078)

(0.156-0.078) 1.0 [0.98]
(2nd preference)
→ (0.153-0.076)

(1.574-1.283) 0.2

(1.574-0.728) 0.05 → (0.079-0.036)
(1st preference)

(0.086-0.046) 1.0 [0.95] → (0.082-0.044)
(2nd preference)

options numbered 111 onwards generate less benefits than the earlier ones, and option 136 *et seq* especially so. This means that we are forced to choose between blocks of options on grounds other than least cost. Here the rule we are given is to choose the least-cost-per-unit-of-benefit, that is, choose those courses of action with the highest value of

$$\frac{\Sigma B}{\Sigma C}$$

On this basis we prefer options 133–135 $\left(\dfrac{1.57}{0.728} = 2.162\right)$

to options 136–145 $\left(\dfrac{0.086}{0.046} = 1.870\right)$

which in turn is better than 128–132,121–127 or 112–120.

Similarly, in year 1, preference is given to options 107–110 (benefit: cost ratio 2.497:1.225 = 2.038) over options 111–145 (ratio 0.161:0.08 = 2.013), but both of these are markedly better than the ratios exhibited by any of the other options. At the end of the day, the 'wait' strategy yields an expected value of (0.204–0.102). If this is compared with the 'buy now' value of (3.497–3.348) it will be obvious that it yields a much better benefit:cost ratio, so the best strategy is to 'wait'.

This example contained a moderate degree of complexity, but there is, of course, far more in the real world. However, if this extra complexity consists merely of further eventualities to be considered, it does not raise different issues of principle. I will therefore confine myself to some key issues that the example takes for granted.

The time horizon is arbitrary in the sense that problems seldom have natural and unique terminal points and it is therefore a matter of judgement where, for analytical purposes, one chooses to truncate the planning horizons. This judgement will need to take note of the timing of any key *related* decisions (for example, the commissioning of new premises), of financial planning or forecasting horizons (for example, the assumption used here that no further purchases would be possible after year 3), the nature of the decision-making or review process, and so on.

The degree of risk is assumed to be reducible to probabilities associated with 'states of the world', and these in turn are supposed to be independent of decisions that 'we' make. It is not supposed that these probabilities are anything other than subjective estimates, though they may be the collective wisdom of 'experts'. The decision–makers are assumed to think in actuarial terms about 'fair gambles' and not to be risk–averse (in the sense that a 0.5 chance of 1000 *plus* a 0.5 chance of 0 is valued the same as the certainty of 500. This is only likely to be true of organisations taking a large number of such decisions, no one of which is sizeable in relation to its total operations. This is likely to be true of the British National Health Service in relation (say) to a particular scanner, but perhaps not to a small hospital faced with the same decision. In this latter case we would need to resolve these 'expected values' into 'certainty–equivalent values', which is a much more complicated form of evaluation.

We have assumed that benefits are reducible to a single unit of value (a QALY) and that these are known (or rather, explicitly estimated) for each alternative. This is often a source of uncertainty in itself and the figure 1000 used here may well be an 'expected value' derived from a probability distribution over many possible benefit levels, depending on the reactions of clinicians, the local incidence of the relevant conditions or problems, the reactions of patients, etc. Moreover, with multi-attribute benefit measures, assessment of gains becomes much more difficult unless dominance relationships emerge amongst the options.

It was assumed in this problem that benefits were not to be evaluated in money terms, though it would have made life much easier if they had. But it might also have changed the decision criterion, because we need then no longer work with benefit:cost *ratios*, but could have turned instead to the theoretically sounder criterion of the *difference* between benefits and costs. For instance, suppose that a QALY is worth £1000, then the benefit figures in the example can be interpreted as in millions of pounds, the same as the cost figures. It will immediately be observed that the 'wait' alternative will now be valued at £0.204 million less £0.102 million (that is, at £0.102 million) whilst the 'buy now' alternative is worth £0.149 million (3.497 *less* 3.348), so the latter is now preferred to the former. This is actually too hasty a conclusion, however, because each choice that was made *within* each strategy needs to be reappraised according to this changed criterion, and the consequences reworked. It so happens that in this example the 'buy now' alternative is unaffected, but in the 'wait' alternative different choices would be made across options 112–145 in year 2 and across all options in year 1. The net effect of these changed preferences is to change the expected value of the 'wait' strategy from (0.204–0.102) to (2.472–1.893), that is, the new excess of benefits over costs is £0.579 million, which is still greater than the net benefits of the 'buy now' strategy (still £0.149 million). The reason for the dramatic change in the 'wait' strategy is that options 98–106 replace options 111–145 as the second preference in year 1 because although the latter has a higher *ratio* of benefits to costs it has a much smaller *difference*.

Thus the outcome can be quite sensitive to the decision rule used, though in this case the 'wait' strategy remains the optimal one.

14. Economics of Coronary Artery Bypass Grafting

INTRODUCTION

The report of a consensus development conference on coronary artery bypass surgery recommended a large increase in the number of such operations in the United Kingdom, to 300 for every million of the population, 'if this represents provision for high benefit patients' (BMJ Editorial 1984). The report acknowledged, however, that such a development would require considerable funds and that 'the problem of assessment of priorities remains. This in turn should take account of estimations of the relative cost-effectiveness of other procedures competing for resources'.

The report went on to say 'We were impressed by one method of measurement combining quality and duration of life. Further development of this approach is recommended so that it can be of help not only in comparison between coronary artery bypass surgery and other priorities but also between the various subgroups of patients whom it is proposed would be treated by coronary artery bypass surgery. Such techniques would also help to identify health service estimates which are being continued despite low benefit'.

This chapter presents the economic analysis given to the panel at the consensus development conference in the hope that this will lead to a better understanding of the methodology and enable better data to be collected and deployed than the rather crude data used here.

THE PROBLEM

The objective of economic appraisal is to ensure that as much benefit as possible is obtained from the resources devoted to health care. In principle the benefit is measured in terms of the effect on life expectancy adjusted for the quality of life. The resources for health care should include not only costs to the service but also costs borne by patients and their families.

Given the amount of unemployment, which is expected to persist in the near future, increases in production that might be associated with employment gains have been disregarded. Procedures should be ranked so that activities which generate more gains to health for every pound of resources take priority over those which generate less; thus the general standard of health in the community would be correspondingly higher.

Coronary artery bypass grafting is one of many contenders for additional resources. Ideally, all such contenders should be compared each time a decision on allocation of resources is made to test which should be cut back and which should be expanded. The central issue before the conference was whether the number of operations for coronary artery bypass grafting should be increased, decreased, or maintained at its present level. To address this problem three factors need to be considered: firstly, which groups of patients stand to gain the most and the least from such operations, secondly, whether any of these groups of patients gain more for every pound of resources than patients awaiting other types of cardiac surgery – for example, transplantation, replacement of valves, insertion of pacemakers and percutaneous transluminal coronary angioplasty and, thirdly, whether other specialties have procedures that are more important than any of these – for example, kidney transplantation, renal dialysis and hip replacement. In an ideal world a better standard of care for the elderly, mentally ill and mentally handicapped, diagnostic methods such as computed tomography, nuclear magnetic resonance – and preventive measures – should also be considered. I shall restrict attention here to the more costly therapeutic technologies.

MEASURING BENEFITS

Generally, clinical trials compare rates of survival at various arbitrarily selected times after treatment has started. For our purposes we need to translate these comparative rates of survival into information on the change in life expectancy, which must then be adjusted for the effects on quality of life: some patients are willing to sacrifice a measure of life expectancy for a better quality of life. This feature is particularly important with respect to coronary artery bypass grafting as the procedure seems to offer a considerable improvement in the quality of life even for patients whose life expectancy has not changed or has even worsened.

To what extent will patients generally exchange duration of life for quality of life? The two principal (crude) components of quality of life in this context are physical mobility and freedom from pain (in other contexts the capacity to perform the activities of daily living and to engage in normal social interaction may be relevant).

Kind *et al.* (1982) based their work on these two factors and it is their work on the valuation of the state of health that is used here to establish profiles of quality of life for the various procedures under investigation. Their classification of the state of disability is as follows: I, no disability; II, slight social disability; III, severe social disability or slight impairment of performance at work, or both, able to do all housework except heavy tasks; IV, choice of work or performance at work severely limited, housewives and old people able to do only light housework but able to go out shopping; V, unable to undertake any paid employment, unable to continue any education, old people confined to home except for escorted outings and short walks and unable to shop, housewives able to perform only a few single tasks; VI, confined to chair or wheelchair or able to move only with support; VII, confined to bed; and VIII, unconscious. Their classification for distress is as follows: A, none; B, mild; C, moderate; and D, severe. They do not claim that these measures exhaust all the features that might be incorporated in a measurement of quality of life.

Table 14.1 shows the actual (median) valuations elicited by Kind *et al.* or each state of health from 70 respondents. Some severe states were regarded as worse than death – that is, had negative valuations – and it was only for those states given a value of below 0.9 (below the line) that the respondents regarded the degree of disability and distress as warranting less than 90 per cent of the score assigned to being fit and well. The 70 respondents included 10 doctors, all of whom appeared to have a much greater aversion to disability and distress than the population at large; they would therefore overvalue reductions in disability and distress compared with the rest of the population.

Table 14.1   Valuation matrix for 70 respondents:
(1 = healthy, 0 = head)

Disability rating	Distress rating			
	A	B	C	D
I	1.000	0.995	0.990	0.967
II	0.990	0.986	0.973	0.932
III	0.980	0.972	0.956	0.912
IV	0.964	0.956	0.942	0.870
V	0.946	0.935	0.900	0.700
VI	0.875	0.845	0.680	0
VII	0.677	0.564	0	– 1.486
VIII	–1.028	*	*	*

* Not applicable

Life expectancy and quality of life can then be joined into a single unit of benefit, the quality-adjusted–life-year. Unfortunately, few clinical studies have attempted a systematic measurement of changes in quality of life in these terms. I therefore asked three well informed cardiologists to give me their judgements on the comparative profiles of health of various patients with angina who had or had not undergone coronary artery bypass grafting. The cardiologists were asked to distinguish cases of severe, moderate and mild angina and within each of these three subgroups to distinguish cases with left main vessel, triple vessel, double vessel and one vessel disease. Figures 14.1 and 14.2 show the expected quality of life profiles obtained from these data. In 67 per cent of patients with disease of the left main vessel and severe angina there would be considerable gains from coronary artery bypass grafting. For 30 per cent the operation would provide no better prognosis than medical management, and for an unfortunate three per cent the operation would prove fatal (Figure 14.1).

Figure 14.1   Expected value of quality and length of life gained for patients with severe angina and left vessel disease

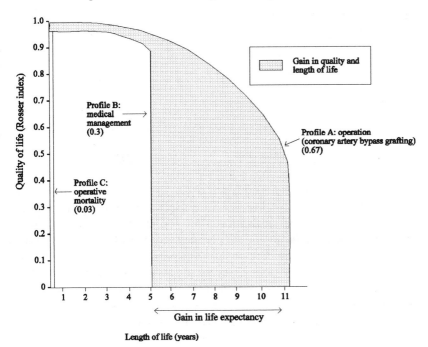

Thus the expected value of coronary artery bypass grafting in this case would be 0.67 of the shaded area minus 0.03 of the unshaded area (representing the quality of life that would have been enjoyed had the operation not been undertaken). In patients with disease of one vessel and severe angina (Figure 14.2) the probabilities would be the same but the outcomes different as coronary artery bypass grafting offers little potential benefit over medical management and, if the operation proves fatal, the patients will have lost the adjusted life expectancy that medical management offers (Wheatley 1984).

Figure 14.2   Expected value of quality and length of life gained for patients with severe angina and one vessel disease

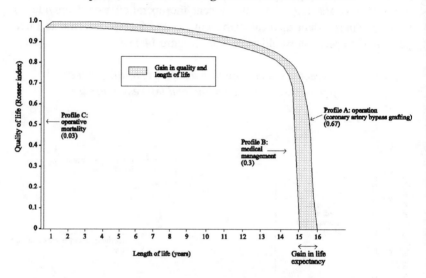

The three cardiologists complained about the difficulty of establishing these profiles with any confidence, which seems to be a serious indictment of the nature of the evaluative work currently carried out, with measurements of the quality of life playing a minor part, so that they were having to rely heavily on their clinical experience. All three cardiologists offered prognoses for the cases of severe angina but one was unable to offer any estimates for the cases of moderate and mild angina. The prognosis for replacement of valves for aortic stenosis were also based on only two respondents and the prognoses for percutaneous coronary angioplasty and pacemakers were from only one respondent. Table 14.2, based on these responses, gives a schedule of the effect on life expectancy adjusted for quality of life.

Table 14.2 Expected value of quality–adjusted–life–years gained
from operation compared with medical management*

| Coronary | Degree of angina | | |
anatomy	Severe	Moderate	Mild
Left main vessel disease	3.5	3	1.5
Triple vessel disease	3	1.5	1
Double vessel disease	2	1	1
One vessel disease (coronary artery			
bypass grafting)	0.5	0.5	
One vessel disease (percutaneous			
transluminal coronary			
angioplasty)	1	1	0.5

* Values are for a 'standard' patient: a 55-year-old man with good left ventricular function and no important concurrent conditions. For other classes of patient gains are probably less – for example, for women, older patients and those with poor left ventricular function, or other important diseases. Gains have been discounted at 5 per cent per annum to maintain comparability with data on cost.

MEASURING COSTS

The resources devoted to diagnosis and treatment include costs to the National Health Service and those falling on patients and their families. As there are few procedures for which much information about private costs is available these have been ignored here. The possibility that some of the rankings might be changed had private costs been included cannot be ruled out.

Several estimates of the service costs of coronary artery bypass grafting have been made in the United Kingdom and the United States of America. A detailed study conducted by the Department of Health and Social Security and the National Health Service in three NHS hospitals in 1980 suggested that the average recurrent cost of bypass surgery, including angiography, was about £3580 at 1983–4 prices, with an allowance of £725 for capital. Outpatient costs were not included. Three other British studies with apparently similar coverage of use of resources, but excluding capital, suggested that the costs ranged from £2500 to £4500 (Wheatley and Dark 1982, Patel 1984). Most studies have assumed that the cost of coronary artery bypass grafting is roughly the same regardless of the number of bypasses performed. American studies have suggested higher costs for bypass grafting at ($17 500) (Weinstein and Stason 1982). Much of the difference is probably explained by the fact that doctors' remuneration and the costs of acute inpatient care are fairly high in the United States of America.

The alternative to bypass surgery is medical treatment. Some studies assume that this costs about the same with and without bypass surgery. Others suggest that medical treatment is considerably reduced after coronary artery bypass grafting (Weinstein and Stason 1982, Coles and Coles 1982). The incidence of myocardial infarction may be reduced after grafting, resulting in further savings in medical costs.

The incidence of repeat operations after coronary artery bypass grafting and late operations after medical management may be similar. Weinstein and Stason suggested that, after working out the difference in the costs of medical care with and without operation and the difference in the cost of treatment of myocardinal infarction, the net cost of coronary artery bypass grafting is about 80 per cent of the surgical cost in cases of severe angina and about 90 per cent in cases of mild angina. When these ratios are applied to the DHSS's estimates of surgical costs of bypass surgery the excess of surgical over medical costs in the United Kingdom is about £2860 for cases of severe angina and £3170 for cases of mild angina. This would suggest a cost of about £3015 for moderate angina. The implication of this British and American hybrid estimate is that the cost of medical management without operation lies between £150 and £70 annually, depending on the severity of angina. The cost of medical care after operation would be about £75 annually.

A report from the Mayo Clinic by Reeder *et al.* (1984) indicates that, owing to the high rate of restenosis, percutaneous transluminal coronary angioplasty is only about 15 per cent cheaper than coronary artery bypass grafting. If this is so in the United Kingdom it would mean excess costs over medical management of between £2400 and £2680.

COST-EFFECTIVENESS

Table 14.3 shows, not surprisingly, that coronary artery bypass grafting offers better value for money in cases of severe angina and left main vessel disease or triple vessel disease and in cases of moderate angina and left main vessel disease than in any other circumstances.

To assess the relative value of coronary artery bypass grafting we need to make comparisons with other forms of expensive treatment such as replacement of valves, implantation of pacemakers and heart transplantation. Thick *et al.* (1978) estimated the cost of inserting a prosthetic valve (Bjork–Shiley) as being £2000, which would be £4540 at 1983–4 prices. This includes the cost of the operation, the valve and

Table 14.3 Coronary artery bypass grafting and percutaneous coronary angioplasty

Degree of angina	Coronary anatomy	Treatment	Present value of extra service costs (£000)	Discounted quality–adjusted–life– years gained	Present value of extra cost per quality–adjusted–life–year gained (£000)
		Severe angina			
Severe	Left main vessel disease	⎫		2.75	1.04
	Triple vessel disease	⎬ Coronary artery	2.85	2.25	1.27
	Double vessel disease	⎪ bypass grafting		1.25	2.28
	One vessel disease	⎭		0.25	11.40
	One vessel disease	Percutaneous transluminal coronary angioplasty	2.4	1.00	2.40
		Moderate angina			
Moderate	Left main vessel disease	⎫		2.25	1.33
	Triple vessel disease	⎬ Coronary artery	3.0	1.25	2.40
	Double vessel disease	⎪ bypass grafting		0.75	4.00
	One vessel disease	⎭		0.25	12.00
	One vessel disease	Percutaneous transluminal coronary angioplasty	2.55	0.75	3.40
		Mild angina			
Mild	Left main vessel disease	⎫		1.25	2.52
	Triple vessel disease	⎬ Coronary artery	3.15	0.50	6.30
	Double vessel disease	⎪ bypass grafting		0.25	12.60
	One vessel disease	⎭			
	One vessel disease		2.68	0.25	10.72
		Percutaneous transluminal coronary angioplasty			

subsequent inpatient care (intensive and general care) but does not include the cost of long term anticoagulant treatment or repeat operations. An estimate of the costs of inserting cardiac pacemakers was made by Barber (1978), which included the costs of implanting, reimplanting and associated check ups based on the experience at two hospitals in the west Midlands. These were revalued to accommodate 1983–4 prices. Initial implantation implies a commitment to future expenditure if the patient survives as replacement pacemakers are required every five years (less often if batteries powered by lithium are used). For heart transplantation Jennett (1984) quoted a figure of £15 000 (November 1982 prices) for initial costs; additional costs for subsequent drugs, etc., need to be included, which I have taken to be slightly higher than those required for kidney transplantation, amounting to an annual figure of about £2000.

For the quality of life I obtained estimates for patients with replaced valves and pacemakers by the same method as for those who had undergone coronary artery bypass grafting. For heart transplantation I used Hellinger's (1982) review of American (mainly from Stanford) experience, which indicated gains in life expectancy of between about two and six years. As techniques have probably improved I took a figure of 5.5, which I assume to be good quality of life, which, with discounting, gave a score of 4.5. Table 14.4 summarises these data and shows that insertion of pacemakers (for heart block) and replacement of valves (for aortic stenosis) are better value for money than coronary artery bypass grafting, though insertion of a pacemaker for the sick sinus syndrome and replacement of valves for mitral problems compare less favourably. Heart transplantation does not seem to be a serious contender. Table 14.5 shows the costs and relative gains in adjusted quality of life for the treatment of end stage renal failure and hip replacement. Interestingly, of all treatments examined so far, hip replacement comes near the top of the league whereas renal dialysis fares less well than heart transplantation.

Table 14.4 Summary of costs and benefits of three cardiac procedures

Procedure	Present value of extra service costs (£000)	Discounted quality–adjusted–life–years gained	Present value of extra service costs per quality–adjusted– life–year gained (£000)
Valve replacement for aortic stenosis	4.5	5	0.9
Pacemaker implantation for atrioventricular heart block	3.5	5	0.7
Heart transplantation	23	4.5	5

Table 14.5   *Summary of cost and benefit of some selected non-*
cardiovascular treatments*

Treatment	Present value of extra service costs (£000)	Discounted quality–adjusted–life–years gained	Present value of extra service costs per quality–adjusted–life– year gained (£000)
Kidney transplantations (cadaver)	15	5	3
Haemodialysis in hospital	70	5	14
Haemodialysis at home	66	6	11
Hip replacement[†]	3	4	0.75

* All costs at 1983–4 prices, including an estimate of annual capital costs. Complications are
   included in costs of end stage renal failure. For hip replacement a 2 per cent rate of failure and
   replacement each year is assumed.
[†] Estimate from DHSS Economic Advisor's Office, November 1984.

DISCUSSION

Before a well informed judgement can be made of whether it is in the
public interest to increase, decrease, or keep constant the number of
operations for coronary artery bypass grafting, reliable comparisons must
be made with other potential users of resources. Such information is not
readily available and the assumptions that I have made are not entirely
satisfactory. Clearly, further research is needed and should be focused
much more on measurement of the quality of life and on costs (both public
and private). Far too much attention has been paid to the rate of survival
which, in the case of coronary artery bypass grafting and many other
therapeutic procedures in which the main benefit is improved quality of
life, is potentially misleading.

Resources need to be redeployed at the margin to procedures for which
the benefits to patients are high in relation to the costs, such as the
insertion of pacemakers for heart block, hip replacement, replacement of
valves for aortic stenosis and coronary artery bypass grafting for severe
angina with left main disease and triple vessel disease and moderate angina
with left main disease. These treatments should take priority over
additional facilities for patients needing kidney transplants and coronary
artery bypass grafting for mild angina with left main disease, moderate

angina with triple vessel disease or one vessel disease and severe angina with one vessel disease, for which the costs per quality adjusted life year gained are higher.

15. Prioritising Waiting Lists

INTRODUCTION

Waiting lists represent improvements in health waiting to be delivered to people. In a well-organised system lists should not include people for whom the potential benefits of health care deteriorate rapidly if they are kept waiting (that is, urgent or acute cases). Nor should people for whom the benefits of earlier treatment are very small get precedence over people for whom the benefits of earlier treatments are quite large. Although routine statistics on waiting lists show the relative numbers of people waiting and the lengths of time they have to wait, nothing is known about the *quality* of waiting lists, that is, whether they represent potentially large benefits which are not being delivered, or only small ones.

It has proved extremely difficult to shift resources from one department to another in the National Health Service because there has been no measure capable of comparing the benefits of one activity with another. The unsatisfactory nature of routine waiting lists data becomes even more apparent (for example, Frankel 1989) when the length of waiting lists, or waiting times, are used as an indicator of need for additional resources. The length of a waiting list is strongly influenced by the behaviour of its 'owner' and can easily be lengthened or shortened by changes in clinical policies, such as shifting the balance of activity between day case and inpatient operative procedures. Thus there is no good reason to suppose that the person with the longest waiting list should have more resources. What needs to be known is the *composition* of the list, for example, a long list of minor cases could signal a lower priority use of additional resources than a short list where the benefits from earlier treatment are large.

METHOD

Five general surgeons at Guy's Hospital agreed to participate in an assessment of the benefits generated by treating patients on their waiting list. The assessment estimated the effects of such treatment on patients' life expectancy and on the quality of their lives. The 'top 22' conditions listed in Table 15.1 accounted for about 90 per cent of the waiting list (after the exclusion of diagnostic referrals and patients who had already declined the offer of treatment).

Table 15.1   The composition of Guy's Hospital general surgical waiting list as at June 1988*

Rank	Case type	Mean age of patients (years)	Number of patients	Cumulative % of waiting list
1	Subcutaneous lumps	41.2	325	28.7
2	Skin lesions	42.4	115	38.9
3	Unilateral VV – female	41.4	107	48.4
4	Unilateral IH – male	50.1	99	57.1
5	Bilateral VV – male	47.4	89	65.0
6	Unilateral VV – male	44.2	50	69.4
7	Ganglion	37.2	37	72.7
8	Piles	47.6	37	76.0
9	Bilateral VV – female	42.2	28	78.4
10	Ingrowing toenail	38.6	24	80.5
11	Circumcision	29.9	20	82.3
12	Anal fissure	31.3	17	83.8
13	Excision of mole	43.0	9	84.6
14	Anal tags	41.4	9	85.4
15	Unilateral IH – female	50.4	9	86.2
16	Hyperhidrosis	23.7	8	86.9
17	Gynaecomastia	30.3	8	87.6
18	Bilateral IH – male	59.1	8	88.3
19	Recurrent IH – male	65.4	7	88.9
20	Anal fistula	33.8	6	89.5
21	Epigastric hernia	42.8	7	90.1
22	Incisional hernia	56.0	6	90.6
	Total		1025	

* Total number on waiting list is 1131
VV = varicose veins
IH = inguinal hernia

Rosser's classification of illness states in Kind *et al.* (1982) was used to assess the pre- and post-operative quality of life of a typical patient waiting for each of these conditions (see Table 15.2). Other quality of life measures could have been used, unfortunately the leading British candidate, the Nottingham Health Profile, does not generate a single index number which is essential for this type of exercise.

The expected changes in life expectancy and in quality of life were evaluated using the valuation matrix of Rosser's 70 respondents (see Table 15.3). Table 15.4 shows the life expectancy data for each of the 22 conditions, together with the average expectations of the five surgeons concerning peri-operative mortality and percentages of treated patients who do not receive symptomatic relief.

Table 15.2   Rosser's classification of illness states

	Disability		Distress
I	No disability	A.	No distress
II	Slight social disability	B.	Mild
III	Severe social disability and/or slight impairment of performance at work	C.	Moderate
	Able to do all housework except very heavy tasks	D.	Severe
IV	Choice of work or performance at work very severely limited		
	Housewives and old people able to do light housework only but able to go out shopping		
V	Unable to undertake any paid employment. Old people confined to home except for escorted outings and short walks and unable to do shopping. Housewives able only to perform a few simple tasks		
VI	Confined to chair or to wheelchair or able to move around in the house only with support from an assistant		
VII	Confined to bed		
VIII	Unconscious		

Table 15.3   Rosser's valuation matrix: all 70 respondents

Disability rating	Distress rating			
	A	B	C	D
I	1.000	0.995	0.990	0.967
II	0.990	0.986	0.973	0.932
III	0.980	0.972	0.956	0.912
IV	0.964	0.956	0.942	0.870
V	0.946	0.935	0.900	0.700
VI	0.875	0.845	0.680	0.000
VII	0.677	0.564	0.000	−1.486
VIII	−1.028	Not applicable		

Fixed points: health = 1 dead = 0
(A negative value indicates that the state is regarded as being worse than being dead.)

Table 15.4   Mean values of five general surgeons' expectations concerning the effects of operating for 20 common conditions

Condition	Peri-operative mortality (%)	% not receiving symptomatic relief	Life expectancy following successful operation (years)
Subcutaneous lumps	0.004	0.2	34.9
Skin lesions	0.002	0.4	33.6
Unilateral VV – female	0.20	10.0	37.3
Unilateral IH – male	0.46	1.4	25.3*
Bilateral VV – male	0.48	10.6	27.7
Unilateral VV – male	0.30	7.0	30.4
Ganglion	0.03	1.8	38.7
Piles	0.41	6.0	28.9
Bilateral VV – female	0.20	14.0	36.7
Ingrowing toenail	0.04	5.4	37.3
Circumcision	0.04	0.2	43.0
Anal fissure	0.06	8.0	44.4
Excision of mole	0.01	0.0	33.1
Anal tags	0.05	9.0	34.6
Unilateral IH – female	0.28	2.2	29.0*
Hyperhidrosis	0.72	6.0	51.8
Gynaecomastia	0.26	1.0	42.6
Bilateral IH – male	0.87	2.8	17.00*
Recurrent IH – male	0.72	3.8	13.00*
Anal fistula	0.82	8.0	38.5
Epigastric hernia	0.42	1.0	33.1
Incisional hernia	1.22	3.6	21.8*

* Condition for which surgery increases life expectancy (according to at least two surgeons)
VV = varicose veins
IH = inguinal hernia

RESULTS

The data generated a table of expected gains in quality–adjusted–life–years (QALYs) for *treatment over no treatment* (see Table 15.5) and for *treatment one year earlier* (see Table 15.7). A description of the methods of calculation can be found in the technical appendix.

The conditions showing greatest benefit from surgery are male bilateral and recurrent inguinal hernia, male and female unilateral inguinal hernia and circumcision (see Table 15.6). A further six conditions show moderate benefit from surgery but benefit for the remaining conditions is low. There is a large inter-consultant variation for some of the conditions, indicated by a wide standard deviation. This is due mainly to one consultant's estimates of quality of life after surgery being consistently more optimistic than the others.

Table 15.5 Benefits from treatment versus no treatment (present value* of the future stream of QALYs)

Condition	QALY gain (mean ± SD)
Subcutaneous lumps	0.28 ± 9.13
Skin lesions	0.26 ± 0.22
Unilateral VV – female	0.29 ± 0.20
Unilateral IH – male	1.34 ± 0.70
Bilateral VV – male	0.26 ± 0.16
Unilateral VV – male	0.22 ± 0.06
Ganglion	0.39 ± 0.25
Piles	0.77 ± 0.40
Bilateral VV – female	0.41 ± 0.24
Ingrowing toenail	0.56 ± 0.25
Circumcision	1.21 ± 1.24
Anal fissure	0.70 ± 0.22
Excision of mole	0.25 ± 0.23
Anal tags	0.20 ± 0.08
Unilateral IH – female	1.12 ± 1.03
Hyperhidrosis	0.69 ± 0.53
Gynaecomastia	0.22 ± 0.10
Bilateral IH – male	1.84 ± 1.69
Recurrent IH – male	1.59 ± 1.80
Anal fistula	0.67 ± 0.41
Epigastric hernia	0.69 ± 0.58
Incisional hernia	0.27 ± 1.10

* Discounted at 5 per cent over the expected lifetime of the patient
VV = varicose veins
IH = inguinal hernia

Table 15.8 lists the conditions showing greatest benefit from *earlier* surgery as male bilateral inguinal hernia, hyperhidrosis and anal fissure. A group of four conditions showing medium benefit can also be identified and again there is large inter-consultant variation for some conditions. All the gains in QALYs are very small, indicating that there is little penalty in terms of QALYs in the event of delay for a year. In two cases, delay of surgery was actually *beneficial*, due to the high peri-operative mortality rate estimates of two surgeons for hernia surgery.

However, these outcome data alone do not give a full picture of the costs and benefits of surgery for these conditions. The benefit data were therefore related to the two major resource constraints which consultants themselves manage, that is, operating theatre time and hospital beds.

Table 15.6 Summary of benefit from surgery

Greatest benefit	male bilateral inguinal hernia
	male recurrent inguinal hernia
	male unilateral inguinal hernia
	circumcision
	female unilateral inguinal hernia
Medium benefit	piles
	anal fissure
	hyperhidrosis
	incisional hernia
	anal fistula
	ingrowing toenail
Large inter-consultant variation	male recurrent inguinal hernia
	circumcision
	excision of mole
	skin lesion
	male bilateral inguinal hernia
	female unilateral inguinal hernia
	incisional hernia
	hyperhidrosis
	female unilateral varicose veins

As part of the resource management initiative at Guy's Hospital, a computer system has been registering the time taken to bring each patient to the theatre suite and the times spent by the patient in the anaesthetic room, the theatre and recovery room.. The typical requirements of every treatment for each of these resources were calculated (see Table 15.9). Treatments were ranked in order by benefit per hour of theatre time (see Table 15.10) and by benefit per inpatient bed day (see Table 15.11). A collation of the actual rank orderings is shown in Table 15.12. It can be seen that the top half of the listed conditions is fairly consistent by all four criteria and compares favourably with Tables 15.6 and 15.8. The longer lengths of stay after surgery for piles and incisional hernia reduces the benefit per bed day for these conditions. Despite the QALY gain for hyperhidrosis being one of the highest, the earlier treatment and the long operating time puts this condition into twentieth place.

Table 15.7 Benefits from immediate treatment versus treatment 1 year later (Present value* of the future stream of QALYs)

Condition	QALY gain (mean ± SD)
Subcutaneous lumps	0.010 ± 0.003
Skin lesions	0.010 ± 0.003
Unilateral VV – female	0.008 ± 0.003
Unilateral IH – male	0.013 ± 0.006
Bilateral VV – male	0.007 ± 0.004
Unilateral VV – male	0.010 ± 0.003
Ganglion	0.019 ± 0.014
Piles	0.030 ± 0.030
Bilateral VV – female	0.010 ± 0.002
Ingrowing toenail	0.033 ± 0.012
Circumcision	0.018 ± 0.007
Anal fissure	0.050 ± 0.037
Excision of mole	0.007 ± 0.004
Anal tags	0.010 ± 0.007
Unilateral IH – female	0.021 ± 0.014
Hyperhidrosis	0.046 ± 0.038
Gynaecomastia	0.014 ± 0.010
Bilateral IH – male	0.068 ± 0.117
Recurrent IH – male	0.039 ± 0.044
Anal fistula	0.020 ± 0.015
Epigastric hernia	0.026 ± 0.022
Incisional hernia	0.014 ± 0.008

* discounted at 5 per cent over the expected lifetime of the patient
VV = varicose vein
IH = inguinal hernia

Table 15.8 Summary of benefit from earlier surgery

Greatest benefit:	male bilateral inguinal hernia
	hyperhidrosis
	anal fissure
Medium benefit:	male recurrent inguinal hernia
	in-growing toe nail
	piles
	incisional hernia
Large inter-consultant variation:	male bilateral inguinal hernia
	hyperhidrosis
	male recurrent inguinal hernia
	piles

Table 15.9 Resource requirements of surgery for each condition

Condition	Operating time* (minutes)	Length of stay (days)
Subcutaneous lumps	32	1.9**
Skin lesions	24	0.4**
Unilateral VV – female	50	2.3
Unilateral IH – male	44	3.5
Bilateral VV – male	54	2.8
Unilateral VV – male	33	4.0
Ganglion	31	1.4
Piles	36	6.0
Bilateral VV – female	62	2.8
Ingrowing toenail	29	1.4
Circumcision	31	1.7
Anal fissure	23	1.8
Excision of mole	28	0.5**
Anal tags	27	1.0
Unilateral IH – female	40	2.9
Hyperhidrosis	56	4.3
Gynaecomastia	38	2.8
Bilateral IH – male	43	4.3
Recurrent IH – male	40	4.0
Anal fistula	23	3.8
Incisional/epigastric hernia	58	5.1

* Excludes anaesthetic time and recovery time
** Day-case surgery included
VV = varicose veins
IH = inguinal hernia

Table 15.10 Benefits per hour of operating time

Treatment versus no treatment		Treatment now versus treatment one year later	
Condition	QALY gain	Condition	QALY gain
Bilateral IH – male	2.567	Anal fissure	0.130
Recurrent IH – male	2.385	Bilateral IH – male	0.095
Circumcision	2.342	Hyperhidrosis	0.049
Unilateral IH – male	1.827	Ingrowing toenail	0.068
Anal fissure	1.826	Recurrent IH – male	0.059
Anal fistula	1.748	Anal fistula	0.052
Unilateral IH – female	1.680	Piles	0.050
Piles	1.283	Ganglion	0.037
Ingrowing toenail	1.159	Circumcision	0.035
Ganglion	0.755	Unilateral IH – female	0.032
Hyperhidrosis	0.739	Incisional hernia	0.027
Incisional hernia	0.714	Skin lesions	0.025
Skin lesions	0.650	Anal tags	0.022
Excision of mole	0.536	Gynaecomastia	0.022
Subcutaneous lumps	0.525	Subcutaneous lumps	0.019
Anal tags	0.444	Unilateral IH – male	0.018
Unilateral VV – male	0.400	Unilateral VV – female	0.018
Bilateral VV – female	0.397	Excision of mole	0.015
Unilateral VV – female	0.348	Epigastric hernia	0.014
Gynaecomasatia	0.347	Bilateral VV – female	0.010
Bilateral VV – male	0.289	Unilateral VV – female	0.010
Epigastric hernia	0.279	Bilateral VV – male	0.008

VV = varicose veins
IH = inguinal hernia

Table 15.11 Benefits per day-bed occupied

Treatment versus no treatment		Treatment now versus treatment one year later	
Condition	QALY gain	Condition	QALY gain
Circumcision	0.712	Anal fissure	0.028
Skin lesions	0.650	Skin lesions	0.925
Excision of mole	0.500	Ingrowing toenail	0.024
Bilateral IH – male	0.428	Bilateral IH – male	0.016
Ingrowing toenail	0.400	Hyperhidrosis	0.014
Recurrent IH – male	0.398	Excision of mole	0.014
Anal fissure	0.389	Ganglion	0.011
Unilateral IH – female	0.386	Circumcision	0.011
Unilateral IH – male	0.383	Recurrent IH – male	0.010
Ganglion	0.279	Anal tags	0.010
Anal tags	0.200	Unilateral IH – female	0.007
Anal fistula	0.176	Anal fistula	0.005
Hyperhidrosis	0.160	Subcutaneous lumps	0.005
Subcutaneous lumps	0.147	Incisional hernia	0.005
Bilateral VV – female	0.146	Piles	0.005
Incisional hernia	0.135	Gynaecomastia	0.005
Piles	0.128	Unilateral IH – male	0.004
Unilateral VV – female	0.126	Bilateral VV – female	0.004
Bilateral VV – male	0.093	Unilateral VV – female	0.004
Gynaecomastia	0.079	Bilateral VV – male	0.003
Unilateral VV – male	0.055	Unilateral VV – male	0.003
Epigastric hernia	0.053	Epigastric hernia	0.003

VV = varicose veins
IH = inguinal hernia

*Table 15.12 Rank ordering of conditions with respect to benefit per
resource requirement*

Condition	Treatment versus no treatment		Treatment now versus treatment one year later	
	Benefit per hour of operating time	Benefit per day bed	Benefit per hour of operating time	Benefit per day bed
Bilateral IH – male	1	4	2	4.5
Recurrent IH – male	2	6	5	9.5
Circumcision	3	1	9	8
Unilateral IH – male	4	9	16	18
Anal fissure	5	7	1	1
Anal fistula	6	12	6	14
Unilateral IH – female	7	8	10	11
Piles	8	17	7	14
Ingrowing toenail	9	5	4	3
Ganglion	10	10	8	6.5
Hyperhidrosis	11	13	3	4.5
Incisional hernia	12	16	11	14
Skin lesions	13	2	12	2
Excision of mole	14	3	18	6.5
Subcutaneous lumps	15	14	15	14
Anal tags	16	11	13	0.5
Unilateral VV – male	17	21	17	21
Bilateral VV – female	18	15	20	18
Unilateral VV – female	19	18	21	18
Gynaecomasatia	20	20	14	14
Bilateral VV – male	21	19	22	21
Epigastric hernia	22	22	19	21

VV = varicose veins
IH = inguinal hernia

DISCUSSION

This first attempt to rate waiting lists by the level of expected benefit to
patients has shown that it is feasible at quite moderate cost. The
composition of the waiting list should be available within the Körner data
set, with each clinician taking about two hours to make the estimates of
pre- and post-quality of life, peri-operative mortality and percentage of
treated patients gaining no symptomatic relief from treatment. The Centre
for Health Economics provided approximately five days of research fellow

time for organising the study and data processing, excluding the three days spent in meetings at the hospital before and during the study. However, there is still a long way to go before the full potential of this approach can be realised. Firstly, a similar exercise needs to be conducted comparing the waiting lists of other specialties, which might then inform resource deployment at the margin. Secondly, it would be better to use 'hard' data to represent patient benefits rather than the informed judgement of the clinicians, although this would be more costly to collect. Thirdly, and also costly in terms of data collection, there are resource costs other than beds and theatre time. Finally, the value matrix derived from Rosser's work could be replaced by more recent and representative views. The calculations should also be repeated using the Quality of Life Indexes to see how sensitive the results are to such substitutions.

TECHNICAL APPENDIX 15.I

The net QALY gains have been calculated by the following equations:

1. *QALY gains by having surgery versus no surgery at all:*
    $$Q = (1 - m)(1 = nr)[Q^*_{T(D)} - Q_{T(D)}] - mQ_{T(D)}$$

    where m =   peri–operative mortality, as a fraction (Curve A)
              nr =   patients receiving *no* symptomatic relief from surgery, as a fraction
              $Q^*T(D)$ =   gross QALY total by having surgery, cumulative over patient survival (Curve B) and discounted at 5 per cent
              $Q^*T(D)$ =   gross QALY total without surgery, cumulative over patient survival (Curve C) and discounted at 5 per cent

Quality of life

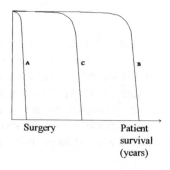

Surgery   Patient
                                           survival
                                           (years)

2. $QALY$ gains by having surgery now versus surgery one year later:
    $$Q = (1 - m)(1 = nr)[Q^*_1 - Q_1]_D - mQ_{1(D)}$$

where   m   =   peri–operative mortality, as a fraction
        nr   =   patients receiving *no* symptomatic relief
                            from surgery, as a fraction
        Q^*_1 =   gross QALY total in the first year, by
                            having surgery
        Q_1 =   gross QALY total in the first year
                            without surgery
        $[Q^*_1 - Q_1]_D$ = difference in QALY totals for the first
                            year, discounted at 5 per cent
        $Q1(D)$ =   gross QALY total in the first year without
                            surgery, discounted at 5 per cent

3. *A worked example*

    Hypothetical basic data

            m   =   .001
            nr   =   .3
            Q^*_1 =   1.00 (and remains constant
                                    in all subsequent *treated* years)
            Q_1 =   0.95 (and remains constant in
                                    all subsequent *untreated* years)
            Life expectancy (T) =   20 years (assumed unaffected by
                                    treatment)

    Note   With a 5 per cent discount rate a delay of one year
             involves a discount factor of .95.   At the same
             discount rate a whole stream at a constant level for 20
             years is worth approximately 13 times the single
             annual value.

    Calculation (1)   *Treatment versus no treatment*
    Formula: Q = $(1 - m)(1 - nr)[Q^*_T(D) - Q_T(D)] = mQ_T(D)$
    In this case: Q = $(1 - .001)(1 - .3)[1.0(13) - .95(13)] - .001$
                              $\times .95(13)$
                     = $.999 \times .7 \times [13.00 - 12.35] - .001 \times 12.35$
                     = $0.45454 - 0.01235 = 0.44219$

Calculation (2)   *Treatment now versus treatment one year later*

Formula: Q = $(1 - m)(1 - nr)[Q^*_1 - Q_1] = mQ_1D$

In this case: Q = $(1 - .001)(1 - .3)[1.00 - .95].95 - .001$
$\times .95 \times .95z$

= $.999 \times .7 \times .05 \times .95 - .001 \times .95 \times .95$

= $0.0332 - 0.0009 = .0323$

16. How Should Information on Cost-Effectiveness influence Clinical Practice?

Information on cost-effectiveness should influence clinical practice through practice guidelines, monitored within clinical audit and reinforced by purchasers through the contractual process. In principle there should be no resistance to this. After all, the objective of cost-effectiveness analysis is to ensure that the limited resources at our disposal are used to bring about the maximum improvement in people's health and that is also the objective of clinical audit and the objective of purchasers. Moreover, various official bodies involved in this area of activity have stated their positions in an encouraging way.

In a recent booklet entitled *The Evolution of Clinical Audit* (NHS Management Executive 1993a) the Department of Health cites one of its earlier publications (NHS Management Executive 1993b) as providing 'the strategic direction for clinical audit': 'Clinical audit involves systematically looking at the procedures used for diagnosis, care and treatment, examining how associated resources are sued and investigating the effect care has on the outcome and quality of life for the patient'.

Unfortunately the booklet then departs significantly from this remit by ignoring resources completely, except for resources devoted to the audit process itself. More recently still, the NHS Management Executive sent round an Executive Letter (NHSME 1993c) which includes a section on the use of clinical guidelines to inform the contracting process. It states that such guidelines will need to be:

- developed and endorsed by the relevant professional bodies
- based on good research evidence of clinical effectiveness
- practical and affordable
- where appropriate, multidisciplinary, and
- take account of patient choices and values

It is well known that the road to hell is paved with good intentions and consensual agreement in principle is no guarantee that things will actually turn out the way that well intentioned people wish. Some of the difficulties are purely practical and the greatest practical difficulty in this field is lack of relevant information. Superficially, there are plenty of data (that is, facts untouched by human thought) about the conduct of clinical practice but most of them cannot be applied systematically even to the issues which the creation of narrowly focused clinical guidelines brings to the fore, let alone to the issues which a cost-effectiveness approach brings to the fore.

There is, however, a prior problem which I fear will misdirect attempts to fill the information void if we are not aware of it. The problem in question is that many of those who pay lip service to the need for clinical practice to be pursued in a cost-effective way do not really appreciate what they are committing themselves to and they tend to shrink from the implications when they realise what they are.

I have written elsewhere about the weaknesses of clinical audit (as typically practised in the United Kingdom) as a mechanism for checking the clinicians behave in a cost-effective way (Williams 1993a). Instead of commenting on the currently rather tentative use by purchasers of the latent power that they have to redirect clinical activity into more cost-effective channels, I will concentrate on the potential role of practice guidelines in this matter.

It has been suggested that purchasers should not be buying 'treatments' but 'treatment protocols' (which I am here calling 'practice guidelines'). The idea is that when a patient presents, the case is identified as falling within a particular guideline and a particular protocol is then followed. This may lead to an expensive treatment being offered or to no treatment being offered, depending on what happens as the protocol is followed. The purchaser pays a fixed sum of money for each presenting case; this takes account of all these possibilities and does not depend on the actual treatment offered to any particular patient. It would be monitored for all presenting cases (by random sampling perhaps) through clinical audit to ensure that the protocol was properly followed. It is against that scenario that I wish to consider the problem of ensuring that these guidelines are drawn up appropriately.

My starting point is a recently published book from the Royal College of Physicians of London (Llewelyn and Hopkins 1993). The introduction states:

> It is important ... to be able to analyse the reasoning process which leads to diagnoses and decisions on clinical management. The discipline which addresses this issue is called 'clinical decision analysis', and draws on a number

of other disciplines, especially probability theory and the assessment of 'utility' or 'value' of outcomes of medical intervention.

A little later an important caveat appears:

> The decision about how to deal with a particular patient may be made from different points of view. It is entirely the patient's point of view which will be considered in this introductory book. In the United Kingdom, most patients do not have to take the cost of care into consideration .... A decision can also be considered from a hospital or departmental manager's point of view. For example, if a limited number of beds is available, a priority decision may have to be made as to who should be admitted. A doctor may have to make a decision which combines both points of view. Decision analysis allows decisions of this kind to be made explicit.

It is on the implications of this last sentence that I wish to elaborate.

AN EXAMPLE

Let us take a very simple clinical decision problem (figure 16.1). For a particular class of presenting signs and symptoms, a diagnostic test will generate two results, A or B. If the result is A, two alternative treatments

Figure 16.1 A simple clinical decision tree

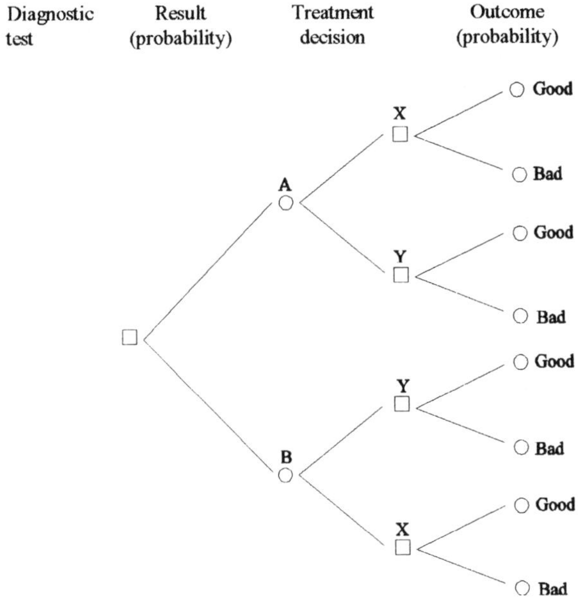

(X and Y) are commonly offered; if the result is B, two alternative treatments (Y and Z) are commonly offered. Each of the three treatments has a 'good' and a 'bad' outcome, with known probability. The clinical problem is: given the test result, which treatment should be offered?

Some basic data are supplied in Figure 16.2, comprising the respective probabilities associated with the two test results and the probabilities associated with each outcome for each treatment. There is also a quantitative indicator of the amount of health gain for each outcome. To keep things simple, this can be thought of as additional years of life but it

Figure 16.2 Expected health gain

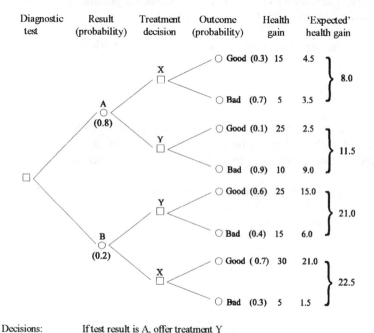

Diagnostic test	Result (probability)	Treatment decision	Outcome (probability)	Health gain	'Expected' health gain

Decisions: If test result is A, offer treatment Y
                If test result is B, offer treatment Z

might be quality adjusted life years or some other measure of benefit appropriate to the situation. When these health gains for each outcome are multiplied by the probability of getting that outcome we arrive at the 'expected' health gain which, when summed for each treatment (that is, across the good and bad outcomes for that treatment), gives us the 'expected' benefit from the treatment. For the alternative treatments on offer when the test result is A, treatment Y offers greater expected benefits than treatment X and should, therefore, be chosen. Similar reasoning will lead to treatment Z being chosen when the test result is B.

THE BASIC APPROACH

Figure 16.2 is an example of the class of decisions analysed in the basic texts on clinical decision-making and it is this same class of decisions that is analysed in Llewelyn and Hopkins (1993). The weakness of this analysis is that it pays no regard to the costs of treatment, being wholly concerned with 'effectiveness'. This renders such a guideline or protocol deeply suspect from a purchaser's viewpoint and is, in my view, irresponsible even from a clinician's viewpoint. As I have argued elsewhere (Chapter 17), clinicians, like purchasers, have responsibilities to a whole group of patients and potential patients, which they can discharge responsibly only by taking into account the consequences for resource use which flow from their decisions (for example the use of the hospital beds or operating theatre time that has been allocated to them). Figure 16.2 ignores that consideration.

There is a temptation at this stage to repair the weaknesses in the 'effectiveness only' approach by throwing in some cost data, almost as an afterthought. In more complex situations than that shown in Figure 16.2, this would be an enormous task if every alternative branch and twig in a dense decision tree had to be properly costed, so what tends to happen is that only the chosen options are costed, just to check that they are 'affordable'. This implies that the decision problem is seen as the one shown in Figure 16.3, where cost per unit of health gain for the two chosen treatments has been added to the previous analysis. If 217 and 222 are 'acceptable' cost levels (if, for example, many other treatments being offered elsewhere in the system exceed this level of cost per unit of health gain) then the clinical seal of approval is fixed to the protocol and it is henceforth claimed to be the cost-effective procedure and recommended on those grounds.

Figure 16.3 Decision tree including costs of preferred options

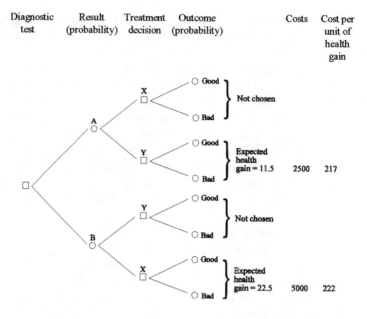

A TURNAROUND

In the more advanced texts on clinical decision analysis these problems are dealt with by integrating the cost data into the decision tree itself, in the manner shown in Figure 16.4 (Weinstein and Fineberg 1980, Williams 1983). This ignores the costs of the test itself, though in a more broad ranging review they would have to be included. The costs that have been added are the costs associated with each treatment option and they have been carefully selected to make a point. When the cost per unit of health gain is calculated for four options, neither of the previous recommendations holds. When the test result is A, it is now treatment X that is recommended rather than treatment Y and when the test result is B it is now Y that is indicated, not Z. So the really cost-effective procedure is the opposite of what was mistakenly claimed at the end of the preceding paragraph.

Let us pause to consider the implications of this turnaround. It means that patients with this condition are not being offered the 'best' treatment if, by 'best', we mean the one that would do each of them the most good. How can this be justified? Let us do some more arithmetic. In the 'effectiveness only' scenario in Figures 16.2 or 16.3, the average cost per presenting patient is 3000 (see appendix). The corresponding figure in the 'cost-effectiveness' scenario in figure 16.4 is 1300.   With resources available for these patients amounting to £300,000 a year, under the 'effectiveness only' scenario 100 patients could be treated each year, whereas under the cost-effectiveness scenario 231 could be treated.   So although each treated patient gets a greater expected benefit under the 'effectiveness only' protocol (13.7 of 10.6), the total benefit is much greater under the cost-effectiveness protocol (2449 of 1370) because many more patients get treated.

Figure 16.4 Decision tree with costs fully incorporated

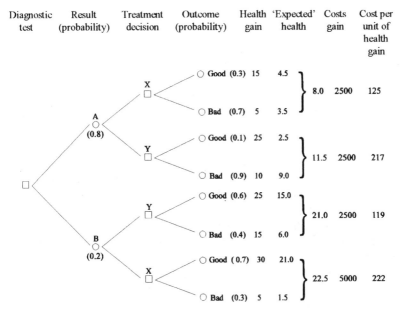

Diagnostic test	Result (probability)	Treatment decision	Outcome (probability)	Health gain	'Expected' health	Costs gain	Cost per unit of health gain
		X	Good (0.3) 15	4.5	8.0	2500	125
			Bad (0.7) 5	3.5			
	A (0.8)						
		Y	Good (0.1) 25	2.5	11.5	2500	217
			Bad (0.9) 10	9.0			
		Y	Good (0.6) 25	15.0	21.0	2500	119
			Bad (0.4) 15	6.0			
	B (0.2)						
		X	Good (0.7) 30	21.0	22.5	5000	222
			Bad (0.3) 5	1.5			

Decisions: If test result is A, offer treatment X
If test result is B, offer treatment Y

CONCLUSION

We are brought back to the question: what objective are we pursuing? Is it 'clinical excellence' (no matter what the costs?) or 'ensuring that the people we serve are as healthy as it is possible for us to make them, given the resources (human and material) available to us?' If we all share the 'healthy' objective, then purchasers, clinicians and decision analysts need to get together to devise clinical guidelines that embody the cost-effectiveness approach to health care. Only in that way will we achieve the ambitions of the august bodies whose views were cited at the outset and who emphasised the need to be systematic and to take into account both resource use and patient outcome (in terms of survival and of quality of life). However, if this is to happen in my lifetime, I fear that a rather more fundamental culture shift is going to be required among clinicians at grass roots level in the United Kingdom than is evident at present, despite the pioneering efforts of the enlightened few. It is in the creation, dissemination, monitoring and purchasing of truly cost-effective treatment protocols that the commitment of the medical profession to this enterprise is about to be tested. I hesitate to predict which way it will go.

APPENDIX

Calculations for results in the test

In Figure 16.3, 80 per cent of patients get treatment A, which costs 2500, and 20 per cent get treatment Z, which costs 5000, so the (weighted) average cost is 0.8 of 2500 (that is, 2000) plus 0.2 of 5000 (that is, 1000), which amounts to 3000 in all. The corresponding figures for the scenario presented in Figures 16.4 are 80 per cent getting treatment X (costing 1000 each) and 20 per cent getting treatment Y (costing 2500 each). This generates a weighted average cost of 0.8 of 1000 (that is, 800) plus 0.2 of 2500 (that is, 500), which comes to 1300 in all.

A budget of 300,000 will thus enable 100 patients to be treated if the policy is as set out in Figure 16.3, but 231 (300,000/1300) can be treated if the policy is as set out in Figure 16.4.

The average benefit from treatment under each scenario can be calculated in a similar manner. In Figure 16.3 it is 0.8 of 11.5 plus 0.2 of 22.5 (which comes to 13.7 in all) whereas in Figure 16.4 it is 0.8 of 8 plus 0.2 of 21 (which comes to 10.6).

Putting together the benefit per patient and the number of patients treated under each scenario, we have, for Figure 16.3, 100 times 13.7 (or 1370) and, for figure 16.4, 231 times 10.6 (or 2449).

PART FOUR

Economics, Ethics and Clinical Freedom

17.   Health Economics: The End of
Clinical Freedom?

Clinical freedom is dead, and no one need regret its passing. Clinical freedom was the right – some seemed to believe the divine right – of doctors to do whatever in their opinion was best for their patients.   In the days when investigation was non-existent and treatment as harmless as it was ineffective the doctor's opinion was all that there was, but now opinion is not good enough. If we do not have the resources to do all that is technically possible then medical care must be limited to what is of proved value and the medical profession will have to set opinion aside.

Clinical freedom died accidentally, crushed between the rising cost of new forms of investigation and treatment and the financial limits inevitable in an economy that cannot expand indefinitely. Clinical freedom should, however, have been strangled long ago, for at best it was a cloak for ignorance and at worst an excuse for quackery. Clinical freedom was a myth that prevented true advance. We must welcome its demise and seize the opportunities now laid out before us. (Hampton 1983)

I think Hampton's rather unkind obituary notice for clinical freedom was in fact premature for, if indeed clinical freedom is dead, it seems reluctant to lie down. It is still frequently paraded before us (rather as with the corpse of El Cid?) in defence of some medical practice which is being challenged by a non-medical person on grounds of its doubtful effectiveness, or its undoubtedly high cost, or both.

A sharply contrasting view to Hampton's was expressed a few years earlier in a letter to the *New England Journal of Medicine* by Loewy (1980).   He wrote: 'Of late an increasing number of papers in this and other journals have been concerned with 'cost-effectiveness' of diagnostic and therapeutic procedures.   Inherent in these articles is the view that choices will be predicated not only on the basis of strictly clinical considerations but also on economic considerations as they may affect the patient, the hospital, and society.   It is my contention that such considerations are not germane to ethical medical practice .... A physician who changes his or her way of practising medicine because of cost rather than purely medical considerations has indeed embarked on the 'slippery slope' of compromised ethics and waffled priorities'.

POLAR POSITIONS

Although prudence suggests that it is better for outsiders like me to remain onlookers when it comes to intraprofessional disputes of this degree of intensity and on topics of this degree of sensitivity, economists are not noted for angelic reticence so I propose to join the other fools from outside the medical profession and rush into the fray.

Before getting too deeply embroiled, however, let me state the issue as sharply as possible so that there is no doubt as to what the fuss is all about. The two polar positions are as follows:

Viewpoint A – It is the doctor's duty to do all he or she can for the patient no matter what the cost.

Viewpoint B – It is the doctor's duty to take costs into account when deciding what course of action to recommend for the patient.

There are various ways forward when faced with this sharp difference of view. The conservative view would be that the quest for efficiency, or cost consciousness, must stop as soon as it begins to breach clinical freedom (which may mean that the quest can never really start). A more radical view would be that if clinical freedom stands in the way of cost-effectiveness then it is clinical freedom that must go (unless of course, as Hampton believes, it has already gone).

A middle of the road view might be that all that is necessary is that the essence of clinical freedom is better understood and, when it is redefined according to that better understanding, the apparent conflict will disappear. This seems to be the stance of Hoffenberg (1987), to whose views I will return later.

Before considering his position I want to develop one of my own, which is essentially that, despite their protestations to the contrary, doctors never have behaved without regard to the costs of their actions. Simply stated, they have been selective and haphazard about what costs they counted and how they counted them. So the issue is not whether or not it is ethical to count costs but rather where the line should be drawn between costs that may be counted and costs that may not.

To capture the essence of the situation it is important to focus on the nature of the relationship between doctor and patient. In theory the patient and the doctor are in a principal–agent relationship in which the doctor is the agent and the patient is the principal. If the doctor were a perfect agent their relationship would then be as follows:

The 'doctor' is there to give the 'patient' all the information the 'patient' needs in order that the 'patient' can make a decision and the 'doctor' should then implement that decision once the 'patient' has made it.

If you find that description of your relationship with your patient a bit odd try reversing the respective roles of the doctor and the patient so that it now reads:

The 'patient' is there to give the 'doctor' all the information the 'doctor' needs in order that the 'doctor' can make a decision, and the 'patient' should then implement that decision once the 'doctor' has made it.

Does that sound more like it actually is? The point is that doctors are not perfect agents and, because of that, they have quite a lot of discretion over what they take into account in exercising their so called 'clinical' judgement, so that the distinction between 'strictly clinical' and 'extraneous' factors is a fuzzy one, as we shall see later when considering Hoffenberg's views.

ECONOMIC CONSIDERATIONS

Now let me illustrate these points with three examples, designed to show that doctors have, quite spontaneously, had regard to 'economic considerations as they may affect the patient, the hospital and society', notwithstanding what Loewy believes. Moreover, they have absorbed these considerations comfortably within what they would still regard as their own clinical judgement rather than as part of any wider 'extraneous' economic or social responsibility.

The first example concerns a patient for whom there are two possible courses of treatment, the safer and more effective of which would require the patient to be off work for three months, entailing a substantial loss of income for him and possibly putting his livelihood in jeopardy altogether. The less safe and less effective treatment entails no such risks. Which treatment do you think most doctors would offer in that situation? Would they regard it as unethical to consider these 'economic considerations as they affect the patient?'

The second example concerns a patient with a stable chronic condition that has already been thoroughly investigated and is being treated conventionally. The patient has seen a TV programme describing some research which seems to suggest that a new drug works wonders for some patients with that condition and the patient wants to be sent for investigation at the nearby regional centre where the TV programme was made. Both the patient and the doctor know that there is only a small chance that the patient will benefit and, in any case, this would be a very expensive course of action for everybody. The doctor also knows that the specialist in question takes a poor view of general practitioners who send him speculative cases lacking clear indications that such referral is justified and that the regional centre is hard pressed to cope with its workload as it is. What do you think most doctors would do in that situation? Would they regard it as unethical to take into account these 'economic considerations as they affect the hospital?'

The final example concerns an elderly woman living with relatives whose daughter has had to switch from full-time to part-time work to cope with the strain of looking after her, who disturbs the neighbours with her odd behaviour and who is costing the health and personal social services a great deal of money in domiciliary support. Although it is acknowledged that the patient's own survival prospects and quality of life will be worse if she is taken into hospital, those around her will certainly be better off and it will release a lot of scarce, domiciliary support for others to use. What do you think most doctors would do in that situation? Would they regard it as unethical to take into account 'economic considerations as they affect society' – that is, people other than the patient?

It is the observation of clinical behaviour in examples such as these that convinces me that doctors do not regard it as unethical to have regard to costs, whether these costs fall on patients, their relatives, the health service, or the community at large. And I do not regard their characteristic behaviour in each of the examples I cited as at all unreasonable. I am sure that we ordinary citizens and, indeed, the courts of law, would support them in their decisions if they were challenged.

GHOST OF CLINICAL FREEDOM

So why is clinical freedom (or, if Hampton is right, the ghost of clinical freedom) invoked when economists argue that clinicians should count costs when making treatment decisions? I suppose it may simply be that insiders are allowed to say things about each other which none of them would allow an outsider to say. After all, if you give these economists an inch they will take an ell and, before you know where you are, there will be a desiccated inhuman calculator in every consulting room.

Time to send for Hoffenberg (1987). I fear that his testimony will not help the scaremongers very much. Early on in his monograph he observes: 'There is no such thing as clinical freedom, nor has there ever been. Nor, for that matter, should there be .... Absolute freedom to make clinical decisions without taking into account the preferences or wishes of the patient could not be countenanced. It is therefore proper that personal, moral, ethical and even legal constraints should be observed. To these must now be added the constraint of limited resources'.

Later he notes that this additional constraint requires doctors to help in rationing access to health care and asks: 'Does acceptance of this function compromise our clinical freedom? It is not easy to identify or isolate the part played by awareness of costs when individual doctors ... decide whether or not to investigate or treat a patient. The open and direct opinion 'This would cost too much', is rarely expressed and, I suspect, not often consciously entertained as a discretionary factor. Even if it were it would be regarded simply as one of the factors that needed to be taken into account in order to establish a course of action. It would not often be seen as detracting from the doctor's clinical freedom. Whether or not he was influenced by costs would be entirely up to him; his autonomy would be unimpaired. A more important question from the point of view of clinical freedom is whether and to what extent doctors are prevented from doing something they wish to do for their patients, by an authoritarian embargo, based on resource considerations, which is in conflict with their professional judgement'.

This comes close to saying that only doctors have the right to determine what the effect of resource constraints should be (which, if true, might explain why clinical freedom is invoked against health economists, who also think they have something to say about how best to respond to resource constraints). But this does not appear to be Hoffenberg's intention, for he goes on to consider prescribing from the limited list and the carefully controlled dissemination of heart transplantation in the United Kingdom as cases where, despite 'the authoritarian encroachment on clinical freedom' there seems to have developed ... a widespread acceptance by the public as well as the profession that resources are not unlimited and that the doctor has a responsibility to Society – through a thoughtful allocation of scarce resources – as well as to the individual patients.'

ENTER HEALTH ECONOMICS

It is here that I want to give health economics a more central role in the debate. It seems to me that much of the public and, indeed, some of the professional discussion of resource constraints, is bedevilled by the feeling that the extension of life, the relief of pain and suffering and the improvement in people's functional capacity, which appropriate health care can provide, should not be limited merely by shortage of money or distorted by purely financial considerations. But in economics (unlike accountancy) there is more to the notion of costs than the spending of money. Economics is based on the principle that a 'cost' is simply a forgone benefit, so when we advocate greater efficiency we are advocating the adoption of two ethical precepts: firstly, to make sure that the sacrifices entailed in pursuing any activity are kept to a minimum and secondly, to make sure that no activity is pursued unless the benefits gained outweigh the benefits forgone.

In applying these general precepts specifically to medical care their implication is that the true cost of treating one patient is the benefit that might have been enjoyed if those same resources had been used to treat other patients. The cost of keeping one patient on hospital haemodialysis for a year is (say) 10 coronary artery bypass grafts forgone. The finance required for each activity is but a superficial mediating variable that should not be allowed to obscure the fundamental choice, which is the one stated. Does 'clinical freedom' include the freedom of the renal physician to deny 10 heart patients their bypass grafts (or vice versa for the cardiologist)? This is the agonising reality that underlies the tension between 'clinical freedom' and 'social responsibility' and I cannot accept that the speed with which this 'social responsibility' permeates the NHS should be limited by what Hoffenberg calls 'clinical autonomy' – that is, by the willingness of the medical profession to accept this wider responsibility (and by their willingness to act accordingly) rather than having to accept it grudgingly or resentfully.

The health economist's notion of 'cost as sacrifice' (rather than cost as financial expenditure) also shows more clearly the ethical issues underlying the 'clinical freedom' *versus* 'health economics' discussion. To return to the earlier example, in deciding whether the treatment of the renal patient should be sacrificed to enable the 10 heart patients to be treated or vice versa, we need a distributive ethic, or an ethic of fairness and this has to be a society-wide principle (or set of principles), preferably deeply ingrained in all citizens and in the professions that serve them.

I think Hoffenberg is right in believing that in Britain we are feeling our way to a large extent unconsciously towards what he calls a 'curious, unstated and ill-defined understanding' of that kind. The health economists are trying to sharpen up and clarify the nature of that understanding, especially through the development of generalised measures of health benefit, such as the quality–adjusted–life–year, leading to the suggestion that treatments for which the 'cost per quality–adjusted–life–year gained' is very high should be given low priority, so long as we have so many patients waiting for treatments where the cost per quality–adjusted–life–year gained is quite low. To put if more crudely, hip replacements are a better buy than heart transplants.

LEGAL CONCEPTS

Recently I noted that even the American courts are moving towards a similar criterion when determining whether it would be proper to initiate treatment or to withdraw treatment once initiated. The key legal concept is that of 'proportionality' and the relevant dictum is: 'Proportionate treatment is that which, in the view of the patient, has at least a reasonable chance of providing benefits to the patient which outweigh the burdens attendant to that treatment' (Ruark *et al.* 1983).

The court suggested that a benefit exists when life sustaining treatment contemplates 'at very least, a remission of symptoms enabling a return toward a normal functioning, integrated existence'. Or, as I would put it, when the expected gain in quality adjusted life years justifies the costs.

In reviewing that difficult territory the Stanford University Medical Center Committee on Ethics made the following general observation, (Ruark *et al.* loc. cit.) which is central to my themes: 'A basic principle of medical ethics is obviously the preservation of life, which is frequently tempered by the second principle, the alleviation of suffering. A third is the injunction that physicians "first do no harm ...." A fourth principle [is] respect for the autonomy of the individual patient .... A fifth fundamental principle is the concept of justice, exemplified by the effort to ensure that medical resources are allocated fairly. The final principle is truth-telling .... Because medical practice often brings these principles into conflict, resolving such conflicts is central to the art of medicine.'

But it is also central to the art of health service management, which operates across a much broader range of choice than that facing any individual clinician and which has a stronger claim to legitimacy than any clinician has when it comes to judging what broad social responsibility entails.

So where does all this leave us as regards the alleged conflict between health economics and clinical freedom? According to Hoffenberg there 'ain't no such animal as clinical freedom in any absolute sense'. According to Hampton clinical freedom died a while back but since it should have been strangled at birth anyway, we do not need to mourn its passing. Yet the symbolic power of this apparent myth seems undiminished and my reading of the situation suggests that it owes this power to its malleability. It is consistent with whatever clinicians choose to make it consistent with.

If I am right in my diagnosis, the immediate task of health economists is to get our notions of efficiency and fairness in the distribution of the benefits of health care so deeply embedded in the clinical consciousness that they come to be thought of as wholly within the realm of clinical autonomy and we outsiders then become redundant. I would be fearful of working myself out of a job by delivering lectures such as this if I thought that all doctors were as receptive to this way of thinking, talking and acting as the Hamptons and Hoffenbergs of this world clearly are. But I have no such fears (or perhaps I should say no such hopes).

ACCOUNTABILITY

I would like now to strike a more positive note and I found one in Hoffenberg's book. As you would expect from someone who has spent most of his professional life simultaneously encouraging doctors to count costs as well as benefits and encouraging managers to count benefits as well as costs, I jumped with joy when I came across this delightful passage in which Hoffenberg manages to sweep away the problem and leave us all on the same side:

> Doctors are best placed to shift the emphasis of managerial enquiry from its pre-occupation with resources and costs to health outcome or patient satisfaction. For this reason alone medical participation in management is imperative. By ensuring that resources are devoted optimally to serve the interests of patients, doctors will find that their own clinical freedom is maximised.

So health economics actually enhances clinical freedom. There seems nothing more to be said. I just wish I could believe it. But I think that Hoffenberg has muddled up two different things. He is right in believing that, if doctors absorb into their realm of the managerial skills needed to minimise the potentially adverse effects of resource constraints, their clinical practice will thereby be improved. However, this does not mean that their clinical freedom has been enhanced; indeed, quite the reverse. I think Hoffenberg has been too concerned to allay the fears of his colleagues and, in resolving this ethical dilemma, he has given too much weight to the relief of distress and too little to truth telling. So let me end by shifting the weight the other way.

Much of the tension between health service managers and the medical profession (and I do not mean between the government and the NHS, which is a different matter) revolves around the issue of who has the authority to decide and give effect to the priority-setting that resource constraints make inevitable. The incursions by management into territory that has hitherto been the preserve of the medical profession are seen as threats to clinical freedom. Demands for greater accountability for the resource consequences (and even for the health consequences) of their clinical activities are still frequently resisted by doctors sheltering behind the banner of clinical freedom. On closer scrutiny, the banner may be threadbare – but its emotional appeal is still strong. Health economics is not the end of clinical freedom but the demand for greater accountability for resource use is a further (and quite proper) limitation of its scope.

A more constructive response by doctors would be to move forward from the key observation that the resolution of conflicting ethical principles is 'central to the art of medicine' and accept that it is also central to the art of health service management and, in that broader forum, many other voices are equally entitled to be heard. Doctors still have an important role in policy making at national, regional, district and unit level. But theirs is not a pre-eminent role, or one which gives them the right to exclude others from any part of the policy discussion in the name of clinical freedom or on any other pretext. It is an agonising realm of discourse which, because of its very nature, cannot be easy intellectually to think through, or comfortable emotionally to implement, so it needs to be approached with great sensitivity and respect for the views of others, remembering all the time that none of us has a hot line to God. And on that celestial note I end this diabolical discussion.

18. Ethics and Equity in the Provision of Health Care

INTRODUCTION

A major purpose in nationalising the provision of health care in the UK was to affect its distribution between people and, in particular, to minimise the impact of willingness and ability to pay upon that distribution. It has never been clear, however, what alternative distribution rule is to apply. There is no shortage of rhetoric about 'equality' and 'need', but most of it is vacuous, by which I mean it does not lead to any clear operational guidelines about who should get priority and at whose expense. The closest we have got so far to such explicit guidelines has been the formulae which determine the *geographical* distribution of NHS funds, the driving force behind which is a notion of need based on relative mortality rates and on the demographic structure. The avowed objective is to bring about equal access for equal need irrespective of where in the UK you happen to be (Daniels 1982, West 1981, Parton 1985, Birch and Maynard 1986).

Equal access for equal need says nothing about the efficiency of the services to which you have access, it being assumed that they will do you good and that the more you have of them the more good they will do you. It also assumes that the range of services offered is the best that can be afforded. A further implication, seldom exposed to public discussion, is that unless a service can be provided equally across all geographical areas it should not be provided at all, for equal access for equal need is quite consistent with no access at all for some needs. Accessibility therefore seems a rather limited conceptual basis for a discussion of the distributive ethic (see also Steele 1981, Mooney 1983, de Jong and Rutten 1983).

The difficulty about going beyond service *provision*, and examining instead service *benefits*, is that we have no *routine* data on the latter and even research findings on the benefit of health care are patchy in both coverage and reliability. Just attempting to define, measure and value such benefits at individual level seems to excite people's passions and what to me seems common-sense propositions (such as that people value *both* improvements in the quality of their lives *and* improvements in their life expectancy and that, therefore, the health service should also attach positive values to both) engender vehement counter-propositions of a quite extreme kind (for example, that increasing people's life expectancy must take absolute priority over improving people's quality of life).

I do not want to go over that well-trodden ground again here (see Kind *et al.* 1982, Williams 1985, Harris 1987) but to concentrate instead on one specific ethical issue, which is: *is a particular improvement in health to be regarded as of equal value no matter who gets it; and, if not, what precisely is its relative value in accruing to one kind of person as opposed to another*?

Before plunging into that deep water, in order to concentrate single-mindedly on the issue posed, the following circumscriptive propositions should be noted.

I assume that benefits have been defined, measured and valued in a manner that is acceptable to everybody. In order to be as neutral as possible on that matter, I will, for expositional simplicity, assume that the health benefit we are talking about is simply one additional year of healthy life expectancy, to be enjoyed between the same dates in calendar time by whoever gets it.

I also assume that the health care system is operating at such a high level of efficiency that it is not possible for it to offer anyone an extra year of healthy life expectancy without depriving someone else of that same prospect.

I further assume that the health care system has at its disposal exactly the right amount of resources, as agreed by the community generally.

In determining the appropriate answer to my question, we are trying to operate from behind the 'veil of ignorance' about our own possible future dependence on the health care system, that is, we are tying to decide what is 'right' in a 'good' society, not what is most likely to be in our own personal interest.

Although the desired distribution rule is to be cast in terms of the distribution of *benefits*, it will in fact also *determine* the distribution of *access*, since the pattern of provision of health care (including its geographical pattern) will follow automatically from the distribution rule once that is linked to the production possibilities open to the health care system.

I am not necessarily seeking a *one-dimensional* distribution rule any more than I seek a one-dimensional concept of benefit) but, if there is more than one dimension that is to have weight, I shall need specific guidance about the 'trade-offs' between one dimension and another in all situations where they are both likely to apply.

Finally, my long-term intention is to conduct a survey of the population at large to find out what *they* think about these issues and I am using this opportunity to explore one possible way in which these issues might be posed to ordinary citizens so that they could respond in an informed way to some carefully designed questions which are as unbiased as it is possible to make them.

THE PERCEIVED IMPORTANCE OF HEALTH AT DIFFERENT LIFE STAGES

During 1985 a survey was conducted in York of 377 people randomly selected from the electoral register, who were interviewed about a variety of matters concerning health. Amongst the many things we were interested in was whether people thought that health benefits were more valuable at some stages in peoples lives than at other stages. So, during the interview, respondents were shown a list of 10 life stages and asked:

> If a choice had to be made between them, in which of the following circumstances would you think it most important to keep people *in general* well?

> Please choose the three sets of circumstances you judge to be most important.

After three had been chosen from the list, respondents were then asked:

> Now from these three sets of circumstances, which would you judge to be most important, and which next most important?

Their answers are set out in Table 18.1.

One of our interests was to test whether or not a consensus exists and this was treated by looking at the relationship between the responses and the sociodemographic data we had collected from respondents. Given the intentional distribution of the 10 life stages selected across the lifespan (from 'infancy' to 'getting very old'), age might be expected to be a major influence on choices. Thus, older subjects could be expected to select later stages of life as the most important to which to ensure well-ness out of a simple egocentrism or self-interest. Similarly, sex might be expected to influence choices due to the relationships between the traditional sex roles and the life stages employed. Thus males might be expected to emphasise school and work-orientated phases of life ('starting school', 'starting work', 'at the peak of earning power', 'retirement') whilst females could be anticipated to emphasise life stages relevant to their role as carer and home-maker ('bringing up children', 'looking after elderly relatives'). Other sociodemographic variables with social role implications (for example, marital status, work status) might also be expected to affect choices.

Table 18.1 York health evaluation survey

'If a choice had to be made between them, in which of the following circumstances would you think it most important to keep people *in general* well? Please choose the three sets of circumstances you judge to be most important'. After being shown the list and choosing three, respondents were asked 'Now from these three sets of circumstances, which would you judge to be most important, and which next most important?'

	Most important		Next most important		Third choice	
	N	%	N	%	N	%
As infants	103	27.3	36	9.5	25	6.6
When starting school	24	6.4	16	4.2	25	6.6
When starting work	14	3.7	37	9.8	42	11.1
When setting up home for the first time	10	2.7	21	5.6	17	4.5
When bringing up children	124	32.9	102	27.1	42	11.1
When at the peak of their earning power	15	4.0	19	5.0	23	6.1
When looking after elderly relatives	19	5.0	44	11.7	38	10.1
When just having retired from work	16	4.2	36	9.5	63	16.7
When coping with the death of a husband or wife	23	6.1	31	8.2	45	11.9
When getting very old	27	7.2	33	8.8	53	14.1
(Unusable responses)	(2)	(0.5)	(2)	(0.5)	(4)	(1.1)
Total	377	100.0	377	100.0	377	100.0

What emerged is that the two life stages 'when bringing up children' and 'as infants', which were by far the most often selected, also showed a high degree of consensus across both sexes and all age groups. If one looks at each life stage separately to see whether there are significant differences in response by any background factor, it turned out that females placed relatively more weight on 'bringing up children' and 'caring for elderly relations' and males on 'setting up home for the first time' and 'when at peak of earning power', thereby reflecting traditional sex roles. Differences in response by other background factors were negligible.

The study was not a suitable vehicle for exploring the next phase of the enquiry, which is to find out *how much more weight* is to be given to the more important life stages compared with the less important. What I need to elicit is something like Table 18.2, which is entirely hypothetical and which purports to indicate, against an average index value of 1.0, how much *more* important the above average life stages are and how much *less* important the below average ones are. One of the problems I shall address later is precisely how one might best elicit such information.

Table 18.2   *Hypothetical data on the relative values*
of good health at different life stages

As infants	2.75
When starting school	0.64
When starting work	0.37
When starting up home for the first time	0.26
When bringing up children	3.31
When at peak of earning power	0.40
When looking after elderly relatives	0.51
When just having returned from work	0.43
When coping with the death of spouse	0.61
When getting very old	0.72

For the time being these data are presented merely to whet the appetite. In the present context they would imply that, in order to give one extra year of healthy life expectancy to an infant (valued at 2.75), it would be worth sacrificing about 10.5 years of healthy life expectancy (each valued at 0.26), which might have been given to someone setting up home for the first time. It is at this level of specificity that the ethical problem of setting up distributional rules needs to be solved if it is going to yield clear operational guidance to those responsible for running the health care system.

POSSIBLE DISTRIBUTION RULES

I earlier explored the possibility that 'life stages' might be a relevant dimension in a distribution rule for health benefits and it did indeed elicit a very clear consensus that health benefits at some life stages are regarded as more valuable than at others. But 'life stages' are an amalgam of age-related elements and role-related elements (and possibly even implicit sex-role factors) so maybe it would be better to attempt a finer discrimination by separating out these different characteristics and dealing with them individually (at first, anyway, even though they may need to be combined later).

What I propose to do next, therefore, is to list, and examine briefly, some possible axes along which discrimination might be considered desirable on ethical grounds (that is, features according to which someone might think that we *should* discriminate between people and we can then consider the likely *consequences* of each such rule (that is, who would benefit and who would lose) compared with some 'standard' rule (with no implication that the 'standard' rule is ethically superior, though it might turn out to be).

The 'standard rule' I shall use is that a unit of health benefit (say 'an additional year of healthy life expectancy') is to be regarded as of equal value no matter who gets it. The data in Table 18.1 indicate that most people do not subscribe to this view but that is of no consequence at present. In terms of Table 18.2, this rule would mean that the numbers shown against each life stage would be 1.0 in each case. Thus, in the distribution of health benefits, the standard case would not discriminate between people on any ground whatever.

Before discussing other possible rules let us first consider the arguments that might be adduced in favour of the 'standard' rule being the rule to be *chosen*, rather than it merely being a reference point. It has a very strong non-discriminatory egalitarian flavour, it is free of judgements about people's worth, or deserts, or influence, or likeability, or appearance, or smell, or manners, or age, sex, wealth, social class, religious beliefs, race, colour, temperament, sexual orientation, or general or particular life-style. It seems prima facie very close to the official medical ethic when it comes to dealing with people.

Before I go any further let me dispose of a red herring which is sure to be drawn across our path sooner or later, namely that the distribution of health benefits should be determined by 'need'. I have argued (Chapter 11) that the most plausible interpretation of what people intend to convey in the health care context by appealing to the notion of need is that someone would be better off *with* the 'needed' treatment than *without* it. This leaves open the issue of whose values determine whether (and to what extent) that person *will* be better off and, more to the point in the present context, how one person's 'needs' (that is, potential benefits) are to be weighed against another's. So all in all an appeal to the notion of need is but a detour which leads us back to the main issue, which is, how a benefit to one kind of person is to be valued relatively to that same benefit to another kind of person.

Let us consider *age* as a fairly straightforward and obvious candidate for consideration. The classic argument is based on the 'good innings' analogy, that is, someone who has had their three-score-years-and-ten has had a good run for their money and precedence should be given to younger people who have not yet had a fair share of the action. The consequences of such a rule, compared with the standard rule, will obviously be an explicit discrimination *against* the old and *in favour* of the young (and, so long as women are living longer than men, it will discriminate *incidentally* against women *as a group*, even though *on an age standardised basis* there would be no such sex discrimination).

Let us next consider *family responsibilities* (for example, looking after young children or elderly relatives) as a basis of a distributive rule. If this were to count, it will favour mainly people in their twenties and thirties (as child carers) and mainly people in their forties to sixties (as carers of the elderly). With present patterns of role division, married (and some single) women will be the main beneficiaries of such a rule. However, is earning money to keep the household going also a family responsibility? If so, the consequences become more blurred, except that single-person households will lose out even more clearly. How narrowly or broadly drawn should this notion of family responsibilities be? Is the motivation for *this* discrimination perhaps the thought that by keeping the 'carer' healthy we are more likely to keep the 'dependants' healthy, so it is really a matter of using the higher value attributed to the benefit to the 'patient' as a proxy for a *sum* of benefits to several people? If so, would it not be better to count the benefits to the different people *directly* instead of tampering with the relative valuations? A final complicating thought here is whether this notion of 'dependency' should be extended outside the family or household, for example, into the workplace, or into any other social setting in which people play a role which is highly valued *by other people*.

This then shades into judgements of *social worth* as a possible basis for a distribution rule. Is a great artist or performer, whose talents are widely appreciated, to be given precedence over an easily replaceable nonentity? How are such judgements to be made? Although in the (admittedly small sample of) writings on ethics I have encountered, this elitist notion is viewed with great suspicion, it is formally entrenched in the health care systems of some countries and, I suspect, more widely practised here than we care to admit and it may well find a ready response in the population at large. I therefore think it has to be considered.

Although *willingness and ability to pay* have been formally rejected as a suitable distribution rule for the NHS, it is a nice question whether this should nevertheless be included in any survey to see whether popular opinion still supports that rejection. Moreover, there is an interesting variant which I think has to be included anyway. It seems to me that it is the differential 'ability' to pay which people find more obnoxious than the differential 'willingness' to pay and we do accept, in the NHS, that people who are willing and able to pay high *time* prices (for example, by waiting in GPs' surgeries or at outpatient clinics in hospitals) get better access than those who are not so willing or able to do so (for example, consider as examples of the former group retired people with cars and as examples of the latter group working single-parents relying on public transport). But this 'willingness to pay' concept has a more subtle non-financial interpretation as *a willingness to make sacrifices for the sake of your health*, for example, by giving up smoking, restraining your enthusiasm for alcohol, eating sensibly, getting regular exercise, and so on.   The distributional issue is then: should the health service value health benefits more highly if they accrue to people who, by their behaviour, clearly value their own health more highly? Put more bluntly, should we discriminate against heavy smokers, heavy drinkers, obese people, freefall parachutists, unclean people, etc. Note that this is a separate issue from the undoubted fact *on the production side* that it is more costly to give an additional year of healthy life expectancy to such people. The question here is should we value that (most costly) benefit differently if people appear *from their own behaviour* not themselves to care much about their own health?

The next possible basis for a distribution rule relies not on how the individual behaves but on the circumstances that the individual is in, these circumstances being assumed to be largely outside that individual's control. An example might be the child born into a poor family with an alcoholic unemployed father and heavy smoking mother with no ideas about proper hygiene or nutrition. Should health benefits for such infants be valued more highly than those for infants generally? At the later stages of life, should someone, who has behaved impeccably (whatever that means substantively) with regard to their own health-related behaviour but has been persistently unemployed and is in poor housing in a poor physical environment, have their health benefits upvalued as some kind of compensatory mechanism for their other deprivations? To go further, does this mean generally that benefits to the poor should be valued more highly than benefits to the rich, in which case have we not reintroduced discrimination by willingness and ability to pay, but now in reverse? Nevertheless, whatever tangles we may get into over this possible dimension of discrimination, I am sure it has to be included in any survey work about possible distribution rules.

There are perhaps other important characteristics by which someone might argue that the NHS should discriminate when allocating health benefits. I have assumed that discrimination by sex, colour, race or creed is unacceptable but if there are others I have missed I would welcome suggestions. For the time being I have been working on the following repertoire:

No discrimination whatever
Discrimination by age (the younger the better)
Discrimination by family responsibilities (the more dependants the better)
Discrimination by social worth (the more talented the better)

Discrimination by own implicit value of health (the more behavioural sacrifices you make for your health the better

Discrimination by socio-economic environment (the more deprived the better)

Discrimination by willingness and ability to pay (the richer and keener the better) [to see what support it commands]

ELICITING PREFERENCES AND TRADE-OFFS

The simplest way into this territory seems to be to begin by asking people whether or not they think there should be discrimination on each individual dimension in turn. For instance, the questions might be posed as in Table 18.3.

Table 18.3 Questionnaire for eliciting preferences

Do you think it right that, in deciding which ill people should be made healthy, the NHS should show any of the following preferences? (Choose one from each group of alternatives.)

1	(A)	The young should get preference over the old	☐	
	(B)	The old should get preference over the young	☐	Tick one
	(C)	Age should make no difference	☐	
2	(A)	People with young children should get preference over those without	☐	
				Tick one
	(B)	Having young children should make no difference	☐	
3	(A)	People looking after elderly relatives should get preference over those not doing so	☐	
				Tick one
	(B)	Looking after elderly relatives should make no difference	☐	
4	(A)	The breadwinner of the household should get preference over the others	☐	
				Tick one
	(B)	Being the breadwinner should make no difference	☐	
5	(A)	Someone who has a lot to contribute to the community should get preference over someone who has little to offer	☐	
				Tick one
	(B)	How talented you are should make no difference	☐	
6	(A)	People who have taken care of their own health should get preference over those who have not	☐	
	(B)	People who have not taken much care of their own health should get preference over those who have	☐	Tick one
	(C)	Whether or not you have cared for your own health should make no difference	☐	

7	(A)	People who are deprived in other ways should get preference when it comes to health	☐	
	(B)	People who are not deprived in other ways should get preference when it comes to health	☐	Tick one
	(C)	Whether or not you are deprived in other ways should make no difference when it comes to health	☐	

8	(A)	People who are willing and able to pay part of the costs should get preference over those who are not	☐	
	(B)	People who are not willing or able to pay part of the costs should get preference over those who are	☐	Tick one
	(C)	Whether or not you are willing and able to pay should make no difference	☐	

| 9 | (A) | Are there any other grounds on which you think the NHS should give preference to one kind of person rather than another?
Yes ☐ No ☐ | |
| | (B) | If yes, who should get preference over whom?
.... should get preference over .... | |

Any respondent who goes for no discrimination in every single case and does not add a write-n candidate, is not asked any further questions, except concerning their personal background (of which anon).

Any respondent who approves of only *one* basis for discrimination, could then be asked:

If we could keep one (state respondents'preferred class of person; for example, 'young person') alive and well for an extra five years by sacrificing that same amount of healthy life expectancy for some (state respondents'non-preferred class of person; for example, 'old people') how many (old people') health would you think it right to sacrifice in order to improve the health of one (young person)?

State number ☐

If a respondent favours discrimination by more than one characteristic, things become much more difficult. Initially they could be asked the above preference intensity question about each nominated characteristic individually. Then we would have to face the trade-off problem since the criteria might be interdependent. Suppose three criteria were nominated, in which the preferred categories were the young, those caring for children and the deprived, treating each characteristic as simply dichotomous (no differences of degree are admissible). Then we might initially seek a rank ordering of eight people with the following stated characteristics:

(A) A deprived young person caring for children
(B) A deprived young person not caring for children

(C)   A deprived old person caring for children
(D)   A deprived old person not caring for children
(E)   A comfortably off young person caring for children
(F)   A comfortably off young person not caring for children
(G)   A comfortably off old person caring for children
(H)   A comfortably off old person not caring for children

For such a respondent Person (A) should be the most preferred and Person (H) the least but it is not clear *a priori* how the others will be ranked. From the earlier question about preference intensity (for each dimension separately) we should get some clues about the relative weights likely to be given to the different elements when combined. However, we cannot assume that this combined weight will be related to the individual weights in any simple way because the combination of being deprived *and* coping with young children may be regarded as attracting much greater priority than the sum of the two characteristics in isolation. I would prefer to test this directly by giving respondents some magnitude estimation task at this stage but have failed so far to come up with one that is readily comprehensible and easily do-able by ordinary people.

Assuming that I eventually find a way through that, the other data I will need from respondents is background about themselves, so that I can explore whether any differences that emerge in people's response patterns are systematically related to anything about them or their situation. For this purpose I need to know their age, sex, household composition and their role in it, whether caring for anyone outside the household, current employment status, current or previous occupation, educational level, religious beliefs, recent experience of illness, smoking and drinking behaviour, main recreation pursuits, housing tenure, household income, attitudes to private health insurance, etc.

SOME PRELIMINARY RESULTS

I had an opportunity, at the conference at which a preliminary version of this chapter was given (essentially sections I–III), to try out my questionnaire on the participants, about two-thirds of whom responded. Later I did the same at meetings of health service manager, hospital doctors and clinical psychologists. In order to get the views of a group of people closer to the ordinary citizenry, I finally tried out my questionnaire (as in Table 18.3), but with a lot of background data also elicited) on a group of secretaries within the university. This pilot study generated the pattern of responses in Table 18.4: that is, 1.3 discriminations per person overall, or

2.1 per person who discriminates at all. It will be noted that just under 40 per cent of the respondents favour no discrimination whatever and, of the five occupational groups with the larger number of respondents, this non-discriminatory view was strongest amongst the secretaries and weakest amongst the doctors and the academics (most of the latter being philosophers of one kind or another).

Table 18.4   Pilot survey respondents

	Number of different discriminations favoured					
	0	1	2	3	4	5
18 clinical psychologists	7	4	4	1	2	0
18 hospital doctors	6	3	4	2	3	0
16 secretaries	8	4	3	0	0	1
12 NHS managers	5	4	1	1	1	0
10 academics	3	1	3	1	1	1
5 nurses	2	1	2	0	0	0
2 others	1	1	0	0	0	0
81 Totals	32	18	17	5	7	2
Total number of discriminations 105 =	0	+ 18	+ 34	+ 15	+ 28	+ 10

But what is rather more interesting is the *choice* of discriminators. This is set out in Table 18.5.

Table 18.5   Choice of discriminators by occupational group

	Frequency of mention							
	Young	With children	Elderly relatives	Bread-winner	Social contribution	Careful of health	Deprived	Will pay
Psychologists	5	6	3	1	1	5	2	0
Doctors	5	6	4	3	3	7	1	0
Secretaries	3	3	6	1	1	1	0	0
Managers	5	1	2	1	1	3	0	0
Academics	6	4	1	0	2	4	1	1
Nurses	1	0	0	0	0	2	1	1
Others	0	0	0	0	0	1	0	0
Totals 105	25	20	16	6	8	23	5	2

It is the first three axes of possible discrimination plus being careful over your own health, which clearly dominate but this frequency count may be rather misleading because it contains five 'votes' from some people and only one from others (see Table 18.4) and none, of course, from the non-discriminators.

If the data in Table 18.5 are reworked to allow each (discriminating) person only one vote (so that if someone votes for two items, each item counts a half and a three item vote counts a third each and so on) then we get the 'adjusted' distribution of votes as set out in Table 18.6.

Table 18.6 Adjusted choice of discriminators by occupational group

	Frequency of mention							
	Young	With children	Elderly relatives	Bread-winner	Social contribution	Careful of health	Deprived	Will pay
Psychologists	3.3	2.8	1.0	0.3	0	2.8	0.8	0
Doctors	1.6	1.9	1.3	0.8	1.3	4.7	0.5	0
Secretaries	1.2	1.2	4.2	0.2	0.2	1.0	0	0
Managers	3.1	0.3	0.6	0.3	0.3	2.5	0	0
Academics	2.3	1.5	0.2	0	0.6	1.8	0.2	0.5
Nurses	1.0	0	0	0	0	1.0	0.5	0.5
Others	0	0	0	0	0	1.0	0	0
Totals 105	12.5	7.7	7.3	1.6	2.4	14.8	2.0	1.0

This obviously changes the rank ordering slightly. It also highlights some other interesting features. For instance, the secretaries are especially concerned with those caring for elderly relatives, the doctors strongly favour those who have taken care of their own health and the psychologists, managers and the academics tend to support discrimination in favour of the young over the old, as can be seen when these same data are reworked as in Table 18.7. To explore these phenomena more closely we need to delve rather more deeply into the background data.

Table 18.7 Adjusted choice of discrimination by main occupational groups standardised so that each now sums to 100

	Young	With children	Elderly relatives	Bread-winner	Social contribution	Careful of health	Deprived	Will pay
Psychologists	30	25	9	3	0	25	7	0
Doctors	13	16	11	7	11	39	4	0
Secretaries	15	15	53	3	3	13	0	0
Managers	44	4	8	4	4	35	0	0
Academics	32	21	3	0	8	25	3	7
Overall*	25	16	15	3	5	30	4	2

* Includes nurses and others, for whom the number of observations are too small to warrant a separate entry.

If we look first at the respondents who support discrimination in favour of *those who have taken care of their own health*, there was no noticeable age or sex effect here, nor did those supporting this kind of discrimination do more things themselves for the sake of their own health than the others did. What *was* very noticeable, however, was that whilst similar proportions of ex-smokers and never-smokers supported this kind of discrimination, not a single one of the 12 current smokers did.

A rather more surprising result from my respondents was that those *favouring the young over the old* were predominantly the old. But those favouring discrimination in favour of *those looking after elderly relatives* did not have more elderly persons in their households, though they were predominantly women and especially women over 40. The group most favouring discrimination in favour of *those looking after young children* were the older *males*, though here there was an even stronger correlation with the number of children in the household (those in favour had 1–2 children per household, those against only 0.6 children per household).

In this pilot survey the self-reported state of people's own health, having had any recent consultations with doctors and whether or not a religious person, had no significant effect on responses. Too few people had low incomes or poor educational qualification for this to be tested as an explanatory variable. It need hardly be said that this sample, being small and unrepresentative, cannot be taken to reflect the views of the Great British Public.

It does, however, indicate the rich vein of material to be mined in this territory, about which we know very little, despite its rather crucial policy importance. It strongly suggests that, although the most common single view may well be that the NHS should not discriminate at all between people when it comes to distributing the benefits of health care, there are just as many people who favour discrimination by more than one criterion (there are 32 of the former and 31 of the latter in Table 18.4). Nor, on this limited evidence, is there much support for the view that doctors (and to some extent the managers), who at present are the ones left to exercise such discrimination as occurs, do accurately reflect the views of the rest of the population. Clearly there is a substantial research task to be undertaken here – and we have not started on the trade-off problem yet.

19. Cost-Effectiveness Analysis: Is it Ethical?

INTRODUCTION

In this chapter I want to tackle two kinds of ethical issue that have been raised by critics of the cost-effectiveness approach. The first is whether cost-effectiveness analysis *per se* is unethical when applied to medical care. The second concentrates on particular assumptions that are usually made within cost-effectiveness studies. I will deal with each issue in turn.

IS COST-EFFECTIVENESS ANALYSIS UNETHICAL *PER SE*?

Many clinicians disapprove of the introduction of economic considerations into priority-seeking in medicine, believing that letting costs influence clinical decisions or policies is simply unethical (see Chapter 17). I shall concentrate on the central issue of whether it is ethical for doctors (and presumably all others involved in patient care) to take costs into account when choosing a course of action.

The key to resolving this conflict is to recall the fundamental meaning of the economists' notion of cost. To an economist 'what will it cost?' means 'what will have to be sacrificed?' and this may be very different from 'how much money will we have to part with?'. So if someone says to me that they must have something *no matter what it costs*, I take them to mean that they must have it *no matter what sacrifices have to be made*. And it is always easier to make such statements if the costs (or sacrifices) are going to be borne by somebody else!

Transferring that little homily back into the field of medical practice, anyone who says that no account should be paid to costs is really saying that no account should be paid to the sacrifices imposed on others. I cannot see on what *ethical* grounds you can ignore the adverse consequences of your actions on other people. You can do so on bureaucratic or legalistic grounds, of course, by saying 'they are not my responsibility' but we all know into what an ethical morass that line of defence leads. The word we normally use to describe people who behave without regard to the costs of their actions is not 'ethical' but *'fanatical'* and I think that fanaticism is just as dangerous in medicine as it is in other walks of life. So I conclude that a caring, responsible and ethical doctor *has* to take costs into account. Indeed, it is *un*ethical *not* to do so!

There is, however, also a somewhat more subtle approach which essentially argues that it is clinical freedom that is being challenged by those pursuing the cost-effectiveness approach and, whereas clinical freedom has a strong moral base in traditional medical ethics, economics has no such moral base and must, therefore, be treated with the utmost suspicion. I have elsewhere (Chapter 17) listed the six basic principles of medical ethics:

1. Preserve life
2. Alleviate suffering
3. Do no harm
4. Tell the truth
5. Respect the patient's autonomy
6. Deal justly with patients

We health economists have no difficulty in accepting that those same six principles should guide our professional activity but it may well be that we put rather more weight on the last one (about dealing justly with patients) than doctors have in the past. Again it comes back to how one takes into account the sacrifices borne by others when deciding what 'dealing justly with patients' actually entails. It clearly requires an appeal to some underlying theory of distributive justice and, in practice, this is usually some egalitarian principle, though typically so imprecisely formulated as not to constitute a very clear guide to action. But since this imprecision is a notable characteristic of all the other five principles too, it simply leaves us where we were before, except that now it is those involved in *the art of health service management* who are resolving the conflicts, rather than the doctors. And since issues of community-wide 'just dealing' between patients will go beyond the scope of any one doctor's realm of action, it could be argued that if the judgements made by a particular doctor (exercising his clinical freedom) clash with those of someone with authority from the community to allocate scarce resources across rival claimants, the clinical freedom of the doctor has the weaker moral claim and can legitimately be constrained accordingly.

ARE PARTICULAR ASSUMPTIONS UNETHICAL? THE MEASURE-MENT OF RESOURCE CONSEQUENCES

If we now move from the general to the particular, it will be useful to consider separately the measurement of resource consequences and the measurement of health consequences.   This section will therefore concentrate on the former and the next section on the latter.

It may come as a surprise that the measurement of resource consequences (for example costs) raises any issues whatever beyond checking for accuracy the boring arithmetic of boring arithmetic people (which can surely be left to other boring arithmetic people) – by which is clearly meant accountants, not economists!).   In fact there are two big ethical issues to consider.   Firstly, should we include amongst the benefits of a treatment any reductions in the indirect costs of illness that are due to that treatment? Secondly, should we ignore the distribution of the costs of the health-care system between different groups in the community?

But before tackling either of those questions, I must make clear the context within which these questions get posed.   In a health-care system guided by market forces (for example, willingness–and–ability–to–pay on the demand side and profit–seeking on the supply side) there need be no ideological unease about who gets the benefits or who bears the costs, provided that the distribution of purchasing power, and the distribution of market power, are both considered ethically acceptable.   No health care system I know of works wholly on that basis, however, while many, including the British National Health Service (NHS), have explicitly rejected it in principle (though it persists to a limited extent in practice). An egalitarian stance pervades these alternative systems though, as we shall see shortly, this egalitarian stance is not sufficiently well specified to offer clear guidance to analysts as to what the distributional policy of the system actually is.   Within this large set of predominantly (but vaguely) egalitarian systems are a smaller number in which it has been decided that the best way to finance them is by taxation (and usually by taxes levied centrally by the national government).   In what follows I shall assume that this is the kind of system we are considering.

So let us return to the first issue, which was whether the reduction of so-called 'indirect costs' should be counted amongst the benefits of a treatment in a cost-effectiveness analysis (ignoring here any technical difficulties that might arise in trying to do so).   The typical situation is as follows: earlier return to work, or less time spent off work, are cited amongst the benefits of a treatment.   We can interpret this phenomenon in several different (but not mutually exclusive) ways.   Firstly, it may be taken as a proxy for improved health status (for example, reduced disability).

Secondly, it may be regarded as the fulfilment of a satisfying social role having value in itself (renewed contact with workmates, a more interesting life, greater self-esteem from doing something useful and valued by others, etc.). Thirdly, it may be significant as a source of income (and hence of better living standards) for the whole household. Finally, it may increase national output and thereby benefit the community generally. It is in this last respect that the ethical problem arises.

It is sometimes claimed that a particular treatment 'pays for itself' because the increase in national output (decrease in indirect costs) that it brings about is larger than the service costs entailed (output used) in providing the treatment. This increase in output is usually measured by the change in the gross earnings of the treated patients (reflecting what their extra input is worth to their employers and ultimately what the consequently increased output is worth to the consumers). A treatment which benefits the unemployable or the retired segments of the population generates no such additional benefits, nor will any such benefits show up for those whose (unremunerated) work is within the home. Even within the working population, this benefit will be greater for the highly paid than for the low paid. So in all these respects the counting of indirect costs seems to run counter to the principle that the provision of treatment should not be influenced by whether you are rich or poor.

We are now on the horns of a dilemma, for ignoring such changes in national output is not the same as denying their existence. If the result of ignoring such changes is to concentrate treatments upon the non-working population, the resources available to provide health care and many other good things in life will be less than they might have been (average real income per head will be lower) and it is well known that health is strongly correlated with real income.

So far the British health economists have got themselves off this hook by a rather neat argument. Observing that the British economy is currently operating with a very high level of unemployment, it is pointed out that if an otherwise employed person is 'off sick', there are plenty of people willing to fill the void, so all that will happen is that there will be some small 'frictional' losses while the system readjusts but, at the end of the day, each sick person 'off work' will have been replaced by some otherwise unemployed person now 'in work' and gross national product will be virtually unchanged. So it is concluded that there are good *economic* arguments for ignoring these indirect costs in the British context. Until such time as full employment returns we need not face the ethical question directly. However, what do people who come from countries which do enjoy near-full employment propose to do?

The second ethical dilemma on the resource side concerns the distribution of the costs of running the health service. At present we ignore this issue completely. We justify this by arguing that, because the NHS is not financed by earmarked taxes, we cannot identify the taxes which are higher because of the NHS. It is quite impossible to answer the question 'who bears the costs of the health service?', so there is no point in asking it!

There are increasingly frequent instances where someone advocates the introduction of some additional levy specifically earmarked for the NHS, because it is well known that the NHS is the one public service for which the majority of the population would be prepared to pay more taxes. The question then arises as to whether we care about the incidence of this *extra* taxation. It would be argued that since we are committed to ignoring the distribution of benefits between the rich and the poor, we should ignore the distribution of costs in the same way. Or we could imbibe the pure milk of the Communist manifesto and say that the guiding principle is 'from each according to his ability, to each according to his need', where 'ability' here refers to ability to pay. The classic alternative taxation principle is that taxes should be proportional to benefits. This is an issue which could become quite hot in the near future and which could greatly complicate the conduct of cost-effectiveness studies, whichever way we turn.

ARE PARTICULAR ASSUMPTIONS UNETHICAL? THE MEASUREMENT OF BENEFITS

It is in the area of benefit measurement that I have encountered the most intense ethical objections to the cost-effectiveness approach. They vary from quite sweeping denials of the right of anyone to sit in judgement on the value of another person's life, to more specific accusations of ageism, racism, sexism, etc. The criticisms seem to reach a particularly high pitch of excitement when the quality–adjusted–life–year (or QALY) is used as the measure of effectiveness in cost-effectiveness studies.

I will not spend much time here on the alleged immorality of one person making judgements about the value of another person's life (or, more correctly, on the value of improvements in another person's health). I think such judgements are inescapable in a system which is expected to behave in a non-capricious manner in discriminating between the well and the ill, between the severely ill and the slightly ill and between those likely to benefit from a particular treatment and those unlikely to do so, in order that some systematic priority-setting can take place in the face of inescapable resource constraints. The supposedly more ethical alternative of making these decisions by lottery certainly has the advantage of irresponsibility (if indeed that is an advantage) but seems to me quite inhuman and uncaring and most people I have spoken to about it find it quite unacceptable. I think it also has a serious internal contradiction which flaws it fatally, which is that lotteries do not spring fully formed from Heaven. They are invented by people. These people have to decide who is eligible to enter this lottery, what the prizes are, how soon and how often you can re-enter the lottery if you fail to win the first time, whether 'tickets' (especially winning tickets) can be traded or given away, and so on. It seems to me to be the *beginning* of a new discussion about discrimination which merely takes the place of the old one but does not get us off that particular ethical hook. Having accepted the inescapability of such judgements, what is of more practical interest is an examination of the ethical implications of those judgements.

The first general point that has to be made is that *every* effectiveness measure implies some value judgement. These frequently go unrecognised, because the effectiveness measure has come to be so widely used that it is conventionally accepted as the appropriate *technical* way of doing things.

Take, for instance, the two-year survival rate as a criterion for choosing between rival treatments. It carries the following implications:

1. To survive less than two years is of no value
2. Having survived two years, further survival is of no additional value
3. It does not matter with what *quality* of life people survive to two years
4. It does not matter who you are

The only one of these implicit assumptions that is acceptable to me is the last one but, as I propose to indicate shortly, even that is not acceptable to everybody. So on ethical grounds the two-year survival rate may well prove to be a totally unacceptable measure of effectiveness! It is a bit worrying, is it not?

The objection that survival rates pay no regard to quality of life can be overcome by adopting the QALY as the effectiveness measure and then investigating quite explicitly the extent to which people are prepared to sacrifice quality-of-life to increase life expectancy, or vice versa. Since people are likely to have different views on this important issue, we have to decide how the different valuations are to be brought together when making collective decisions. A typical technical solution is to take the arithmetic mean, which in the case of a skewed distribution gives a lot of weight to the upper extreme. This can be avoided by using median or modal values, or some more complex mean (for example, the geometric). What we are in fact doing here is deciding whose values shall count for how much in whatever policy issue is being addressed by the study (and this is equally true of clinical trials as for cost-effectiveness studies). This has quite important implications for the ethical problem of 'dealing justly with patients', so maybe that boring arithmetic is not as unimportant as it looks!

I noted earlier the strongly-held – but vaguely-articulated – egalitarian notions which supposedly guide many health-care systems. When aggregating benefits across people we have to know what it is that is supposed to be equally valued across people, otherwise aggregation is impossible. So the analysts are having to fill the void (wittingly or unwittingly) by adopting a precise stance on matters on which society is not offering clear guidance. There are many potential targets for such an egalitarian ethic. Even if attention is restricted to outcome measures (and some people argue that process measures are equally relevant) the list of candidates is quite long. A few obvious ones are a life–year, a QALY, whether the person has already had 'a fair innings', the 'rest of your life' (favoured by some philosophers because it minimises the scope for judgements about length and quality), or some other differentiating characteristic to do with who is getting the benefit. Like all other analysts, in order to get the work done I have filled that void in my own field, by explicitly assuming that a QALY is a QALY is a QALY no matter who gets it. I do not feel easy about this, so I have been trying to find out what the general public thinks about these matters and what those responsible for priority-setting in health care think about these things. The difficulty with such empirical work is, of course, that most of the time they do not think about these things at all, so it is quite difficult to elicit their views in a systematic and reliable way. Although I do not pretend to have cracked this problem yet, let me offer you some preliminary results to think over.

Whether a unit of benefit should have a different value depending on who will get it pervades much of the discussion about distributive ethics. The characteristics that are frequently mentioned in these debates are age, sex, marital status, whether with or without children, occupation and whether the person has cared for his or her own health. A few years ago we conducted a survey amongst nearly 400 randomly-selected adult citizens of York asking them at what stages in life they considered it most important to be healthy. The life stages considered were:

As infants	When at peak of earning power
When starting school	When looking after elderly relatives
When starting work	When just having retired from work

When setting up home for the first time   When coping with death of spouse

When bringing up children   When getting very old

The results surprised me. There was a very clear consensus that the most important time to be healthy was when bringing up children, which is a time in people's lives when they are in fact usually quite healthy but which is obviously also a time when people feel extremely vulnerable if they are not healthy. Running a close second was when you are an infant, the reason usually given here being that a healthy start in life is a good investment for the future. All other life stages were far behind these two in importance.

I later tried out a somewhat different approach (see Chapter 18), asking a convenience sample of some 80 people on what grounds they thought the NHS should discriminate between different sorts of people when determining priorities. The largest single group (40 per cent of respondents) thought there should be no discrimination whatever. In the other 60 per cent the preferred bases of discrimination were according to whether people had or had not cared for their own health (which was particularly prominent amongst doctors and health service managers), next in importance being a preference for the young. Roughly equal, in third place, were those looking after children (except amongst health service managers) and those looking after elderly relatives (especially amongst secretaries, who were mostly middle-aged women).

A much larger survey has been carried out in Cardiff on just over 700 people selected randomly from the electoral register. They were asked to choose between people with different characteristics when treatment could only be given to one of them. The results indicated a strong preference for the young over the old, except for the very young, where an eight-year-old would be given precedence over a two-year-old. Incidentally, the older respondents also manifested this preference for the young over the old, typically adducing by way of justification the 'fair innings' principle. In addition, married people were preferred to single people, non-smokers to smokers and light drinkers to heavy drinkers. No clear view emerged on the other dimensions tested (sex and occupation).

I later did a similar survey on the senior members of an English health authority with strikingly similar results. Again the young were to be given precedence over the old, non-smokers over smokers and light drinkers over heavy drinkers. We tested in addition the dimension of having children of school age versus having no children, eliciting a very strong preference for the former. Again, sex and occupation made no difference. These data have already convinced me that there is a very strong consensus in Britain concerning discrimination by age, by whether someone has young children and by smoking and drinking habits. It probably influences treatment priorities almost unconsciously at local levels. But should *we* be taking all this into account in clinical trials and in cost-effectiveness studies? If so, how are we to do so? At present I am sticking uneasily to a QALY is a QALY is a QALY, pending further clarification of just *how much* extra weight is to be given to the favoured categories over the unfavoured ones.

CONCLUSIONS

I hope that I have now convinced you not only that the evaluation of health–care activities is an ethical minefield, strewn with explosive material not easily detected by the naked eye, but also that bringing this material out into the open and analysing it (both by logical discourse and by empirical enquiry) is an important extension of the analyst's role. I think it is our duty to rush in where others fear to tread, even if in the process we find ourselves being maligned as insensitive troublemakers and even if the misguided criticise our analytical techniques because they require quite strong ethical assumptions to be made. It is not that any of these analytical techniques are ethical or unethical *per se*, it is more a matter of ensuring that their particular ethical assumptions are appropriate in the context in which they are being used. To do that requires *us* to be clear about the ethical assumptions built into our studies, but it also requires our 'clients' to be clear about what ethical assumptions are appropriate in *their* worlds.

I suspect that we will make faster progress with our task than they will with theirs, but perhaps it is precisely through our questioning that their position will become clearer, both to themselves and to everyone else. And with increasingly insistent demands for greater professional and political accountability in the provision of health, that must be a good thing!

20. QALYs and Ethics: A Health Economist's Perspective

OBJECTIVES

The purpose of this chapter is to examine some of the arguments that have been put forward to support the claim that it is unethical to use QALYs in priority-setting in health care. I shall first describe the essence of the QALY concept and then set out my views on the nature and purpose of ethical discourse. The core of the chapter then follows, in which I shall categorise those who object to QALYs *on ethical grounds*[1] as follows:

1. those who reject *all* collective priority setting as unethical;
2. those who accept the need for collective priority setting but believe that it is contrary to medical ethics;
3. those who accept the need for collective priority;
4. those who accept the need for collective priority setting in principle, but are unwilling to specify how it should be done in practice.

By 'collective priority setting', I mean priority setting intended to guide the use of *public* resources devoted to health care.[2]

I will mostly be concerned with such priority-setting at a 'policy' or 'planning' level, but when the occasion demands I will also consider its implications at a clinical level (for example, in influencing decisions made by individual doctors about individual patients).

COLLECTIVE PRIORITY-SETTING AND THE QALY

In the presence of scarcity, resources devoted to the health care of one person will be denied to some other person who might have benefited from them.[3] Clinicians are quite used to this phenomenon with respect to the allocation of their own time and of any other resources that they control as practice managers. They are trained to discriminate between those who will benefit greatly from treatment and those who will not and, by this means, 'clinical priorities' are established, which are based on some broad assessment of risks, benefits and costs. The role of costs here is crucial, because they represent sacrifices made by other potential patients who did not get treated. Thus the economists' argument that medical practice should concentrate on those treatments that are known to be cost-effective, is designed to ensure that the benefits gained by those to whom treatments are offered are greater than the benefits sacrificed by those who are denied treatment. That is what 'doing as much good as possible with our limited resources' means.

Priority setting is inevitably painful and its consequences are bound to be unfortunate for someone or other. It is therefore understandable that many people cling to the romantic illusion that, if only more resources were devoted to health care, they can escape from the process altogether.[4] But when more resources *are* made available, we still have to decide which are the highest priority uses to which these additional resources should be put. This too requires us to think carefully about collective priority-setting, so it is not really an escape route at all.

Collective priority setting requires us to be able to compare systematically the benefits of different kinds of health care, provided in different setting, by different clinicians, for patients with different characteristics, suffering from different conditions at different levels of severity. This requires a benefit measure which is extremely versatile and which has interval scale measurement properties (so that we can compare the size of differences in levels of benefit between treatments). Any measure which fails to fulfil these rather stringent requirements will be inadequate *in principle* as an aid to the priority-setting.

QALYs were designed to serve that purpose[5] but they require one further prior commitment, namely that the benefits of health care relate to *both* a person's length of life *and* a person's quality of life. We know that people value *both* of these fundamental attributes of life, so we need a measure of benefit which incorporates both and which reflects the fact that most people are willing to sacrifice some quality of life in order to gain some additional life expectancy and vice versa.[6]

If some health care activity would give someone an extra year of healthy life expectancy, then that would be counted as 1 QALY. If the best we can do is provide someone with an additional year in a rather poor state of health, that would count as less than 1 QALY and would be lower the worse the health state is. Thus, the QALY is to be contrasted with measures such as 'survival rates', commonly used as the sole success criteria in clinical trials, which implicitly assume that only life expectancy is of any concern to people. The essence of the QALY concept is that effects on life expectancy and effects on quality of life are brought together in a single measure and the bulk of the empirical work involved in making the concept operational is concerned with eliciting the values that people attach to different health states and the extent to which they regard them as better or worse than being dead. Conventionally, being dead is regarded as of zero value to everybody and being healthy is regarded as being worth 1 to everybody. This convention has important ethical implications because it represents a strong and quite specific egalitarian position. We will see later, however, that it can be replaced with something more complex if some other ethical position is preferred.

A QALY measure can, in principle, embrace any health-related quality–of–life characteristic that is important to people. The particular measure with which I am most familiar (the Euroqol measure) (Williams 1995) covers mobility, self-care, usual activities, pain/discomfort and anxiety/depression. In that classification system 'usual activities' are whatever the individual's usual activities are and are not restricted to work activities. So, although developed primarily by economists, the QALY is not a measure of people's economic worth but a measure of whatever aspects of people's health-related quality–of–life they themselves value.

THE NATURE AND PURPOSE OF ETHICAL DISCOURSE

This is a large and forbidding topic and I embark upon it only to make clear to the reader my own perceptions. I see ethics as being about what is right and wrong in a moral rather than in a factual sense. I see all ethical principles as 'relative' (that is, no one principle 'trumps' all others). The ethical principles that are of greatest interest in the context of collective priority-setting are those concerned with distributive justice. When applied to publicly-financed health care, they typically have strong (but poorly defined) egalitarian foundations[7] and in a democratic society they also rest on a notion of 'consent', if one expects unanimity, for justice is an 'essentially contested' concept.[8] By this is meant that 'there are competing conceptions of justice, all of which have respectable arguments in their favour' (Chadwick 1993).

What then is the role of 'ethical discourse?' I take it to be clarificatory. It enables us to explore the implications of any principle or choice we might advocate, so as to ensure that we really do understand and accept them. For if we do not, then the principle is insufficient (and perhaps even totally inappropriate) or the choice is wrong (in the sense of not being in accordance with the principles we believe we hold). This clarificatory role is very important and to be welcomed wholeheartedly.

I observe, however, that 'ethical discourse' is also being used for a different purpose, which is to argue (or often simply to assert or imply) that one ethical position is superior to another. Thus, what is engaged upon is not an ethical discourse but a political discourse (that is, an attempt to persuade people to support one position rather than another) by labelling one 'ethical' and the other 'unethical', whereas the truth is that each is 'ethical' according to its own associated conception of right and wrong. It is, of course, perfectly proper, indeed important, that those versed in the discipline of philosophical argument should engage the public in discussions about the basic principles underlying health care policy. However, this requires some analytical detachment if 'expertise' is being claimed, otherwise the role of the 'expert' becomes a cover under which special pleading is made for the kind of health care system that that individual (as a citizen) would like to have, or for the class of person which that individual (as a citizen) would prefer to have favoured.

It is, of course, very difficult for any of us to maintain that detachment and we doubtless all slip up at times and I see one of the benefits of ethical discourse as being to point out to the unsuspecting that they are being asked to subscribe to a position that they might not wish to hold. Much of the ethical critique of QALYs has had precisely this objective in mind and the resulting discussion has been very enlightening. My criticisms of it will be based on two general propositions. The first is that most commentators seem to believe that the ethical commitments that the QALY concept requires are narrower than they really are. This may be partly due to the fact that the empirical work that has so far been done has had to be based on certain simplifying assumptions, which may have led commentators to believe that these simplifying assumptions are intrinsic to the *concept*, when in fact they are not. The second is that these critiques have typically been conducted in a rather unsystematic way. To conduct a systematic critique requires the setting of prior criteria of judgement (which, because of the 'essentially contestable' nature of the subject matter, would need to be quite broad-ranging) and then the comparison (in principle) of QALY-based priority-setting with non-QALY-based ('Brand X') priority setting. I would then expect that QALYs would show up better than 'Brand X' on some criteria but worse than 'Brand X' on other criteria. It would then be left to the reader (as a citizen) to decide which of the respective merits and demerits are more important. If this level of detachment is too much to hope for in such contentious territory, one might at least expect the *analysis* to be clearly separated from the *advocacy*.[9]

IS COLLECTIVE PRIORITY-SETTING UNETHICAL *PER SE*?

Those who reject *all* collective priority setting as unethical typically assert that it is immoral for one person to sit in judgement on the worth of other people's lives, which is what collective priority-setting appears to require us to do.[10]   However, since they accept the fact of scarcity, they acknowledge that some people must be denied the benefits of health care. However, they want that done in a manner which is free of any interpersonal judgements of relative worth. They believe that this can be done by recourse to a lottery.[11] The trouble with this supposed solution is that lotteries do not fall like manna from heaven but have to be devised and run by people, who have to determine who shall be eligible, when and under what conditions, for each and every treatment that is on offer.[12]   So recourse to a lottery simply brings us back to the very same priority-setting issues that it was supposed to avoid. Ultimately someone has to make a conscious decision on how best to discriminate between people when confronted with scarcity. The problems then simply reappear in a different context.

Instead of seeking to avoid the making of interpersonal judgement of life's value, it seems more fruitful to try to achieve as much detachment as possible when making them. An entirely different sort of lottery could have an important role to play in that process. What I have in mind is the thought experiment involved in approaching collective priority setting from behind the 'veil of ignorance' (Rawls 1972). We have to imagine ourselves outside the society of which we are members and then choose that set of rules for collective priority-setting which would be most likely to achieve the distribution of health benefits we think best for our society. Then, and only then, will we be assigned, *by lottery*, an actual place in that society. We may find ourselves favoured by our rules, or we may be one of the unfortunate people who are disadvantaged by them, but we would have achieved a set of rules which we would have to accept as fair.

IS COLLECTIVE PRIORITY-SETTING CONTRARY TO MEDICAL ETHICS?

My second group of objectors are those who accept the need for collective priority-setting but believe that it is contrary to medical ethics. In the extreme, such people believe that it is the doctor's duty to do everything possible for the patient in front of him or her, no matter what the costs (Loewy 1980). In a resource-constrained system 'cost' means 'sacrifice' (in this case the benefits forgone by the person who did not get treated).

Thus 'no matter what the costs' means 'no matter what the sacrifices borne by others'. If medical ethics include an injunction to deal justly with patients,[13] then there *has to be* some weighing of the benefits to one person against the sacrifices of another. So I think that this supposed ethical conflict between the economists' argument that costs (that is, sacrifices) must be taken into account *in every treatment decision* and the precepts of medical ethics, is non-existent. Medical ethics does *not* require everything possible to be done for one patient no matter what the consequences for any of the others.

WHY MIGHT THE QALY APPROACH TO COLLECTIVE PRIORITY–SETTING BE UNACCEPTABLE?

My third group of objectors consists of those who accept the need for collective priority-setting and do not believe that it is contrary to medical ethics but cannot accept the QALY approach to it on other ethical grounds. There seem to be four distinct ethical issues raised here. Firstly, whose values should count? Secondly, how should we move from individual values to group values? Thirdly, should we not be concerned with the distribution of the benefits of health care across different people, as well as with the total amount of such benefits? Fourthly, are there other benefits from health care which QALYs do not pick up? I will tackle each of these in turn.

Is The QALY approach unacceptable because it uses the wrong people's values?

In principle the QALY concept is extremely accommodating in that it can accept anybody's views about what is important in health-related quality of life and anybody's views about the trade-off between length and quality of life. In practice, the early empirical work was based on professional judgements (mostly those of doctors). More recent work has been based on the views of patients and the general public, while my own work has concentrated on the latter because I am anxious to find out whether the values of the practitioners, their patients and the general public coincide. What the QALY concept does, quite properly, is to bring this question to the fore and point up the difficulties that are likely to arise if the priorities of a particular group of patients differ from those of their doctors or of the wider society of which they are part. In principle, since every treatment decision entails benefits to some and disbenefits to others, in a democratic society the views of *all* affected parties should count. Since the sacrifices involved in treating particular groups of patients will be widely spread and difficult to identify with any precision, this points inexorably to the general public as the most appropriate reference group. Some people have advocated using the values of a particular reference group as the collective view (for example, the views of the most disadvantaged, or of people with particular moral, legal or political authority). This alternative position raises other obvious concerns about elitism in a supposedly democratic society. Adoption of the QALY approach does not require you to take sides in this particular dispute but it does require you to be explicit about what the values are that are being used and where they came from. This is an important contribution to the ethical discourse.

Is The QALY Approach unacceptable because of the way it moves from individual to group values?

Once again, there is nothing in the QALY approach which requires aggregation to be conducted in any particular way. However, collective priority setting does require a collective view[14] so *some* method of aggregation has to be adopted and, whatever method is used, it will have strong ethical implications. The simplest method is to postulate that everybody's views, however extreme, must be given equal weight and a simple average is then taken to represent the collective view. A somewhat more complicated position is involved in taking the median view as the collective view. The median view is the one that would command a simple majority in a one-person-one-vote system. With a skewed distribution of values (which is what is commonly found) it gives less weight to extreme views than would the taking of a simple average. But whichever position is taken on this issue, the QALY approach has the great advantage that it is not possible to hide what you have done, so it is quite easy for others to tease out the ethical implications. The implications of a non-populist stance are that some people's views count for nothing, which itself would seem to require an ethical justification, since everybody's welfare will be affected by collective priority-setting.

Something more has to be said here about the conflict between the clinical standpoint and the societal viewpoint, because many clinicians believe that it is unethical for them to replace the values of each individual patient with some collective set of values. That it creates an ethical dilemma for them cannot be denied but it is one they have lived with for a long time.[15] Only in a purely private market (with no charity and no insurance) have doctors been in a position where they could do whatever the patient demanded (that is, wanted and was willing to pay for). In all other circumstances doctors have been constrained by somebody else's willingness to pay (for example, insurers, friendly societies, voluntary hospitals, taxpayers) and by their own willingness to put in more of their own resources without being paid. So it seems to me that such protests are more political than ethical, by which I mean that they are a protest against increasing public demands for accountability in areas which previously were ones where the clinicians had unchallenged private discretion. The ethical grounds for resistance are in fact weak.

Is The QALY approach unacceptable because it ignores the interpersonal distribution of health gains?

My theme here is the same as before. There is nothing in the QALY approach which requires QALYs to be used only in a maximising context, although it is QALY maximisation that is the natural interpretation of the drive for efficiency in health care. QALYs will also have a role in more complex rules and more complex rules will almost certainly be needed if collective priority setting is to reflect the views of the general public.

The simplest and commonest use of QALY calculations at present is based on the assumption that a year of healthy life expectancy is to be regarded as of equal value to everybody. Note that this does not say that it *is* of equal value to everybody. That is unknowable. What it says is that if that social judgement is appropriate, then what follows from it will be appropriate. If it is not, then what follows will be irrelevant. A strong egalitarian case could be made for that assumption, since it implies that it does not matter at all who the beneficiary is. Like justice, it is blind.[16] It pays no regard to race, sex, occupation, family circumstances, wealth or influence.[17] In this respect it follows precisely the assumptions underlying the use of the more conventional outcome measures used in clinical trials, which simply count the number of people with the specified outcome characteristic. Following hallowed tradition may not, however, carry much weight if a sizeable majority of the general public would prefer some discrimination between potential beneficiaries according to their personal characteristics or circumstances.

For instance, there is evidence (Charny *et al.* 1989) that most people (including the elderly) would give extra weight to benefits accruing to young people over the same benefits accruing to old people. There is a similarly widespread view that people with young children should have some priority over their childless contemporaries. It is quite possible to build these differential weightings into QALY calculations, the implication being that instead of maximising *unweighted* QALYs, we need to weight them according to the relative priority assigned to the particular characteristics of the beneficiary. Note that if this populist view is accepted, [18] it would count *in favour* of the simple QALY approach that it discriminates incidentally against the elderly and not be an *unethical* implication at all.[19]

It has also been argued that the interpersonal distribution of health gains should be influenced by how healthy people will be at the end of the process. For instance, not on health gains but on the level of health itself. That statement itself requires clarification as to whether it is people's current health status that is to be the focus of concern, or their prospective health (for example, disability-free life-expectancy). The former may be taken to mean the current 'quality–adjustment' part of QALYs and the latter the present value of a person's expected future flow of QALYs. In either case QALY measurement will have a role to play, even though QALY maximisation is no longer the (sole) objective. The rationale for some equalisation of achieved health is typically a variant of the 'fair innings' argument, namely that it would be wrong to devote resources to improving the health of those who have already had a long and healthy life when those resources could be used to improve the health of someone who otherwise will have a shorter and/or more unhealthy life. In weighing the benefits of health gains between people of different ages this consideration almost certainly lies behind having equity–weights which favour the young, as discussed in the previous paragraph. However, the intention now appears to be to go further and advocate a more extreme form of discrimination against the older and healthier members of the community, by establishing an absolute (or at least a very strong) priority in favour of the more unfortunate over the less unfortunate in the manner of Rawls (1972). It requires the eliciting of a coefficient representing society's degree of aversion to such inequality (which will tell us how great a sacrifice in the health of others we are willing to accept in the interests of this kind of equity).

Something similar seems to be implied by the 'double jeopardy' argument.[20] The proposition is that being in a poor state of health, or having a past history of being in poor health, generates a moral claim which is independent of what can be done to improve (Nord 1993) has found some evidence of support for this belief in a Norwegian population. If this belief proves to be widespread, the role of the QALY measure will be twofold. First, it will be necessary to estimate how bad each person's health state with and without treatment, in order to identify the most beneficial treatments. Then it will be necessary to look at how bad the states are that people have been in, so as to decide how best to arrange the distribution of health *gains* to make the consequential distribution of *levels of health* more acceptable. To do this will require the measurement in QALYs of each person's lifetime experience of health and the generation of a coefficient representing society's aversion to inequality in individual experiences of lifetime health. Once more QALYs play a key role, even though QALY maximisation is not the (sole) objective.

Is the QALY approach unacceptable because there are benefits from health care other than improvements in health?

There obviously are such benefits, in that the provision of health care generates a livelihood for millions of people. What people usually have in mind here is more to do with the benefits to the consumers of health care rather than to its providers.

A more relevant claim in the current context is that people get satisfaction from health care in ways which do not show up as improved health. A key issue in clarifying what is at stake here is what is meant by 'health'. In the QALY context it is whatever aspect of health-related quality–of–life that is of value to people and the length of that life. This is to be sharply differentiated from a narrow clinical definition of health, as concerned with whatever biomedical measures of physiological (or psychological) abnormality are of clinical relevance. Clearly the QALY will pick up many benefits from health care which may not show up in any narrow clinical assessment (though with the broadening of clinical criteria to include many quality–of–life issues, an increasing number of clinicians do in fact adopt a QALY-type approach to benefit measurement) (Williams 1985b). Thus, providing information or reassurance to people about a clinically incurable condition could well be evaluated within a QALY framework for its potentially favourable impact, as could 'palliative' care.[21] So the question is, are there other benefits which in principle the QALY concept will not pick up?

It is important to recall here that the present discussion is centred on priority-setting in a public health care system, so we are not specifically concerned with the broader issue of why one might prefer a public to a private system (which is usually based on the argument that the former determines priorities according 'need' and the latter according to ability to pay, with all the distributional implications that have been discussed above). Most of the arguments about group participation, caring externalities and altruism relate to that choice rather than to the ones we are now considering (McGuire *et al.* 1988, ch. 4). They may also bear just as heavily on how equitable and efficient is the tax system which finances health care as on the distribution of the health care itself (Van Doorslaer *et al.* 1993).

Where the crunch comes for collective priority-setting is whether concern about the *process* of providing and receiving health care is simply derived from any of the more fundamental concern about outcomes which have already been discussed, or whether it is a legitimate focus for ethical concern in itself. To put the matter as starkly as possible, if we had in place a system which satisfied all our efficiency and equity goals *as they relate to any aspect to do with people's health*, would we have any additional ethical concerns about the impact of health care upon the welfare of its consumers?[22]

The usual focus of concern here has been accessibility: is there a legitimate ethical concern requiring people to have access to health care even if it will not improve their health (as measured in QALY terms)? The argument seems to turn on the extent to which it is right to divert health care resources away from improving people's health and to devote them instead to reinforcing people's sense of security about care being available to them when they need it. If 'need' means 'capacity to benefit' and what is available will benefit people and what is available is available on equal terms to everybody, what more ought to be done? All of these conditions would be fulfilled by a simple QALY-maximising system and, if the implied argument is that some people are more deserving than others, the only difference would be that we would need to use 'equity-weighted' QALYs, or some coefficient of aversion to inequality, to determine that terms on which health care is offered. Whichever of these systems is espoused, they require only that the distribution of health *outcomes* be taken into account. No extra rules about accessibility are required.

There is a more broader argument that might also be considered here, namely that respecting the autonomy of the patient implies maximising the patient's satisfaction from health care. The counter argument is that priorities are set by the whole society and that, if what would give an individual satisfaction (for example, getting a medical certificate to say that he is unfit for work when he isn't) clashes with what society requires of the system, then it is the social norms which should prevail.

Is collective priority–setting necessary in principle but unacceptable in practice?

This brings me to my final set of people: those who accept the need for collective priority-setting in principle but are unwilling to specify how it should be done in practice. A typical stance is to point out all the difficulties involved with some particular approach and then to sit on the fence waiting for the next candidate to come by and then do the same. This would be fine if the implied ideal method were available to us, or if we could suspend all health care decision-making until it were. But there is no perfect system on offer and we cannot wait. As with a well-conducted clinical trial, the new has to be compared systematically, according to preselected criteria, with what already exists. If the same criteria as are used to criticise the QALY approach were used *in an even-handed way* to criticise current practice, or any feasible alternative to it, how would these alternatives make out? It is irresponsible to do less.

What is the nature of the 'social contract' in the field of health care?

I must, finally, venture into the territory of the political philosophers and make a few remarks about the supposed nature of the 'social contract' between the citizens and their (public) health care system. In doing so, we must bear in mind that the problem of collective priority-setting is central to that supposed contract, because it arises from the tension between each citizen's desire to have the best health care possible when he or she needs it and the conflicting desire to have as good a length of quality of life as possible in all other respects at all other times. Thus the tensions evident in the public debate merely reflect the tensions we all experience in our private lives, as patients on the one hand and as taxpayers on the other.

I see a social contract as the informal understanding reached between the members of a community as to what are their responsibilities towards their fellow citizens and what are the responsibilities of their fellow citizens towards them. Some of these responsibilities are to be discharged directly between one individual and another, others through various collective institutions set up for this purpose (for example, a public health care system and its associated tax system). Once more it must be stressed that there are no overriding rights or responsibilities in this multidimensional world.[23] Different societies at different times will work within different 'understandings' and not everyone in a given society at a given time will accept the general understanding that the majority has accepted. We are back to 'essentially contested' concepts.

Harris (1987) argues that the social contract has at its heart 'the idea that we treat each person with the same concern and respect' and then, somewhat surprisingly in view of his advocacy elsewhere of lotteries, goes on to quote Dworkin approvingly as observing that 'I do not show equal concern if I flip a coin to decide' whether to give the only dose of a drug to someone who will otherwise die rather than to someone who is merely uncomfortable. He says that treating people equally is not the same as treating them the same. This presumably goes back to Aristotle and the notion that equals are to be treated equally and unequals unequally in proportion to the relevant inequality. The former is what is commonly called horizontal equity and the latter vertical equity. In this formulation of the social contract, everyone is to be treated with the same concern and respect *in relation to what is relevant in the circumstances* and the role of the social contract is to define what is relevant. It may be that age is held to be relevant but not race. It may be that capacity to benefit is held to be relevant but not income or wealth. A decision not to offer someone a particular treatment is ethical and in accordance with the social contract if everyone in the same relevant circumstances will also not be treated, provided that each individual's circumstances have been properly assessed and the social contract was drawn up in a legitimate manner. There is nothing in the QALY approach which prevents us from treating each person with equal concern and respect in this sense. Indeed, careful measurement of health status in QALY terms might help to ensure a better assessment of an individual's circumstances and a more acceptable and openly accountable way of deciding who shall and shall not be offered which treatments.

This brings us to the nub of the matter as far as the social contract idea is concerned, namely how it is to be determined in a legitimate manner. Paul Menzel (1990) argues that if collective priorities are based on the informed views of the general population about the prospective risks they are willing to take (as from behind the veil of ignorance?) with their health, then the likely QALY gains from different treatments will inevitably be part of that process. Note, however, that he believes that 'informed consent' implies using valuations only from people who have actually experienced the health states in question, which may well restrict severely the number of people who are allowed to play a role in some parts of the process. One could go a step further and ask people about which discriminations they consider 'fair' and which 'unfair' and here it is difficult to think of any good reason why any competent person should be excluded. This will probably expose an overwhelming consensus in favour of the young and in favour of those with young children when it comes to distributing the benefits of health care. Within the context of a contractarian approach to priority-setting, why should this be resisted (especially since the older members of the community share in that view)?

One further point needs to be mentioned and that is the view that surveys, whether conducted by questionnaire or by individual interview, fail to capture the deliberative and reflective attributes intrinsic to any socially acceptable collective priority-setting. Would it not be better to rely on wide-ranging and deeply penetrating discussion amongst detached, well informed and responsible people? Quite apart from how this panel is to be chosen (clearly not a random sample of the general population) two problems arise. Firstly, what evidence does this panel need from outside itself to ensure that it is in touch with the community on whose behalf it is acting? Secondly, what evidence does the community need from the panel to ensure that whatever conclusions it comes to were arrived at in a manner that is acceptable to the wider community? Dealing with the first problem will surely require the kind of data the QALY approach throws up and the second will require the panel to be explicit about how it has resolved all the issues which the QALY approach highlights. So it seems to me that the negotiation of a social contract regarding collective priority-setting in health care *requires* the adoption of a QALY approach, for this is the approach which maximises the use of evidence, which maximises participation, which maximises openness and which offers the most comprehensive framework of thought for tackling the many ethical issues involved.

CONCLUSION

The essence of my argument in this chapter is that, although the QALY-based approach to collective priority-setting in health care is 'consequentialist' (meaning that it emphasises the importance of measuring the impact of health care upon people's health), it is not necessarily 'utilitarian' (in the sense of requiring the objective to be the simple *maximisation health gains*). QALYs have a key role to play in *any* system of health care priority-setting in which the impact of health care upon people's health is a relevant consideration. It is difficult to conceive of any ethical system of collective priority-setting for health care in which this would not be the case. Indeed, in most public systems it appears to be the *dominant* consideration (Calman 1994). It therefore seems to me that the QALY approach will become the dominant approach.

NOTES

1.   There are, of course, other possible grounds for objecting to QALYs, such as the practical problems in calculating and applying them. These practical problems will not be addressed here, since the ethical objections are prior. If they are valid, the practical problems become irrelevant.

2.   I am not here concerned with the problem of priority setting in private health care systems, which I have examined elsewhere. See Williams 1988e, 1992.

3.   Throughout this chapter I shall assume that each and every health care activity is being provided efficiently, in the sense that it does not use more resources than are absolutely necessary. I do not believe that this is actually the case in *any* health care system and I also believe that such inefficiency is unethical, in that it avoidably deprives people of beneficial health care. I wish here to concentrate on a more difficult problem, namely how best to choose between efficiently produced health care activities.

4.   John Harris (1985), for instance, after devoting 20 pages to discussing the pros and cons of different principles which might be used in priority-setting, finally comes to the section headed 'What should we do?' and concludes lamely 'The most moral and the most

honourable way of dealing with the difficultie ... is to try to ensure that we have sufficient resources to devote to postponing death, wherever and whenever we can, whether for long or for short periods, so that we do not have to choose between people invidiously'. And when all of our resources have been devoted to giving everyone one more gasp ...? More recently, and in a more pragmatic vein, Christine Hancock (1993) writes: 'Given the difficulty of finding a basis for rational rationing decisions and of commanding consensus in the community ... we need to consider whether the present system could be re-examined and managed more efficiently and with greater fairness'. She goes on to suggest that this be done 'Assuming that the demand for health care is not infinite; that it does not outstrip resources; and that it may not in fact require a huge injection of resources ....' One must applaud the desire to re-examine the efficiency and equity of the present system, which is precisely what QALYs are designed to assist but, if the demand for health care does not outstrip resources, the problem no longer exists, so that does not seem to be the best basis for such a re-examination. There is no evidence whatever from international experience to suggest that a minor increase in the proportion of GNP devoted to health care will eliminate the need for priority-setting. Even La Puma (1990) seems anxious to evade the issues raised by his critics by recourse to the notion that 'physicians and policymakers should work together to understand each other's ideas and to serve patients *without rationing the care that they need*' emphasis supplied). If discussion alone could resolve the problems, we would have solved them by now.

5. See Williams (1993b). I do not here distinguish between QALYs and allegedly rival concepts such as Healthy Year Equivalents (HYEs) since they are best seen as different variants of the same general approach (see, for instance, Symposium on QALYs and HYEs (1995) and they raise the same ethical issues as are considered here.

6. Explicit recognition of this as a trade-off problem immediately creates problems for those who believe that any and every life-threatening condition should take absolute priority over any and every non-life-threatening condition. Harris (1987) himself seems uncertain about his position here. He starts by arguing that it is lives that have to be saved, not life years. He then criticises the QALY approach for trying to maximise life years, observing that with that objective 'the best thing we can do is to devote our resources to increasing the population'. Without any apparent evidence for this proposition, he goes to assert that 'Birth control, abortion and sex education come out very badly on the QALY scale of priorities', forgetting that what QALY maximisation is about is *quality-adjusted* life years, by which criterion one could doubtless make out a very good case for all three of those activities. A little later some doubts seem to creep in however, for he concedes that 'while it maybe that life saving should not *always* have priority over life enhancement, the dangers of adopting QALYs which regard only one dimension of the rival claims ... as morally relevant should be clear enough.' As will be argued later, QALYs can embrace any dimension of health-related quality–of–life that people think important and can be modified to reflect a wide range of equity concerns too.

7. The crucial unresolved problem being to specify what it is that is to be held to be equal between people. See Williams (1994) and Elhauge (1994).

8. See Chadwick (1993) 1955–6.

9. I have not found any such commentaries. If you know of any that come at all close to meeting my desiderata, please send them to me!

10. Strictly speaking it requires us only to sit in judgement on the value of *changes* in health-related aspects of people's lives. For some people, however, any interpersonal comparison of worth is unacceptable (for instance, Harris 1985), especially pp.16–17. I have argued for a long time that since such comparisons are unavoidable, the best strategy is to make them explicit (Williams 1981).

11. See, for instance, Harris (1985), p. 88. Harris's advocacy of lotteries has been endorsed by O'Donnell (1986) p. 59 who also believes them to be 'a much more practical solution than the economists'. My own views were set out in a letter to the editor (Williams 1986). Evans (1993) also appears to regard this as a viable solution.

12. This has been further elaborated by Elhauge (1994).

13.   As appears to be the case, see, for instance, Gillon (1985)

14.   Elhauge (1994) confronts this problem openly but then seeks to resolve it by recommending that we have different health care systems for different people. This merely resuscitates the equality problem and brings to the fore the conflict between equality and freedom of choice.

15.   This point has been acknowledged by the BMA who say 'All health care systems ration care by some method' and the important thing is to ensure that 'rationing should be done openly'. This suggests to me that we should start with full disclosure of how it has been done hitherto so as to test whether the more explicit QALY-based methods are more or less ethically acceptable than the old implicit ones. See chapter 12 of BMA (1993).

16.   An anonymous reviewer of an earlier version of this chapter objected to this analogy on the ground that 'the *law* is blind because it describes a process, rather than outcomes'. But the analogy I am drawing does not relate to outcomes, but to who the recipient of a given outcome might be. The corresponding ideal of 'blindness' in a system of justice would be concerned with not having one law for the rich and another for the poor, or one for women and another for men. This seems to me a peculiarly apt analogy.

17.   It may nevertheless involve *incidental* discrimination against a particular sex, race or socio-economic group who, in some specific circumstance, stand to benefit less from health care than do the others. But before this can be castigated as unethical, what is required is a further principle which reflects a different ethical concern, for instance, the principle that there should be greater equality in (quality adjusted) life expectancy within the community, which, in turn, will discriminate against some other subset of the community. Whilst it is important to note who will be disadvantaged by a rule or set of rules, to note that someone or other will be disadvantaged is not an *ethical* objection against such a rule or set of rules. Kappel and Sandee (1992) argue that 'either we should go for equality; and in that case QALYs are unfair because they haven't got enough of an ageist bias' or 'we should go for consequentialism; and in that case QALYs have just the right sort of ageist bias' (p. 316).

18.   Since this is one of the 'essentially contestable' aspects of distributive ethics, there will, as always, be a significant minority who disagree with the majority view. Elhauge, for instance, argues strongly against allowing such personal characteristics to play any role in the allocation system, which would bring us back to unweighted QALYs. See Elhauge (1994), especially pp. 1518 *et seq.*

19.   Contrary to the argument of Harris (1987) p. 119, the arguments put forward by Evans (1993) for favouring the elderly are more complex. After arguing impeccably that 'We should aim to acquire enough information to be able to work out for each person, regardless of age, what the cost–benefit ratio of a contemplated intervention would be', he concedes that 'age may be associated, on average, with differences in outcome from treatment', which implies that 'on average' his cost–benefit ratios will (incidentally) discriminate against older people. But he then shrinks from accepting this, saying that 'cost-effectiveness underlies the ethics of the purveyor rather than the customer', seemingly forgetting that 'effectiveness' is precisely about the benefits to 'customers', and that costs are about the sacrifices imposed on other potential 'customers'. He then simply asserts that 'Other things being equal, the 30-year-old and the 80-year-old will be equally grateful for having their lives saved', even though that conclusion is unknowable and the relevant matter is what the two individuals and the rest of society regard as *just*. His final conclusion that 'there is no more equitable or open form of rationing than a good old fashioned orderly British queue' depends, of course, for its validity upon how people became eligible to join the queue, who decided the priorities between them and on what grounds. Queues are like lotteries in that they are the result of someone's decisions about priorities, they are not a way of avoiding those decisions.

20.   As promoted, for instance, by Harris (1987) and challenged by Singer *et al.* (1995).

21.   In this respect La Puma and Lawlor (1990) are mistaken when, after observing that 'Patients often seek physicians for attention, information, reassurance, encouragement and permission, not just prescriptions and procedures' they go on to assert that 'In QALYs there is no attempt to integrate the therapeutic value and outcome of talking with patients or their families'. But talking to patients and their families *is* a procedure, and its therapeutic

outcome upon the length and quality of patients' lives (and those of their relatives) is precisely what QALYs are designed to measure!

22. This is how the issue is posed by Mooney (1989). Mooney asserts that 'equality of access ... is not considered relevant in QALY maximisation' but this is not true. If 'need' means capacity to benefit and then 'equal access for equal need' will be required for QALY maximisation and the greater the need the more important it will be to ensure access.

23. For instance, it is not the case, in principle or practice, that 'there are no financial constraints on society's obligation to attempt to ensure equality before the law' (Harris (1987), for that would imply that the sacrifices borne by others would not be taken into account, which in turn implies a lack of concern or respect for *their* welfare.

21. Conceptual and Empirical Issues in the Efficiency–Equity Trade-Off in the Provision of Health Care or, If we are going to get a fair innings, someone will need to keep the score!

BACKGROUND

The inescapable imbalance between the technical capabilities of the health care system and the resources available to exploit them make priority-setting in health inevitable. This can be conducted in a variety of ways, each of which reflects a particular ideology (Williams 1988f). Here I shall concentrate on priority-setting which is informed, in part at least, by some kind of egalitarian ideology.

When analysing the problem of priority-setting, economists routinely accept that there is a conflict between equity and efficiency objectives, but rarely go on to estimate the actual trade-off between them. This reluctance may be partly explained by the fact that, before attempting to do so, it would be necessary to define both equity and efficiency in terms that were policy-relevant, unambiguous and in a manner that made measurement possible. The purpose of this chapter is to explore what is entailed in that quest.

In the case of priority-setting *for health and health care* there is an immediate complication in that the debate is frequently conducted in the language of 'need'. Thus the issue is seen as determining which needs shall be met and which shall be left unmet. Unfortunately two distinct needs tend to be confused here, namely the 'need for health' and the 'need for health care'. Thus, when people argue that priority should be given to 'those in greatest need', it is not clear whether they mean the person who is in the worst health state, or the person who will benefit most from health care, or the person who will be left in the worst health state after treatment.

The first and last of these are about people's need for *health*, concerning which there may be nothing whatever that the *health care* system can do. Indeed, it needs constantly to be borne in mind that the main determinants of health – and the best ways of improving it – may lay outside the purview of the health care system entirely. Nutrition, genetics, lifestyle (including occupation) and physical and social environment are all likely to be more important than health care in this respect. So demonstrating a 'need for health' has no one-to-one relationship with a 'need for health care'. It may indicate a 'need for *effective* health care' but this is a health-research prioritising issue rather than a health-care prioritising issue, and will not be pursued further here.

THE CULYER/WAGSTAFF APPROACH

Culyer and Wagstaff (1993) and Culyer (1995) approach these matters with commendable clarity and are careful to distinguish the different meanings that can be given to the concepts used in the common rhetoric about efficiency and equity. They have laid out analytically the complex relationships between them and the reasons why some of these concepts may not lead where their propounders imagine (or wish). They also show that in the *provision* of health care,[1] it is possible to define equity and efficiency (quite plausibly) such that no conflict occurs between them. If 'efficiency' simply means operating somewhere on the production possibility frontier for health (given the resources and technologies available), then any equity criterion that is applied in order to choose a particular distribution of health on that frontier will automatically satisfy the efficiency criterion too. All that remains is a conflict *between* different equity objectives (and there has to be at least one, otherwise the system is underdetermined, in the sense that we would not know where on the frontier we want to be). So is this all much ado about nothing?

I think not. Suppose we take a different view as to what efficiency means in the provision of health care, namely maximising health gain as measured in some standardised way such as in life-years or in quality-adjusted-life-years (QALYs). This is a different interpretation of 'health maximisation' from that used by Culyer.[2] For him health maximisation simply means getting to the frontier. With the type of production possibility frontier that is normally assumed, standardised health-gain maximisation will define a single point on such a frontier, where it is tangential to a Social Welfare Function in which all units of health are equally valued no matter who gets them.[3] From this perspective, the equity-efficiency trade-off becomes the estimation of the number of (equally-valued) units of health that are sacrificed by moving from that point to any other on the frontier. As Culyer shows, several different 'other points' may be chosen, each of which could be justified by a specific equity argument.

Figure 21.1 is a modified version of the NE Quadrant of the Culyer diagrammatics, in which the production possibility curve represents what *health services* can do to affect the health of A and B. I have drawn it so as to represent the view that the scope for bringing about change through health care alone is rather limited. Figure 21.2 enlarges the segment where

*Figure 21.1   Health production possibilities
with social welfare function*

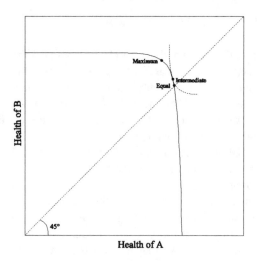

Health of A

the action is likely to be and identifies three key points to which I shall return later. The point labelled 'equal' is where the health of A and the health of B are identical (lying as it does on the 45° line from the origin). It is the preferred position for a pure egalitarian. The point labelled 'maximum' is where the aggregate health of the two individuals (measured, say, in *n* life years) is as great as possible. This would be the preferred position of a pure efficiency maximiser (in the meaning of efficiency that I am using).[4] For those wishing to pursue *both* efficiency *and* equity, some 'intermediate' position will be preferred, depending on the precise shape of their social welfare function. It is this 'intermediate' position that is the focus of interest in this chapter.

Any argument for departing from the strict maximisation of equally-valued units of health, adducing some personal characteristic of the beneficiary as a relevant consideration, I regard as an equity issue. A careful distinction needs to be drawn here between personal characteristics of the beneficiary which affect the *quantum* of benefit to be expected from a treatment and personal characteristics which affect the *value attached to any given quantum* of benefit. For instance, being old may reduce the number of extra life years gained from an intervention (a *quantum* effect related to age). The quantum effect will influence the shape and position *of the production possibility frontier*. The valuation effect will influence the

*Figure 21.2 Health production possibilities
with social welfare function:
enlarged segment*

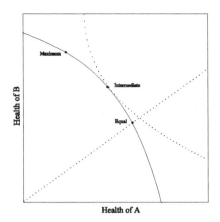

Health of A

shape *of the social welfare function*. I regard the former as an efficiency issue and the latter as an equity issue. Those who do not like the implication that *in the efficiency calculus* the quantum effect may discriminate against some group whom they regard as particularly deserving (for example, the old, or the poor, or smokers, or those with serious concurrent medical conditions) are, of course, free to pursue some countervailing argument *as an equity argument* (which is what typically already happens in the literature).

SOME POSSIBLE EQUITY PRINCIPLES

It is impossible within the short span of a single chapter to provide a comprehensive account of all the equity principles that have been, or might be, put forward as relevant for priority-setting in health care. I will not even consider in detail all those reviewed by Culyer op cit. (1995). My selection is guided partly by the results of some surveys of popular opinion and partly by a personal judgement that ultimately it is going to be *equity concepts related to outcome* (that is, health) which will be decisive, not concepts which focus on process or resources.[5] In various surveys that have been conducted to elicit people's views as to who should be given priority over others, there are some recurring themes. One is that the young should in general be given priority over the old (though not infants over slightly older children). Another is that those looking after young children should have priority over those without that responsibility. Many people also believe that priority should be given to those who have cared for their own health over those who have not (for example, by smoking, drug abuse or heavy drinking). These views have also come through clearly in opportunistic surveys I have done over the past five years with health professionals from a wide variety of backgrounds and from many different countries (see Table 21.1).

Table 21.1 Who should be given priority if only one can be treated?

	Percentage distribution of respondents within each Choice					
Choice	First strongly preferred to second	First slightly preferred to second	Unable to choose to first	Second slightly preferred to first	Second strongly preferred	Row totals (per cent)
5 year old vs 70	58	22	13	5	2	100
35 years vs 60	29	42	28	2	0	100
2 years vs 8 years	4	5	69	17	5	100
Sngl vs Marr	3	3	69	21	4	100
Smokr vs N–Smk	6	8	43	31	12	100
H Drinkr vs Lt	5	6	23	41	24	100
Female vs Male	5	4	91	1	0	100
Unemp vs Empl	3	3	84	8	2	100
Dir vs Unskld	6	7	85	2	0	100
Drivr vs Tchr	3	1	87	7	2	100
Care of Chldrn	28	37	31	2	2	100

NUMBER OF OBSERVATIONS: 519

Composition of respondents: Health economics students 130, Swedish hospital staff 127, overseas course members 88, NZ public health students 78, York citizens 39, English Health Authority members 38, Junior hospital doctors 19.

I have also noted that the literature has generated another recurring theme, namely that those who have had a hard life (either specifically to do with health, or more broadly to do with life in general) should not be further penalised in health care priority-setting by applying to them the full rigours of the efficiency calculus. Rawls' (1972) advocacy of the rule that policy should be guided solely by its effects on the worst off member(s) of a community is the most extreme manifestation of this kind of argument. It also manifests itself in a slightly less extreme form as a 'double jeopardy' argument (see Harris 1987, Singer *et al.* 1995), where it is said that those who have already experienced significant misfortune should not have further tribulations imposed upon them simply because they are not good candidates (within the efficiency calculus) for the receipt of health care. This in turn seems related to a similar but more pervasive argument concerning people's entitlement to a 'fair innings'.[6] Later in this chapter I intend to follow the implications of this last argument in some detail but, for the present, suffice it to say that it reflects the feeling that everyone is entitled to some 'normal' span of health (usually expressed in terms of life years, for example, 'three score years and ten'). The implication is that anyone failing to achieve this has in a sense been cheated, whilst anyone getting more than this is 'living on borrowed time'. Thus whereas the 'double jeopardy' argument directs attention only to the down-side, the 'fair innings' argument considers both downside and up-side.

I have found it useful to classify equity arguments into those concerned with 'desert', and those concerned with Aristotelian notions of horizontal and vertical equity. Arguments based on desert (that is, one person *deserving* special consideration, for good or ill, because of something they have done or not done) include those mentioned earlier which related to having or not having children and caring or not caring for one's own health. They touch a strong moralistic strain in many people and clearly influence the behaviour and policies of many health care professionals, so they cannot be lightly dismissed. They also raise a raft of difficult issues concerning the boundaries between personal and social responsibilities, which I do not propose to address here (see Schwartz 1995). I shall therefore not pursue *these* equity notions further on this occasion.

The notion of horizontal equity concerns the equal treatment of equals, the problem being to determine which are the relevant characteristics upon which horizontal equity is to be brought to bear (Williams 1994). It is common to assert that characteristics such as wealth, position, race, or religion should not influence the value attached to the benefits derived from health care (and hence to the value attached to access those benefits). Sometimes location is added to this list, meaning that people should have equal access to the benefits of health care no matter where they live but again we need to remember that, even if any given quantum of benefit is *valued* identically whichever of these groups of people get it, each of these same distinguishing characteristics may affect the *quantum* of benefit derived from any treatment. It must also be remembered that *denying access altogether* is perfectly equitable (from the viewpoint of horizontal equity) provided that *everyone* in like circumstances is denied access. There may also be some overlap with the preceding discussion about 'desert' when people assert that (say) smoking status should not be a relevant consideration when valuing benefits (any more than race or religion are). In taking such a position they are in fact asserting that this is a matter of horizontal equity and not a matter of 'desert' (or 'contributory negligence'). Different perspectives generate different conclusions!

It is however, the vertical equity issue about 'treating unequals unequally in proportion to the relevant inequality' upon which I shall concentrate for the rest of this chapter. It requires us to embark upon three voyages:

(a)   to decide what is a relevant personal characteristic by which to classify people for policy purposes;
(b)   to decide how we are going to measure health;
(c)   to decide how we are going to measure a health inequality.

With this information before us, we then have to decide just how averse we are to the inequality so described, that is, what sacrifices in the original efficiency maximand we would accept to achieve a specified reduction in the policy-relevant inequality.

Unfortunately, almost every conceivable categorisation of personal characteristics is likely to be relevant for some policy decision or other. Policies are made at levels ranging from the World Health Organisation and the World Bank, through public bodies at national, regional and local levels, to hospitals, clinics and individual clinicians. Some of these policy-makers specialise only in health or health care matters; some have to weigh the claims of health and health care against the claims made for other ways of improving the length and quality of people's lives, such as social care, education and the relief of poverty. Out of this welter of material I have selected a few items which seem to me to be important at a fairly aggregate level, to serve as examples of how policy analysis of health inequalities might proceed. I leave it to the imagination of the reader to identify ways in which these ideas might be applied more widely, with such modifications as may be necessary to make them relevant in other contexts. I will also make a few suggestions myself at the end indicating where I think things should be going.

QUANTIFICATION

Debates about equity are on the whole not cast in quantitative terms. Much more typical is the philosophical argument, designed to persuade the reader in principle to adopt or abandon a particular position, often being conducted around illustrative case studies. These present problems to be resolved, which are usually inconclusive. This is because ethical principles are 'essentially contestable', by which is meant that 'there are competing conceptions of justice, all of which have respectable arguments in their favour'.[7] Most commentators observe that such principles inevitably conflict with each other and none 'trumps' the others, so that no principle is absolute. In the presence of such unanimity about the *relativity* of ethical values, it is surprising that there has been almost no attempt to establish empirical trade-offs (or even to assert a personal opinion as to what they should be).

If the nature and implications of particular positions are to be clarified in a policy--elevant way, this discussion has to move on to seek quantification of what are otherwise often quite vacuous assertions (for example, that access should be determined by need, without 'access' or 'need' being defined in an operationally meaningful manner and without any attempt to estimate the costs, in terms of other benefits forgone). Only with some quantification will it be possible to devise rules that can be applied in a consistent manner with a reasonable chance of checking on performance (that is, holding people accountable). At present, although reassurance is frequently offered that equity considerations have been taken into account, there is no way of establishing what bearing, if any, those principles actually had upon the outcomes. Judging by the persistence of health inequalities and the almost universal agreement that they are deplorable, it is tempting to conclude that the rhetoric is not matched by any real commitment to do anything effective. Quantification thus has potential for clarification, for performance measurement, for accountability and for policy analysis and reappraisal. The quest for greater quantification of equity considerations seems worth pursuing on those grounds alone, despite the hostility it is likely to engender from those who mistakenly equate precision with lack of humanity.[8]

One of the few issues in the equity field about which there seems to be an overwhelming consensus is that the young should have some priority over the old (a view held by the old as well as by the young). Unfortunately this remarkable consensus is likely to evaporate when the quantitative issue is addressed as to *how much* priority the young should have, since I doubt whether many people would interpret this priority ranking to mean that everyone over a certain age should be denied *all* health care.

Figure 21.3 Value of a year of life
Relative value of a year of life at age x   (World Bank)

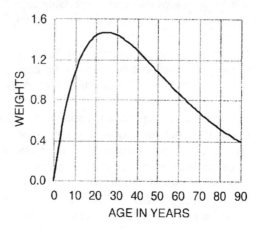

The World Bank has clearly taken this on board in a very explicit way, by publishing[9] the set of 'age–weights' reproduced here as Figure 21.3 and using these weights when measuring the 'burden of disease' in different countries. The authors of the *World Development Report 1993* (World Bank 1993) observe that 'Most societies attach more importance to a year of life lived by a young or middle-aged adult than to a year of life lived by a child or an elderly person'. So in their table of age–weights note the zero weight attached to newborns, the peak in the late twenties and the decline below 1 in the early fifties. These age–weights are important social judgements, which may well be appropriate to the conditions in the countries with which the World Bank mainly deals in the field of health. Thus, too high a birth rate may justify the low weight given to infants; the peak productivity of manual workers in countries living near subsistence may justify the high weights in the range 15 to 40; while the low life expectancy at birth may justify the weight less than 1 given after the age of about 50. In richer countries things may be different. For example, with a low birth rate the starting values will be higher, if the productivity of a largely non-manual work force peaks later, the peak will be later while, with life expectancy at birth much higher, the decline below an age–weight of 1 will come later.

Thus one simple route into the equity-efficiency trade-off is to use weights such as these in the maximisation process, so that instead of maximising life years or QALYs, it is age-weighted life-years or QALYs that are maximised.[10]   It would only be by coincidence that the maximisation of age-weighted health gains also maximised unweighted health gains. The 'efficiency loss' should be calculable from the difference between them.

A powerful part of the rhetoric about equity in health care employs the notion of equality applied to *people's whole lifetime experiences*, not just their current situation. Adopting that broad perspective and looking first at life expectancy, we find considerable variation *both* between countries *and* within countries.

Whereas those living in the poorest 20 countries in the world have a life expectancy at birth of less than 50 years, those living in the richest 20 have over 75 years (World Bank 1993). How averse are we to this enormous discrepancy and what, if anything, could we do about it? By inference from the volume of resources transferred in cash or kind from rich to poor countries, it seems that we in the richer countries have only a weak aversion to this kind of inequality and to the extent that we do respond, we are more likely to devote aid to improving nutrition, water supply and disposal and simple disease prevention measures, than to investing heavily in personal medical care. In that respect we are probably pursing the right strategy.

We could get some clues as to what the quantitative potential of improved health care might be by estimating the extra gains in life-years that could be achieved if health care systems operated at their frontier. The difficulty here is that there is a different 'frontier' according to the volume of resources available to the system, so the level of resource commitment needs to be taken into account too. A promising approach at an international level is to compare real per capita expenditure on health with feasible but unexploited per capita increases in life expectancy. It seems reasonable to expect that those countries which spend a lot on health care have a smaller amount of unexploited health gain potential than those which spend a little. Figure 21.4, which attempts this calculation is based on data in Towse (1995) and shows that this is not exactly the case. Those countries nearest the origin in that figure are closest to their respective frontiers, but, as usual, the USA is clearly well inside its frontier with a large volume of unexploited gains despite very high expenditure per head. If the US health care system were to become more efficient, it would make the international distribution of life expectancy even more unequal, though greater efficiency might do something to reduce inequalities *within* the USA. From an international equity viewpoint we need to explore the potential for concentrating resources where the poorest countries will be best able to exploit this untapped potential for generating additional life years. I will return to this point shortly.

Figure 21.4 Technical efficiency of health care:
a comparison of systems

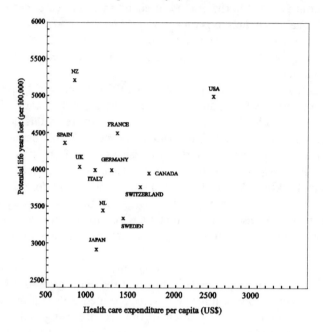

Figure 21.5 Survival rates: males versus females

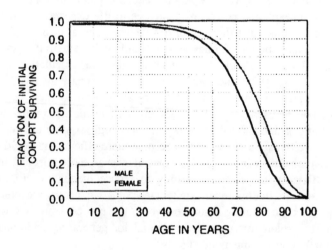

Figure 21.6 Survival rate from initial cohort males: by social class groupings

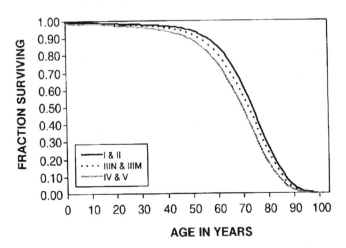

There are similar problems *within* most countries, with life expectancy differing significantly by sex, social class, region etc. For instance, in most advanced countries women live longer than men and, for the UK, their respective survival rates are shown in Figure 21.5. If we take life expectancy at birth as defining a 'fair innings' within any society, then in the UK in 1980 this would have been about 74 years, with females getting three years more than their fair share and males getting three years less. On equity grounds some redistribution of health care (and other) resources seems called for. This is also true in the UK for the different occupational groups, where the differences in male survival rates between the professional and managerial groups (social classes 1 and 2) on the one hand and the semi-skilled and unskilled manual workers (social classes 4 and 5) on the other, is shown in Figure 21.6. Life expectancy at birth between the two groups has been estimated to differ by about 5 years (72.5 compared with 67.7).[11] Again the equalisation of life chances in terms of life expectancy seems to require some changes in public policy, though not wholly confined to health care. The (limited) contribution of health care to reducing these inequalities could be exploited more fully by weighting additional life-years gained from the various health care activities according to whether they accrue to males or females and according to the social class of the potential recipient, just as the Bank has done with age–weights.

THE 'FAIR INNINGS' ARGUMENT APPLIED TO QUALITY-ADJUSTED LIFE EXPECTANCY

If the concept of a 'fair innings' is to capture the full flavour of this kind of thinking, it needs to be extended beyond *crude* life expectancy to *quality-adjusted* life expectancy. Otherwise it will not be possible to reflect the views of those who consider that a lifetime of poor quality health entitles a person to special consideration in the current allocation of health care, even if life expectancy is unaffected.

Once more the *World Development Report 1973* provides a lead.[12] In measuring the burden of disease in various countries, an estimate was made not only of effects on life expectancy but also of the disabling effects of each disease upon those suffering from it whilst still alive. These various disabilities were then valued (by a panel of experts) on a scale where dead = 1 and healthy = 0, and these weights were then used to measure disability-adjusted life years (DALYs), which are a primitive kind of QALY. The estimated burden of disease, measured in DALYs on a per capita basis, varies fivefold between the richest and poorest regions of the world (World Bank 1993). However, the weakness of these data is that they measure a 'need for health' rather than a 'need for health care' and, to assess the potential contribution of *health care* itself, only those DALY losses that are feasibly avoidable are relevant. Supposing that it were only communicable diseases that fell into that category, then although the scope for improving life expectancy through health care is reduced by more than half and the differences in potential per capita DALY gains between regions of the world actually increase from fivefold to more than thirtyfold, the greatest gains are in the poorest regions. To keep this in perspective, the total elimination of communicable diseases will increase DALYs per capita in the poorest region by only .4 (and it would have a negligible effect in the richer regions), so the gap between rich and poor regions will still be very wide. Clearly policies related to health care are *not* the major factor.

More sophisticated measurement of health-related quality–of–life is now under way in several countries[13] and, although I shall here draw only on UK material, the data from these other countries tells a similar story about the unequal distribution of health within them, by population characteristics that are relevant to their own health care policies. However, they are all fairly rich countries and it is *possible* that comparable data from much poorer countries will tell a very different story – though I doubt it.

The actual UK data on health-related quality–of–life that I shall use are from a survey representative of the adult population living in their own homes, conducted by the Research Group on the Measurement and

Valuation of Health at York, in collaboration with Social and Community Planning and Research in London.[15] The survival data are drawn from Bloomfield and Haberman (1992).

Health-related quality-of-life by age for men and women from our data is shown in Figure 21.7. These are based on self-reported health states at each age, using the EuroQol EQ–5D descriptive system (see Table 21.2), valued by a set of weights derived from the whole population by time–trade–methods using the methods of Chapter 10. The data shown in these two figures can be combined with the survival data presented earlier (in Figure 21.5), to yield an estimate of *quality-adjusted* life expectancy (QALE) at birth for the UK. For the whole population this is just under 60 QALYs, with males expecting about 57 and females nearly 62. But more interesting is the implication that, to reach a 'fair innings' of 60 QALYs, an average male would have to survive to the age of 72, while the average female would need to survive only to the age of 69. The punch line is that only 56 per cent of males will manage to survive long enough, whereas 79 per cent of females will do so! This difference of 23 per cent in the

Figure 21.7 Health-related quality-of-life compared: males and females

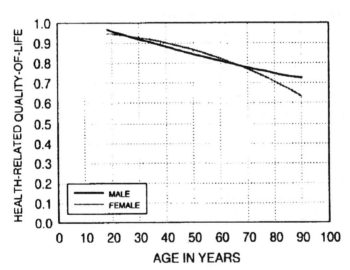

proportion of the two populations enjoying a 'fair innings' is a good indicator of the pervasiveness and magnitude of the problem. The limited contribution that health care might make in reducing these inequalities might again best be tapped by applying a differential valuation of QALY gains according to whether they accrue to men or women.

It is clear from other data which we have (which confirm already well-established findings for the UK) that surviving members of social classes 4 and 5 have noticeably worse health than their contemporaries in social classes 1 and 2, especially once they are past 40. This is shown in Figure 21.8 (which is comparable to Figure 21.7 above). When these data are

Table 21.2 Health-related quality-of-life descriptors
as presented in EuroQol EQ-5D

By placing a tick in one box in each group below, please indicate which statements best describe your own health state today.

Mobility
I have no problems in walking about   ☐
I have some problems in walking about   ☐
I am confined to bed   ☐

Self-care
I have no problems with self-care   ☐
I have some problems washing or dressing myself   ☐
I am unable to wash or dress myself   ☐

Usual activities *(for example, work, study, housework,*
                     family or leisure activities)
I have no problems with performing my usual activities   ☐
I have some problems with performing my usual activities   ☐
I am unable to perform my usual activities   ☐

Pain/discomfort
I have no pain or discomfort   ☐
I have moderate pain or discomfort   ☐
I have extreme pain or discomfort   ☐

Anxiety/depression
I am not anxious or depressed   ☐
I am moderately anxious or depressed   ☐
I am extremely anxious or depressed   ☐

combined with the differences in survival rates (shown earlier in Figure 21.6) we find that the quality-adjusted life expectancy at birth of someone in social classes 1 and 2 is about 66 QALYs but for someone in social classes 4 and 5 it is only about 57 QALYs.[14] To achieve the mean value of about 61.5 QALYs (a 'fair innings') they would need to live to be 65 and 71 years old respectively, a feat achieved by about 76 per cent of social classes 1 and 2 but by only 46 per cent of social classes 4 and 5.

Figure 21.8 Health-related quality-of-life compared: males, social classes 1/2 versus 4/5

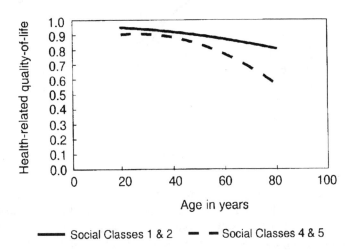

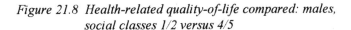

How should we respond to such data, assuming that they are approximately true? Clearly that depends on (a) how convinced you are that the 'fair innings' argument is a good basis for making equity adjustments in the allocation of health care, (b) how convinced you are that QALE at birth is a good indicator in any country of what constitutes a 'fair innings' (and of departures from it) and (c) on how far you are willing to have the overall level of health of the community reduced in order to reduce inequalities in the distribution of health. So far I have not considered the latter point in detail and this is the time to do so.

Figure 21.2 was a close-up of the segment of the health production possibility frontier that is relevant to policy-making about health, indicating three points on it which might be objectives of policy. These were the pure efficiency point ('maximum'), the pure equity point ('equal') and one in between ('intermediate'), the latter being the point of tangency with a social welfare function. In Figure 21.9 these abstractions are given substance (indeed they had it before but I concealed it!). In Figure 21.9, health is measured as life expectancy at birth and 'A' and 'B' have become UK social classes 4 and 5 and 1 and 2 respectively. The point labelled 'intermediate' is the actual current distribution of life expectancy at birth

*Figures 21.9   Life expectancy at birth
by social class*

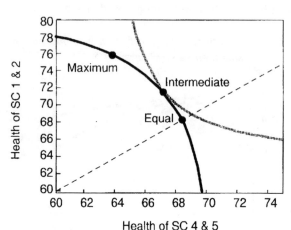

between the two, namely 67 for social classes 4 and 5 and 72 for social
classes 1 and 2. The production possibility frontier and the social welfare
function are hypothetical and, as drawn, charitably represent the current
situation as being the best we can do on efficiency grounds and where we
would choose to be on equity grounds, given the circumstances in which we
find ourselves. Neither of these propositions may be true.

Henceforth I am going to ignore the production possibility frontier,
because my argument does not need it. Even if we were inside the frontier
rather than on it (which is almost certain to be the case), if we still had
equity objectives and would still prefer to be closer than we are to the locus
of points representing equal health for all, we would want those views to
influence resource allocation in health care. It would not be a wise strategy
to say 'let us first of all get to the frontier and *then* we will worry about
equity'. Firstly, we are unlikely ever to get to the frontier and, secondly,
even if we did, we might find ourselves in such an extremely inequitable
situation that there would be no way of remedying it. It is better to accept
that inefficiency and inequity are both endemic in all systems and both
have to be worked on simultaneously.

Striving to reduce inequity will not, however, be costless in the sense that
whatever is stopping us pushing out towards the frontier, is also likely to be
resisting our attempts to shift towards greater equality. So I am going to
assume that moves towards greater equality can be achieved only by
reducing the level of overall health. The equity-efficiency trade-off then

consists in discovering the answer to the question 'how big a sacrifice in the overall health of the population would you be prepared to accept in order to eliminate the disparities in health between A and B?' Different As and Bs could then be specified and the current disparities between them would be the actual comparator in the above question.

Returning to social class differentials between males in the UK, suppose people were prepared to sacrifice six months of life expectancy at birth, in order to eliminate the disparity of five years. This is the situation represented in Figure 21.10, which starts from the same 'current UK situation' (previously called 'intermediate') but which also shows the mean life expectancy (69.5 years) and the equal life expectancy that people would settle for if the five-year disparity could be eliminated ('equal at 69'). This last point, the current situation and the obverse of the current situation (where the respective situations of A and B are reversed), are all on the same social welfare contour. By fitting a suitable function to them (that is, one that is symmetrical about the 45° line from the origin), we can draw the relevant social welfare contour.[15] This is in fact the one that has been there from the beginning. From it we can compute the gradient at the current situation. It is approximately −2, meaning that we should attach *twice the weight* to improving the life expectancy at birth of people in SC4&5 as we do to doing so for SC1&2. As and when the disparity reduces, these relative weights will decline as we move along the social welfare contour towards the locus of points of perfect equality. Obviously, the smaller the efficiency sacrifice that people are willing to make, the less will be the curvature of the contour, and the smaller the differential weight at any point in this space.

Pursuing a similar procedure[16] for three other cases and pulling everything together, we get the results shown in Table 21.3. The reason why the degree of curvature ('r') is so much lower in the last case is that the six-month sacrifice is offered in exchange for a reduction in a much larger discrepancy and hence shows a less strong aversion to inequality than in the other cases. It is, of course, quite likely that people will regard the different discrepancies with differing degrees of concern according to their size, nature and likely causes, so that my assumption of a constant six-month trade-off in efficiency terms must be regarded as purely illustrative.

There are two respects in which it might be desirable to 'fine tune' the crude version of the 'fair innings' argument that I have presented so far. One would be to make it more dynamic and the other would be to individualise it. I will sketch out each possibility in turn.

*Figure 21.10 Inequalities in life expectancy
by social class*

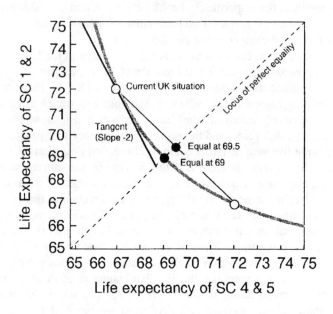

Table 21.3 Summary of inequality findings

Variable	Ha	Hb	discrepancy	r	Gradient
Life expectancy at birth	Males	Females	6 years	7.5	– 2.0
Life expectancy at birth	SC4&5	SC1&2	5 years	10.4	– 2.1
Quality-adjusted life expectancy at birth	Males	Females	5 QALYs	7.4	– 1.8
Quality-adjusted life expectancy at birth	SC4&5	SC1&2	9 QALYs	2.0	– 1.8

In the Culyer diagrammatics, the situation in which A and B find themselves is not 'at birth', but some way into their respective lives.[17] In the present context this raises the issue as to whether it is right that the fair innings is defined *as at birth*, or whether it should be recalculated for survivors according to where they are at present. For instance, for those surviving to the age of 20, 40, 60 and 80 the expected age of death and the expected lifetime QALY totals increase as shown in Table 21.4. The changes are not dramatic until the later ages, when survival is a more distinctive achievement and survival with good HRQOL even more so (especially for social classes 4 and 5). But I think it would be wrong in principle to 'renegotiate' the notion of a fair innings in that way, because in raising it for survivors we would delay the point at which anyone gets penalised for having more than their fair share, which means redistributing resources away from those less likely to survive. So I think the 'fair innings' is best defined with respect to prospects at birth.

That, however, leaves open the question of having the weights change over a person's lifetime as they approach and perhaps eventually exceed the predetermined 'fair innings'. The situation for males and females is set out in Figures 21.11 and 21.12, the former relating to life expectancy (or rather 'expected age at death') and the latter to quality–adjusted–life–expectancy (or rather 'expected lifetime QALY total'). It will be noted that, since females are always expected to be on the 'up-side', the equity weights applying to them are always low but get still lower for survivors as their prospects slowly improve still more. Males start out on the 'down-side', with high weights (around 2.0) initially, which decline with age and reach 1.0 for those who survive into their early sixties and thereafter become less than 1 when their *prospects* are that they will exceed the 'fair innings'. Thus there would be generated a set of age weights to apply *within* each relevant subgroup of the population.

Table 21.4 UK health expectancy at different ages

AGE	EXPECTED AGE AT DEATH			EXPECTED LIFE-TIME QALY TOTAL		
	Male	Female	Diff	Male	Female	Diff
At birth	71.5	77.5	6.0	57.0	61.8	4.8
At 20	71.9	77.7	5.8	57.4	62.0	4.6
At 40	72.4	78.1	5.7	57.9	62.3	4.4
At 60	74.0	79.1	5.1	59.1	63.1	4.0
At 80	81.8	84.0	2.2	64.4	65.8	1.4
	SC1&2	SC4&5	Diff	SC1&2	SC4&5	Diff
At birth	72.0	67.2	4.8	65.8	57.1	8.7
At 20	72.3	67.7	4.6	66.0	57.5	8.5
At 40	72.8	68.5	4.3	66.5	58.2	8.3
At 60	74.1	71.0	3.1	67.6	60.3	7.3
At 80	81.7	81.0	0.7	73.9	66.9	7.0

Figure 21.11 Survival and gradient of SWF
'fair innings' = expected death at 74.5

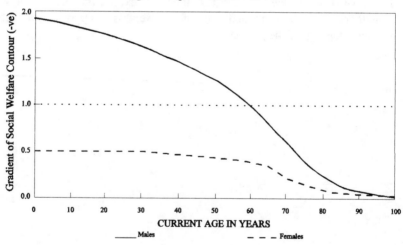

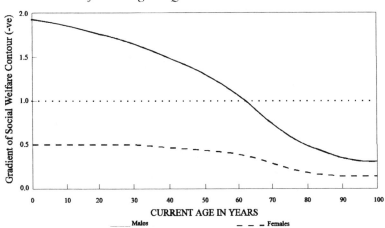

Figure 21.12   Survival and gradient of SWF
'fair innings' = QALYs at death 59.4

It would be possible to go a step further and identify other groups whose health is of special policy relevance, for instance, the permanently disabled. This is rather complex territory, but to indicate the possibilities, in Figure 21.13, I have sketched out four simple cases where, from birth to death, the disabled person is in one of four stable states which, in health-related quality-of-life terms, are rated at .5, .6, .7 or .8. No deterioration with age has been assumed and survival rates are taken to be those for males in the UK. The social welfare function used is the one that was generated for the social class differences in quality–adjusted–life–expectancy (r = 2). It will be seen that for the most severe disability considered, the equity weights would be extremely high if we took the view that the disabled are entitled to the same 'fair innings' as everyone else (in this case rounded to 60 QALYs over their lifetime). In the most severe case this is just not deliverable and, even at levels .6 and .7, it is rather unlikely and would be even more unlikely if deterioration with age were taken into account. There is much to ponder there.

What is the relationship between the age–weights considered in the preceding section and the age–weights promulgated by the World Bank? Do we need both, or are they substitutes for each other? I think we need both, because they are picking up different things. The World Bank's age–weights are designed to reflect the view that people's years of life (whether quality-adjusted or not) have a different (social) value according to the individuals age (which in turn reflects their likely life stage). The years between 20 and 50 are particularly valuable to society because those are

Figure 21.13   Equity weights for the disabled:
'fair innings' expected QALYs at
death = 60

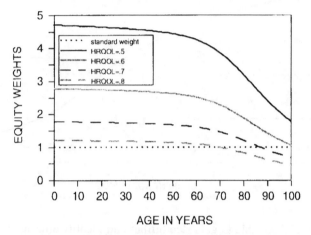

ASSUMPTIONS: Constant HRQOL over entire lifetime, Typical
male survival rates, Degree of curvature of SWF = 2
(as for social class differences in quality–adjusted life expectancy)

the years of procreation and child-rearing. They are also the years in which
the rest of society gets the maximum economic return from earlier
investments in an individual's education and training etc.

The age–weights relating to the 'fair innings' argument have a different
rationale. They are concerned with the likelihood that, over your lifetime,
you will achieve a certain target number of life years or QALYs. So what
ought to be happening is that the World Bank age–weights are applied
before a 'fair innings' is calculated, though I have not done so here because
of my wish to highlight one thing at a time and not take on board
everything at once.

POLICY AND RESEARCH IMPLICATIONS

This consideration of equity issues, alongside the pursuit of efficiency, was
motivated by two different desires. The first was to relate discussions of
social justice, as typically conducted by non-economists, more closely to
ways of thinking that are natural to economists so as to encourage other
economists to pick them up and run with them. The second was to impose
some quantitative rigour upon the the assertions made by non-economists
about what is equitable so that, whenever it is argued that more weight
should be given to one class of persons, it has to be acknowledged that this

means that some specified other class of person is going to suffer. There is a regrettable tendency for equity arguments to be conducted within a rhetorical framework in which it appears possible to 'do good' at no opportunity cost whatever. It generates a great deal of righteous self-satisfaction for the romantic escapists, it puts economists back in the role of the dismal scientists always stressing the sacrifices, but it does not help the hard-pressed decision-makers, who grapple with the issues in real–life every day.

This is a first attempt to take one such equity principle, the 'fair innings' argument, and subject it to empirical manipulation. The term 'manipulation' is used advisably, because the empirical data that are employed are stretched to their limits (some would doubtless say 'beyond their limits') in order to stimulate thought and to indicate what could be done if we chose to go down that route. This sort of manipulation is essential if we are to bring home the fact that giving priority to one group inevitably disadvantages others, a consequence which many advocates of particular equity principles fail to make clear (and they rarely state who will be called upon to make what sort of sacrifice even when they do acknowledge this implication). I have not attempted a philosophical critique of the 'fair innings' argument itself, which I must confess I found intuitively appealing at first sight. I still find it intuitively appealing at second sight but, considering what an apparently simple idea it is, it seems to lead us into deep water rather quickly. Nonetheless, it does seem to capture a great deal of the concerns that people express when resisting the single-minded pursuit of efficiency notions in health care.

Subject to what might emerge from further exploration of these ideas, however, I would like to end with some observations which rest on the tentative conclusion that I have got the broad picture right, even though the detail may be suspect.

For a long time I have been of the view that the best way to integrate efficiency and equity considerations in the provision of health care would be to attach equity–weights to QALYs. QALYs measure benefits of health care in standard units and equity–weights allow benefit valuation to become person-specific to the extent that that is policy-relevant. There is, however, a danger that such weights become arbitrary and capricious – used to fudge outcomes in ways that would not be acceptable if their basis were exposed. One safeguard against this is to have some underlying (or 'overarching'?) general principle enunciated, which can be confronted with evidence, so that its various implications can be explored in a quantitative way. The fact that it is possible to move from a general notion of a 'fair innings' to specific numerical weights, reflecting the marginal social value of health gains to different people, seems to me to be an important breakthrough.

It also sparks off an interesting research agenda of both a theoretical and an empirical kind. Most of the theoretical work will be interdisciplinary, requiring (some) economists to get to grips with both the philosophical concepts and the sociological propositions concerning the relevant categorisations of social structures that would be relevant to defining a 'fair innings'. For instance, should we be, or are we, equally averse to health inequalities between social classes as between the sexes? Or, more broadly, should we be concerned about inequalities between ethnic groups, regions, urban and rural communities, etc.? Should there be just one 'fair innings' for everyone served by a particular health service, or is there one for men, another for women, and variants within each of these for social class, race, location and so on? This is not something that economists can decide but it is something we can force out into the open and seek to have clarified.

At an empirical level this enterprise will be quite data hungry. I have been surprised at the paucity of life expectancy data for different subgroups within the community. I already knew that there was a dearth of data on health-related quality-of-life, which I and my colleagues have been busy remedying over the past few years. We know little or nothing about the willingness of people to sacrifice life expectancy (especially their own?) in order to reduce health inequalities in their own communities (or even in other communities). It would also be interesting to discover whether the countries we generally regard as strongly egalitarian (for example, the Nordic countries) do in fact manifest a stronger aversion to inequality in health than those we regard as of a more libertarian ideology (for example, the USA). Countries which already have disaggregated mortality and HRQOL data might try to imitate some of my speculative calculations to see whether this sheds any useful light on their own problems, while countries which do not have such data but which do have a policy interest in this sort of inequality should set about collecting the relevant data.

Finally, I think that the focus on outcomes-with-equity-weights-attached has great advantages over other approaches to the reduction of health inequalities, because it rules out the giving of priority to things that do no good, which is the danger with equalisation policies activities that concentrate on process (including access) or on resources. It also sets clear limits on how far we are willing to go in helping the disadvantaged, for a weight of 2 indicates that we are willing to spend twice as much as the norm to provide 1 extra QALY for such a person – but that is the limit. This is far preferable to the vague notion of 'priority groups' whose champions are left to compete on unclear terms with more powerful competitors in the annual scramble for resources.

So I commend the 'fair innings' argument as a promising way forward for economists interested in the equitable provision of health care. If we are to know whether or not we are getting a 'fair innings', however, somebody has to keep the score! That is a challenge to researchers and practitioners alike, and one that will involve collaboration between many disciplines.

ACKNOWLEDGEMENTS

The general stimulus for this chapter has come from the many non-economists with whom I have argued about the rival moral claims of equity and efficiency over the years. In preparation itself I owe a considerable debt to Paul Dolan, who not only helped me by extracting the relevant data on HRQOL by social class from the survey work undertaken by the MVH Group but, in his characteristically argumentative way, helped me to focus more sharply on the measurement issues involved. Without the HRQOL data generated by the MVH Group in York, in collaboration with SCPR in London, I could not have done this work, and the long-term support of the UK Department of Health in London in financing it is gratefully acknowledged. At key moments when I was stumbling over the mathematical and computing demands made by my approach, I drew on the crucial expertise of Arnold Arthurs and Minoru Kunizaki. Whether any of these people will actually approve of what I have done as a result is quite another matter! They should remain blameless.

NOTES

1. Like me, they ignore the efficiency and equity issues that attend the *method of financing* of that provision. Van Doorslaer and colleagues have demonstrated that that is also important to any overall judgement about the equitability of any health care system (especially between rich and poor). See Van Doorslaer *et al.* (1993).
2. On page 18 of Culyer (1995) he identifies 'the treatment of *efficiency* as being concerned with identifying the health frontier, which I have termed 'health maximisation', and the treatment of *equity* as being concerned with the selection of a point on it'. One such point is the case where an equal social weight is attached to each unit of health (be it a life-year or a QALY) at the margin, which he refers to in his text on page 16 of the paper, but does not show on any of his diagrams. This point represents an alternative view of what 'health maximisation' means and it is this view that I have adopted here. *Not* attaching equal social weight at the margin to each unit of health, but allowing this value to vary according to the characteristics of the recipient, then becomes the equity issue. In this conceptualisation of the situation, there clearly *is* a conflict between efficiency and equity.
3. In Culyer's diagrammatics, such a Social Welfare Function generates iso-welfare contours which are straight lines at 45° to each axis.
4. The fact that the point 'maximum' lies above the point 'equal' indicates that the production possibilities have here been assumed to be such that it is possible to do more for the health of B than for the health of A.

5. There are plenty of commentators who favour a prominent role for process measures *alongside* outcome measures. Much of the practical work on equity has been concerned with bringing about a more equitable distribution of *resources*, in the hope that this will lead to a more equitable distribution of *outcomes*. In the UK the most recent manifestation of this approach was Carr-Hill et al. (1994), which was only partially implemented by the UK Government. I will comment later on why I think a more direct focus on outcomes may be a better long term strategy than these indirect approaches.

6. This argument is expounded in Harris (1985) where it appears on pages 91–94. He seems to have some difficulty in accepting it, however, as may be gained from the following passage: 'What the fair innings argument needs to do is to capture and express in a workable form the truth that while it is always a *misfortune* to die when one wants to go on living, it is not a *tragedy* to die in old age; but it is on the other hand both a tragedy and a misfortune to be cut off prematurely.' Without telling us what are the key distinctions between a 'misfortune' (a person's judgement about themselves?) and a 'tragedy' (a social judgement that separates one misfortune from another?), he ends up accepting 'a reasonable form of the fair innings argument' as being one in which 'people who had achieved old age or who were closely approaching it would not have their lives further prolonged when this could only be achieved at the cost of the lives of those who were not nearing old age'.

7. For a more detailed account of this notion, see Chadwick (1993) who cites Gallie (1955–6).

8. Two very important, easily quantifiable, variables are going to be totally ignored in what follows, however, and they are (a) people's attitudes towards risk and (b) their feeling about time preference. It is quite possible that aversion to inequality is as much motivated by risk aversion as by any sense of justice or solidarity. And if views about time preference are very different between different groups whose health prospects I consider later, this will greatly complicate the rather simple welfare implications I am trying to draw. But things are complicated enough as it is and it seems a better strategy not to tackle everything at once!

9. See *World Development Report 1993*. Box figure 1.3 on page 26 and further discussion on page 213.

10. Too many people have unthinkingly assumed that QALYs can only be used in a QALY-maximising context. I challenged this assumption in Williams (1994) and have dealt with it more fully in Williams (1996).

11. See Bloomfield and Haberman (1992). These survival rates by social class are based on different data from the survival rates by sex and are not comparable with each other. The problem of comparability is discussed in Bloomfield and Haberman, but it is not a central issue here since I am not using the two data sets together.

12. See Appendix B, and especially page 214 of that Report.

13. Not all of these results have yet been published, but a general account of the research programme is given in Chapter 10. We have no data from our study relating to the quality adjustments for those under the age of 18, so they have been assumed to be equal to those of the 18-year-olds. We also have only very limited data for over-80s, so for them the trends already evident in the data have been extrapolated.

14. Obviously this assumes that a person remains in the same social class throughout their lives. Movement between social classes will make these differences less clear-cut at individual level.

15. The function used was of the CES type, which allows for considerable flexibility in the key parameters. The basic equation is: $W = q[bHa^{\wedge} - r + (1-b)Hb^{\wedge} - r]^{\wedge} - (1/r)$.
W is an index of the overall level of welfare, here assigned the value it takes at its intersection with the perfect equality locus. Thus in Figure 21.J it is 69. The coefficient 'q' translates units of health into units of welfare, and is here assumed for simplicity's sake to be 1. The coefficient 'b' determines the weight given to the health of each party, which is here assumed to be equal, so the coefficient takes the value 5. Ha and Hb are the levels of health of the two parties (measured in identical units, in this case 'years of life expectancy at birth'). Finally, the coefficient 'r' is the parameter determining the degree of curvature of

the contours, which here represents the extent of social aversion to inequality. It is this parameter that is estimated from the data. Its value on each occasion is shown in Table 21.3.

16. More specifically, by assuming that society would sacrifice six months of life expectancy (or, as appropriate, of quality–adjusted–life–expectancy) to eliminate each of the stated discrepancies and keeping all the other parameters (except 'r') at the values set out in the preceding footnote.

17. This later starting point for the analysis is identified in Culyer's diagrams as the situation 'PS' (present situation) in which A and B have already enjoyed some life years or QALYs, and the problem is what to do from here on.

References

Anonymous (1984). Consensus development conference: Coronary artery bypass grafting. (Editorial) *British Medical Journal*, **289**, 1527–9.

Ansoff, H.I. (ed.) (1969), *Business strategy*, Penguin Modern Management Readings, Harmondsworth, Middlesex.

Anthony, R.N. (1965), *Planning and Control Systems: A Framework for Analysis*, Harvard University Press, Boston.

Avorn, (1984), 'Benefit and Cost Analysis in Geriatric Care: Turning Age Discrimination into Health Policy' *New England Journal of Medicine*, **310**, 1294–1301.

Banta, (1979), International Workshop on the Evolution of Medical Technology, Stockholm September 18–19th, S.P.R.I. Report, 156.

Barber, P.R. (1978), *The Economics of Cardiac Pacemakers*, University of York, MSc dissertation, York.

Barry, B.M. (1970), *Sociologists, Economists and Democracy*, Collier Macmillan, London.

Barry, B. (1965), *Political Argument*, Routledge and Kegan Paul, London.

Baumol, W.J. (1967), *Business Behavior, Value and Growth*, revised edition, Harcourt, Brace and World, New York.

Birch, S. and Maynard, A. K. (1986) *The R.A.W.P. Review*, Centre for Health Economics Discussion Paper, **19**, University of York, York.

Black, D. (1958), *The Theory of Committees and Elections*, Cambridge University Press, Cambridge.

Blades *et al.* (1986) *The International Bibliography of Health Economics*, Wheatsheaf, Brighton.

Bloomfield, D.S.F. and Haberman, M.A. (1992), 'Male social class mortality differences around 1981: An extension to include childhood ages', *Journal of the Institute of Actuaries*, **119**, 545–559

Bone, M.R. *et al.* (1995), *Health Expectancy and its uses*, H.M.S.O., London.

Bowen, H. (1948), *Toward social economy*, Rinehart, New York.

Bradshaw, J. (1972), 'A taxonomy of social need' in McLachlan (1972), 69–82.

Bråkenheim, C.R. (1990), 'Vårdens På Lika Villkar' in Calltorp, J. and Bråkenheim, C.R. (eds.) *Vårdens Pris*, Verburo, Stockholm, 35–54.

British Medical Association (1993), *Medical Ethics Today: Its Practice and Philosophy*, British Medical Association, London.

Buchanan, J.M. and Tullock, G. (1962), *The Calculus of Consent*, University of Michigan Press, Ann Arbor.

Butterfield, W.J.H. (1968), *Priorities in Medicine*, Nuffield Provincial Hospitals Trust, London.

Calman, K.C. (1994), 'The ethics of allocation of scarce health care resources: A view from the centre', *Journal of Medical Ethics*, **20**, 71–74.

Canvin, R.W. and Pearson, N.G. (eds.), (1973), *Needs of the Elderly*, University of Exeter, Exeter.

Card, W.I. Rusenkiewicz, M and Phillips, C. I. (1977), Utility estimates of a set of states of health, *Methods of Information in Medicine*, **16**, 168–176.

Carr–Hill, R.A. *et al.* (1994), *A Formula for Distributing N.H.S. Revenues Based on Small Area Use of Hosipital Beds*, York, Centre for Health Economics.

Chadwick, R. (1993), 'Justice in priority–setting' in *Rationing in Action*, London, British Medical Journal Publishing Group, 85–95.

Chambers, M.L. (January 1973), 'Construction of a utility function for a police force', paper presented at an R.E.S. Specialist Conference on Decision Analysis, Lancaster.

Charny, M.C. *et al.* (1989), *Choosing who shall not be treated in the N.H.S.*, *Social Science and Medicine*, **28** (12), 1331-1338.

Churchill, D.N., Morgan, J., Torrance, G.W. (1984), Quality of life in end–stage enal disease, *Peritoneal Dialysis Bulletin*, **4**, 20–3.

Cochrane, A.L. (1972), *Effectiveness and Efficiency*, Nuffield Provincial Hospitals Trust. London.

Coles, J.G. and Coles, J.C. (1982), The cost–effectiveness of myocardial revascularization. *Canadian Journal of Surgery*, **25**, 123–6.

Cooper, M.H. (1974), 'The economics of need: The experience of the British health service', in Perlman (1974), 89–107.

Cooper, M.H. and Culyer, A.J. (1973), *Health Economics*, Penguin, London.

Culyer, A.J. (1971a), 'The nature of the commodity 'health care' and its efficient allocation', *Oxford Economic Papers*, **23**, 189–211.

Culyer, A.J. (1971b), 'Medical care and the economics of giving', *Economica*, **38**, 295–303.

Culyer, A.J. (1974), *Economic Policies and Social Goals*, Martin Robertson, London.

Culyer, A.J. (1976), *Need and the National Health Service: economics and social choice*, Martin Robertson, London.

Culyer, A.J., (1978), 'Needs, values and health status measurement' in Culyer, A.J. and Wright, K.G. (eds.), *Economic Aspects of Health Services*, Martin Robertson, London, 9–31.

Culyer, A.J. (ed.), (1983), *Health Indicators*, Martin Robertson, London.

Culyer, A.J. (1995), *Equality of what in health policy?*, Discussion Paper, **142**, Centre for Health Economics, York.

Culyer, A.J. Lavers, R.J. and Williams, A, (1971) 'Social indicators: health' *Social Trends*, **2**, 31–42.

Culyer, A.J., Lavers, R.J. and Williams, A. (1972, 'Health indicators', in *Social Indicators and Social Policy*, (eds.) Shonfield, A. and Shaw, S., Heinemann, London.

Culyer, A.J. and Wagstaff, A. (1993), 'Equity and equality in health care', *Journal of Health Economics*, **12**, 431–445.

Curry, R.L. Jr, and Wade, L.L. (1968), *A Theory of Political Exchange*, Prentice Hall Contemporary Political Theory Series, Englewood Cliffs, N J.

Cyert, E.M. and March, J.G. (1963), *A Behavioral Theory of the Firm*, Prentice Hall, New York.

Daniels (1982), 'Equity of access to health care: some conceptual and ethical issues', *Millbank Memorial Fund Quarterly*, **60**, 51–81.

Dasgupta, A.K. and Pearce, D.W. (1972) *Cost–Benefit Analysis: Theory and Practice*, Macmillan, London.

De Jong, G. and Rutten, F.F.H. (1983), 'Justice and health for all' *Social Science and Medicine*, **17**, 1091.

Diesing, P, (1962), *Reason in Society*, University of Illinois Press, Urbana.

Doll, R. (1973), 'Monitoring the National Health Service', *Proceedings of the Royal Society of Medicine*, **66**, 729–740.

Donabedian, A, (1971), 'Social responsibility for personal health services: An examination of basic values', *Inquiry*, **8**, 3–19.

Dorfman, R. (1966), *Operations Research, Surveys of Economic Theory: Vol III – Resource allocation*, Macmillan, London.

Downs, A. (1957), *An Economic Theory of Democracy*, Harper and Bros, New York.

Dror, Y. (1967), 'Policy analysts: A new professional role in government service', *Public Administration Review*, **27**, 200–201.

Drummond, M.F. (1980), *Principles of Economic Appraisal in Health Care*, Oxford University Press, Oxford.

Drummond, M.F. (1981), *Studies in Economic Appraisal in Health Care*, Oxford University Press, Oxford.

Elhauge, E. (1994),'Allocating health care morally', *California Law Review*, **82**, 1492–1510.

Emery, F.E. (ed.), (1969), *Systems Thinking: Selected Readings*, Penguin Modern Management Readings, Penguin Education, Harmondsworth, Middlesex.

Enthoven, Alain.C. (1985), *Reflections on Improving Efficiency in the National Health Service*, Nuffield Occasional Paper, Nuffield Provincial Hospitals Trust, London.

Evans, J.G. (1993), 'This patient or that patient?', *Rationing in Action*, British Medical Journal Publishing Group, London, 118–124.

Fanshel, S. and Bush, J.W. (1970), 'A health status index and its application of health status outcomes', *Operational Research*, **18** (5), 1021–1066.

Feldstein (1967), *Economic Analysis for Health Service Efficiency*, North–Holland, Amsterdam.

Flagle, (1963), 'Operational Research in the Health Services', *Annals of the New York Academy of Science*, **107**, 748–759.

Florey, C. du V., Weddell, J.M. and Leeder, S.R. (1976). 'The epidemiologist's contribution to medical care planning and evaluation', *Australian and New Zealand Journal of Medicine*, **6**, 74–78.

Frankel, S. (1989), 'The natural history of waiting lists: Some wider explanations for an unnecessary problem', *Health Trends*, **21**, 56–58.

Fuchs, V. (1974), *Who Shall Live?, Health, Economics and Social Choice*, Basic Books, New York.

Garrard, J. and Bennett, A.E. (1971), 'A validated interview schedule for use in population surveys of chronic disease and disability', *British Journal of Preventive and Social Medicine*, **25**, 97–104.

Gater, R., Kind, P. and Gudex C. (1995), 'Quality of life in liaison psychiatry: a comparison of patient and clinician assessment' *British Journal of Psychiatry*, **166**, 515–520.

Gillon (1985), *Philosophical Medical Ethics*, Wiley, Chichester.

Grossman (1972), *The Demand for Health*, N.B.E.R., New York.

Gruenberg, E.M. and Brandon, S. (1966), 'Measurement of the incidence of chronic severe social breakdown', *Millbank Memorial Fund Quarterly*, **44**.

Gudex, C. (1986), *QALYs and Their Use by the Health Service*, Discussion Paper, **20**, Centre for Health Economics, University of York, York.

Gudex, C. (1995) 'Quality of life in end–stage renal disease', *Quality of Life Research*, **4**, 359–366.

Gudex, C., Williams, A., Jourdan, M., Mason, R., Maynard, J., O'Flynn, R., and Rendall M. (1990), 'Prioritising waiting lists', *Health Trends*, **22**, 103–108.

Hamilton, M. (1960), 'A rating scale for depression', *Journal of Neurology and Psychiatry*, **23**.

Hampton, J.R. (1983), 'The end of clinical freedom, *British Medical Journal*, **287**, 1237–1238

Hancock, C. (1993), 'Getting a quart out of a pint pot', in *Rationing in Action*, British Medical Journal Publishing Group, London, 15–24.

Harris, A.I. (1971), *Handicapped and Impaired in Great Britain*, H.M.S.O., London.

Harris, J. (1985), *The Value of Life*, Routledge & Kegan Paul, London.

Harris, J. (1987), 'QALYfying the Value of Life', *Journal of Medical Ethics*, **13**, 117–123.

Hauser, M.H. (ed.), (1972), *The Economics of Medical Care*, Allen and Unwin, London.

Hawood, J. (1969), *Inverse Linear Programming as a Means of Establishing Librarians' Benefit Criteria*, Mimeo, University of Durham.

Head, J.G. (1966), 'On merit goods', *Finanzarchiv*, **25**, 1–15.

Hellinger, F. (1982) 'An analysis of public program for heart transplantation', *Journal of Human Resources*, **17**, 307–13.

Hoffenberg, R. (1987), *Clinical Freedom*, Nuffield Provincial Hospitals Trust, London.

Holland, W.W. (1977), 'A general view', in Holland, W.W. and Gilderdale, S. (eds.), *Epidemiology and Health*, Holland, Henry Kimpton, London, 11–28.

Holland, W.W. *et al.* (eds.) (1979), *Measurement Levels of Health*, W.H.O. European Regional Studies, **7**, W.H.O., Copenhagen.

Hutton, J. and Williams, A (1988) 'Cost–effetiveness analysis of diagnostic technology: a decision framework applied to the case of MRI', in Duru, G., Engelbrecht, R., Flagle, C. D. and van Eimeren, W. (eds.), *System Science in Health Care, Vol 2. Health Care Systems and Actors, Fourth International Society for System Science in Health Care Conference, Lyon 4–8 July 1988*, Masson, Paris, 493–496.

Isaacs, B. and Walkley, F.A. (1963), 'The assessment of the mental state of elderly hospital patients using a simple questinnaire', *American Journal of Psychiatry*, **120**, 173–174.

Jennett, J.B. (1984), *High Technology Medicine*, Nuffield Provincial Hospital Trust, London.

Johnson, E.M. and Huber, G.P. (1977), 'The technology of utility assessment', *Transactions on Systems, Man and Cybernetics*, S.M.C., **7**, 311–325.

Kappel, K. and Sandee, P. (1992), 'QALYs, age and fairness', *Bioethics*, **6**, 297–316.

Katz, S *et al.* (1963), 'Studies of illness in the aged: The index of independence in activities of daily living', *Journal of the American Medical Association*, **185**, 914–919.

Kemp, D.A.M., Kemp, M.S. and Havery, R.O. (1967), *The Quantum of Damages*, Sweet and Maxwell, London.

Kind, P. and Gudex, C. (1994), 'Measuring health status in the community: a comparison of methods', *Journal of Epidemiology and Community Health*, **48**, 86–91.

Kind, P., Rosser, R. and Williams, A (1982), 'The valuation of quality of life: some psychometric evidence', in Jones–Lee, M.W. (ed.) *The Value of Life and Safety*, North Holland, Amsterdam, 159–170.

Kind, P. and Sims, S. (1987), *C.T. Scanning in a District General Hospital*, Research Report, Centre for Health Economics, University of York, York.

Kirkwood, T.B.L. (1981), 'Repair and its evolution: survival versus reproduction', in Townsend, C.R. and Calow, P. (eds.), *Physiological Ecology: An Evolutionary Appraisal, Blackwell, Oxford.*

Klarman, H.E. Francis, J O'S. and Rosenthal, G. D. (1968), 'Cost effectiveness analysis applied to the treatment of chronic renal disease', *Medical Care*, **6**, 48–54.

La Puma, J. (1990), Letter to Editor, *Journal of the American Medical Association*, **264**, 2503–2504.

La Puma, J. and Lawlor, E.F. (1990), 'Quality–adjusted Life Years: Implications for Physicians and Policymakers', *Journal of the American Medical Association*, **263**, 2917–2921.

Layard, R. (ed.) (1972) *Cost Benefit Analysis: Selected Readings*, Penguin Modern Economics Readings, London.

Liefman–Keil, E. (1974), 'Consumer protection, incentives and externalities in the drug market', in Perlman (1974), 117–129.

Lindblom, C.E. (1959), 'The science of muddling through', *Public Administration Review*, **29**, (1), 79–88.

Linder, F.A. (1965), 'National health interview surveys' in World Health Organisation, *Trends in the Study of Morbidity and Mortality*, Public Health Papers, **27**, Geneva.

Lindsay, C.M. (1973), 'Medical care and Equality', in Cooper and Culyer (1973), 75–89.

Llewelyn, H. and Hopkins, A. (1993), *Analysing How We Reach Clinical Decisions,* Royal College of Physicians, London.

Loewy, E.N. (1980), 'Cost should not be a factor in medical care' *New England Journal of Medicine*, **302**, 697.

Machlup, F. (1965), 'Comments' on a paper by B. A. Weisbrod, in: R. Dorfman, (ed.) *Measuring Benefits of Government Investments*, Brookings Institution Studies of Government Finance, Washington DC,149–157.

McGuire, A. *et al.* (1988), *The Economics of Health Care*, Routledge and Kegan Paul, London.

McKean, R.N. (1968), 'The use of shadow prices', in: S. B. Chase, Jr (ed.), *Problems in Public Expenditure Analysis*, Brookings Institution Studies of Government Finance, Washington, DC, 33–77.

McLachlan, G. (ed.), (1971), *Portfolio for Health*, Oxford University Press, London.

McLachlan, G, (ed.), (1972), *Problems and Progress in Medical Care*, Seventh Series, Oxford University Press, London.

McLure, C.E. (1968), 'Merit wants: a normatively empty box', *Finanzarchiv*, **27**, 474–483.

Magdelaine, M. *et al.* (1967), 'Unindicateur de la morbidité appliqué aux données d'une enquête sur la consommation medicale', *Consommation*, **2**, 3–41.

Mancini, P.V. (1984) *Costs of Treating End–Stage Renal Failure*, Department of Health and Social Security, London.

Marshall, T.H. (1973), 'The philosophy and history of need', in Canvin and Pearson (1973).

Martin, F.M. (1977), 'Social medicine and its contribution to social policy', *Lancet*, **2**, 1336–1338.

Matthew, G.K. (1971), 'Measuring need and evaluating services', in McLachlan (1971), 27–46.

Maynard, A. and Williams, A., (1984), 'Privatisation and the National Health Service', in Le Grand, J. and Robinson, R. (eds.), *Privatisation and the Welfare State*, Allen and Unwin, London, 95–110.

Meade, T. (1975), 'Epidemiology, health and health services', in *Specialised Futures: Essays in Honour of George Godber*, Nuffield Provincial Hospitals Trust, London, 267–295.

Menzel, P. (1990), *Strong Medicine: The Ethical Rationing of Health Care*, Oxford University Press, Oxford.

Mishan, (1981), *Economic Efficiency and Social Welfare*, Allen and Unwin, London, Chapters 14 and 16.

Mooney, (1994), *Key Issues in Health Economics*, London, Harvester Wheatsheaf.

Mooney. (1983), 'Equity in Health Care: Confronting the Confusion', *Effective Health Care*, **1**.

Mooney, (1989), 'QALYs: are they enough? A health economist's perspective', *Journal of Medical Ethics*, **15**, 148–152.

Moser, C. A. (1970), 'Some general developments in social statistics', *Social Trends*, **1**, London, H.M.S.O., 7–11.

Musgrave, R.A. and Peacock A.T. (eds.), (1958), *Classics in the Theory of Public Finance*, Macmillan, London.

Mushkin, S, (ed.) (1972), *Public Prices for Public Products*, The Urban Institute, Washington, DC.

Muurinen, J–M. (1982), 'Demand for health', *Journal of Health Economics*, **1**, 1–28.

Neuhauser, D. (1979), *International Workshop on the Evaluation of Medical Technology*, (SPRI Report), Stockholm, 13–16.

NHS Management Executive, (1993a), *The Evolution of Clinical Audit*, Leeds, NHSME, Leeds.

NHS Management Executive, (1993b), *Clinical Audit: Meeting and Improving Standards in Health Care*, Leeds, NHSME, Leeds.

NHS Management Executive, (1993c), *Improving Clinical Effectiveness*, NHSME, Leeds.

Nord, E. (1993), 'The relevance of health state after treatment in privatising between different patients', *Journal of Medical Ethics*, **19**, 37–42.

O'Donnell, M. (1986), 'One man's burden', *British Medical Journal*, **293**, 59.

Office of Health Economics (1971), *Prospects in Health*, OHE, London.

Olson, M. (1965), *The Logic of Collective Action*, Harvard University Press, Cambridge.

Packer, A.H. (1968), 'Applying cost–effectiveness concepts to the community health system', *Operations Research*, **16**, 227–253.

Patel, M. (1984), *Health economics/community medicine approach to planning of facilities for coronary artery bypass surgery at the NHS district level*, paper presented to the Health Economists' Study Group meeting at the University of Aberdeen.

Parton, C. (1985), *The Policy of Resource Allocation and Its Ramifications: A Review*, Nuffield Provincial Hospitals Trust Occasional Paper, **2**, NPHT, London.

Patrick, D.K., Bush, J.W. and Chen, M.M. (1973), 'Methods for measuring levels of well–being for a health status index', *Health Services Research*, **8**, 229–244.

Pazner, E.A. (1972), 'Merit wants and the theory of taxation', *Public Finance*, **27**, 460–472.

Perlman, M, (ed.) (1974), *The Economics of Health and Medical Care, Proceedings of a Conference held by the International Economic Association at Tokyo*, Macmillan, London.

Piachaud, D. and Weddell, J.M. (1972), 'The Economics of treating varicose veins', *International Journal of Epidemiology*, **1**, 287–294.

Pole, J.D. (1968), 'Economic aspects of screening for disease', *Screening in Medical Care: Reviewing the Evidence. A Collection of Essays*, Oxford University Press for The Nuffield Provincial Hospitals Trust, London, 141–158.

Pole, J.D. (1971), in: McLachlan, G. (ed.), *Problems and Progress in Medical Care: Essays on Current Research*, 5th ser. pp 45–55, Oxford University Press for The Nuffield Provincial Hospitals Trust, London.

Pole, J.D. (1972), 'The economics of mass radiography', in Hauser (1972), 105–114.

Raiffa, H. (1968), *Decision Analysis*, Addison–Wesley, Reading Massachussetts.

Rawls, J. (1972), *A Theory of Justice*, Oxford University Press, Oxford.

Reeder, G.S., Krishan, I., Nobrega, F. *et al.*, (1984), 'Is percutaneous coronary angioplasty less expensive than bypass surgery?' *New England Journal of Medicine*, **311**, 1157–1162.

Roskill, Commission, (1970), *Commission on the Third London Airport, Papers and Proceedings*, **7**, (Parts 1 and 2), HMSO, London.

Roskill, Commission, (1971), *Commission on the Third London Airport*, Report, HMSO, London.

Ross, J.M. (1968), 'The Management Consultant's Lament', *O and M Bulletin*, **23**, 1.

Rosser, R. (1979), (Subsequently published; see next entry) (Rosser, 1983).

Rosser, R. (1983), 'Issues of measurement in the design of health indicators: A review', in Culyer (1983), 34–81.

Rosser, R. and Kind, P. (1978), 'A scale of valuations of states of illness: Is there a social consensus?' *International Journal of Epidemiology*, **7**, 347–58.

Rosser, R and Watts, V. (1971), (Subsequently published – see next entry).

Rosser, R. and Watts, V. (1972), 'The measurement of hospital output', *International Journal of Epidemiology*, **1**, 361–368.

Rosser, R. and Watts, V. (1975), 'Disability – a clinical classification', *New Law Journal*, **125**, 323–328.

Rosser, R. and Watts, V.C. (1978) 'The measurement of illness', *Journal of Operational Research*, **29**, 529–40.

Royal College of Physicians of London (1989), *First Report on Medical Audit*, R.C.P., London.

Royal College of Surgeons of England (1989), *Guidelines to Clinical Audit in Surgical Practice*, R.C.S.E., London.

Ruark, J.E., Raffin, T.A. and Stanford University Medical Center Committee on Ethics, (1988) 'Initiating and withdrawing life support: principles and practice in adult medicine', *New England Journal of Medicine*, **318**, 25–30.

Sackett, D.L. and Torrance, G.W. (1978), 'The utility of different health states as perceived by the general public', *Journal of Chronic Disease*, **31**, 697–704.

Schultze, C.L. (1968), *The Politics and Economics of Public Spending*, The Brookings Institution, Washington DC.

Schwartz, R.L. (1995), 'Life style, health status and distributive justice', in Brubbs, A. and Mehlman, M.J. (eds.), *Justice and Health Care*, Wiley, London, 225–250.

Self, P. (1970), 'Nonsense on stilts: the futility of Roskill', *Political Quarterly*, **41**, 249–260, (reprinted in New Society, 2 July 1970).

Sellin, T. and Wolfgang, M.E. (1964), *The Measurement of Delinquency*, Wiley, New York.

Sen, A.K. (1985), *Commodities and capabilities*, North–Holland, Amsterdam.

Shonfield, A. and Shaw, S. (eds.) (1972), *Social Indicators and Social Policy*, Heinemann, London.

Shoup, D.C. Mehay, S. L. (1971), *Program Budgetry for Urban Police Services*, Institute of Government and Public Affairs, University of California Los Angeles.

Sibley, R.M. and Calow, P. (1986), *Physiological Ecology of Animals: An Evolutionary Approach*, Blackwell, Oxford.

Simon, H.A. (1955), 'A behavioural model of rational choice', *Quarterly Journal of Economics* **59**, 99–118.

Simon, H.A. (1956), 'Rational choice and the structure of the environment', *Psychological Review*, **63**, 159–138.

Singer, P. *et al.* (1995), 'Double Jeopardy and the use of QALYs in health care allocation', *Journal of Medical Ethics*, **21**, 144–157.

Smolensky, E., Tideman, T.N. and Nicholes, D. (1972), 'Waiting time as a congestion charge' in Mushkin (1972), 95–108.

Sommerhoff, G. (1950), *Analytical Biology*, Oxford University Press, London.

Sommerhoff, G. (1969), 'The abstract of characteristics of living systems', in Emery (1969), 147–202.

Spek, J.E. (1972), 'On the economic analysis of health and medical care in a Swedish health district', in Hauser (1972), 261–268.

Steele, R. (1981), 'Marginal met need and geographical equity in health care', *Scottish Journal of Political Economy*, **28**, 186–205.

Stevens, S.S. (1971) 'Issues in psychophysical measurement' *Psychological Review*, **78** (5), 426–450.

Stocking, B. and Morrison, S.L. (1978), *The Image and the Reality*, Oxford Publishing Press, London.

Suchman, E.A. (1967), *Evaluative Research: Principles and Practice in Public Services and Social Action Programmes*, Russell Sage Foundation, New York, 75–77.

Sugden, R. L. and Williams, A. (1978), *The Principles of Practical Cost-Benefit Analysis*, Oxford University Press, Oxford.

Sullivan (1965), *Conceptual Problems in Developing an Index of Health*, Office of Health Statistics Analysis, Department of Health, Education and Welfare, Washington DC.

Symposium on QALYs and HYEs (1995), *Journal of Health Economics*, **14**, 1–45.

Thick, M.G., Lewis, C.T., Weaver, E.J.M. *et al.* (1978) 'An attempt to measure the costs and benefits of cardiac valve surgery at the London Hospital in adults of working age', *Health Trends*, **10**, 58–60

Thorner, R.M. and Remein, Q.R. (1967), *Principles and Procedures in the Evaluation of Screening for Disease*, US Dept. of HEW, Public Health Monograph, **67**, Washington DC.

Thouez, J.P. (1979), 'Health Measurement Bibliography', *Social Science and Medicine*, **13D**.

Torrance, G.W. (1976), 'Social preferences for health states: An empirical evaluation of 3 measurement techniques', *Socio–Economic Planning Science*, **10**, 129–136.

Torrance, G.W. (1976), 'Towards a utility theory, foundation of health status index models', *Health Services Research*, **11**, 349–369.

Torrance, G.W., Sackett, D.L. and Thomas, W.H. (1973), 'A utility maximisation model for evaluation of health care programs', *Health Services Research* **7**, 118–33.

Towse, A. (ed.), (1995), *Financing Health Care in the UK: A Discussion of NERA's Prototype Model to Replace the NHS*, Office of Health Economics, London.

Tribe, L. (1972), 'Policy science: analysis or ideology?', *Philosophy and Public Affairs*, **2**, 66–110.

Tunstall, J. (1966), *Old and Alone*, Routledge and Kegan Paul, London.

Van Doorslaer, E.K.A. (1987), *Health Knowledge and the Demand for Health Care*, University of Limburg, Maastricht.

Van Doorslaer, E.K.A. *et al.* (1993), *Equity in the Finance and Delivery of Health Care: An International Perspective*, Oxford University Press, Oxford.

Wager, R. (1972), *Care of the Elderly – An Exercise in Cost Benefit Analysis. Commissioned by Essex County Council,* The Institute of Municipal Treasurers and Accountants, London.

Wagstaff, A. (1986), 'Demand for health: theory and application', *Journal of Epidemiology and Community Health*, **40**, 1–11.

Wall, W.D. and Williams, H.L. (1972), *Longitudinal Studies and the Social Sciences,* Heinemann, London.

Wasfie, T.J., Brown, J.H. (1982) The present role of coronary artery surgery, *Practitioner*, **225**, 739–44.

Weinstein, M.C. and Fineberg, H.V. (1980), *Clinical Decision Analysis,* W.B. Saunders, Philadelphia.

Weinstein, M.C., Stason, W.B. (1982) 'Cost effectiveness of coronary artery bypass surgery', *Circulation*, **66** (suppl III), 56–66.

Weisbrod, B. (1968), 'Income redistribution effects and benefit–cost analysis', in: S.B. Chase Jr, (ed.) *Problems in Public Expenditure Analysis,* Brookings Institution, Washington DC, 177–222.

West (1981), 'Theoretical and practical equity in the NHS in England', *Social Science and Medicine*, **15**, 117–122.

Wheatley, D.J. (1984) *Surgical Prospects,* Paper presented to consensus development conference on coronary artery bypass grafting. Kings Fund, London.

Wheatley, D.J., Dark, J.H. (1982) 'The present role of coronary artery surgery'. *Practitioner*, **226**, 435–40.

Whitmore, G.A. (1973), 'Health state preferences and social choice', in Berg, R.L. (ed.) *Health Status Indexes,* Hospital Research and Educational Trust, Chicago, 135–155.

Wiggins, D. and Dirmen, S. (1987), 'Needs, Need, Needing', *Journal of Medical Ethics*, **13**, 63–68.

Wilcock, G.K. (1978) 'Benefits of total hip replacement to older patients and the community', *British Medical Journal*, **1**, 37–9.

Wildavsky, A. (1966), 'The political economy of efficiency: cost benefit analysis, systems analysis and program budgeting', *Public Administration Review*, **26**, 292–310.

Wildavsky, A. (1969), 'Rescuing policy analysis from PPBS', *Public Administration Review*, **29**, 189–192.

Williams, A. (1970), 'Cost benefit analysis', in A. Cairncross, (ed.), *The managed economy,* Blackwell, Oxford.

Williams, A. (1973), 'Cost–benefit analysis: bastard science? And/or insidious poison in the body politick', in Wolfe, J. N. (ed.) *Cost Benefit and Cost Effectiveness: Studies and Analysis,* Allen and Unwin, London, 30–60.

Williams, A. (1974a), 'Measuring the effectiveness of health care systems', in Perlman, 196–202.

Williams, A. (1974b), 'Need' as a demand concept (with special reference to health), Culyer (1974), 60–76.

Williams, A. (1979), 'One Economist's View of Social Medicine', *Journal of Epidemiology and Community Health*, **33**, 3–7.

Williams, A. (1981), 'Welfare economics health status measurement' in Van der Gaag, J. and Perlman, M. (eds.), *Health, Economics and Health Economics*, North Holland, Amsterdam, 171–281.

Williams, A. (1983), The role of economics in the evaluation of health care technologies. In: Culyer, A.J., Horisberger, B. (eds.) *Economic and Medical Evaluation of Health Care Technologies*, Springer, Berlin, 47–80.

Williams, A. (1984), *Medical Ethics*, Nuffield/York Portfolio, Folio 2 Nuffield Provincial Hospitals Trust, London.

Williams, A. (1985a), 'The economics of coronary artery bypass grafting', *British Medical Journa*, **291**, 326–9.

Williams, A. (1985b), 'The nature, meaning and measurement of health and illness: an economic viewpoint' *Social Science and Medicine*, **20**, 10, 1023–1027.

Williams, A. (1986), 'QALY's or short straws', *British Medical Journal*, **293**, 337–338.

Williams, A. (1987a), 'Health Economics: The cheerful face of the dismal science', in Williams, A. (ed.), *Health and Economics*, Macmillan, Houndmills, 1–11.

Williams, A. (1987b), 'The importance of quality of life in policy decisions' in Walker, S.R. and Rosser, R.M. (eds.), *Quality of Life: Assessment and Application. Proceedings of the Centre for Medicines Research Workshop held at the CIBA Foundation, London, March 3rd 1987*, MTP Press, Lancaster, 279–290.

Williams, A. (1988a), '... Makes a man healthy, wealthy and wise! (Or from folklore to system science)'. In Duru, G. et al. (eds.) *System Science in Health Care Vol 2 Health Care System and Actors*, Masson, Paris, 57–60.

Williams, A. (1988b), 'Applications in mangement', in Teeling Smith, G. (ed.), in *Measuring Health: a Practical Approach*, Wiley, Chichester 225–243.

Williams, A. (1988c), 'Ethics and efficiency in the provision of health care' in Bell, J.M. and Mendus, S. (eds.) *Philosophy and Medical Welfare*, Cambridge University Press, Cambridge (Royal Institute of Philosophy Lecture Series **23**, Supplement to 'Philosophy'), 111–126.

Williams, A. (1988d), 'Health economics: the end of clinical freedom?' *British Medical Journal*, **297**, 1183–6.

Williams, A. (1988e), 'Economics and the rational use of medical technology' in Rutten, F.F.H. and Reiser, S.J. (eds.) *The Economics of Medical Technology*, Springer Verlag, Berlin, 109–120.

Williams, A. (1988f), 'Priority setting in public and private health care systems: A guide through the ideological jungle', *Journal of Health Economics*, **7**, 173–183.

Williams, A. (1992a), 'Cost effectiveness analysis: Is it ethical?', *Journal of Medical Ethics*, **18**, 7–11.

Williams, A. (1992b), 'Priority setting in a needs-based system' in Gelijns, A.C. (ed.), *Technology and Health Care in an Era of Limits*, Academy Press, Washington DC, 79–95.

Williams, A. (1993a) 'Quality assurance from the perspective of health economics', *Proceedings of the Royal Society of Edinburgh*, **101B**, 105–14.

Williams, A. (1993b), 'The importance of quality of life in policy decisions', in Walter, S.R. and Rosser, R.M. (eds.), *Quality of Life: Assessment and Application*, (second edition), MTP Press, Lancaster, 427–439.

Williams, A. (1994), 'Economics, society and health care ethics', in Gillon, R. (ed.), *Principles of Health Care Ethics*, Wiley, Chichester, 829–841.

Williams, A. (1995) *The Role of the Euroqol Instrument in QALY Calculations*, Discussion Paper, **130**, Centre for Health Economics, University of York, York.

Williamson, O.E. (1964), *The Economics of Discretionary Behaviour: Managerial Objectives in a Theory of the Firm*, Prentice Hall, New York.

Wilson, J.M.G and Jungner, G. (1968), *Principles and Practice of Screening for Disease*, Public Health Paper, **34**, WHO, Geneva.

World Bank (1993), World Development Report, World Bank, Washington, DC.

Wright, K. (1974), 'Alternative measures of the output of social programmes: The Elderly', in Culyer (1974), 239–272.

Wright, S.J. (1986), *Age, Sex and Health: A Summary of Findings from the York Health Evaluation Survey*, CHE Discussion Paper, **15**, University of York, York.

Index

Abdalla, Mona 159
Avorn 84

Banta 203
Barber, P.R. 245
Barry, B. 180–81, 184
Barry, B.M. 25n
Baumol, W.J. 25n
Behavioural studies
  nature of 41
Bennett, A.E. 104n
Biological model of health 65, 66–9
  adaptation 66–9
  damage, repair of 68–9
  disposable soma theory 68–9
  economic model compared
    at individual level 75–9
    at population level 80–85
  energy 67–8
  fitness 67, 68–9, 75
  genotype 66, 77
  natural selection 66
  phenotype 66–7
  variation 66–7
  see also Evolutionary physiology
Birch, S. 282
Black, D. 25n
Blades 87n
Bloomfield, D.S.F. 335, 348n
Bowen, H. 25n
Bradshaw, J. 186
  on need 182–3
Bräkenheim, C.R. 84
Brandon, S. 104n
Buchanan, J.M. 25n
Budgets
  activity 201
    meaning of 195
    organisation distinguished 195–6
  annual nature of 196–7
  cash control 194, 196, 197

characteristics of 196–7
decision-making 195, 197
elements of 194–6
evaluation of outcomes 195, 197
financial targets, setting of 194
input orientation of 196, 197–9
monitoring of performance 194, 195,
  197, 200
organisations 198, 199
  activity distinguished 195–6
  meaning of 195
output budgeting 199
planning and 197–8
programme budgeting 199
recording of decisions 195, 197, 200
shadow costs and 200
transmission of information 195, 200
Bush, J.W. 103n, 111
Butterfield, W.J.H. 184

Calman, K.C. 318
Calow, P. 67, 87n
Card, W.I. 111
Carr-Hill, R.A. 348n
Carstairs 103n
Chadwick, R. 307, 319n, 348n
Chambers, M.L. 104n
Charny, M.C. 84, 312
Chen, M.M. 111
Chevalier, Maurice 24
Chronic renal failure
  treatments of, cost–benefit analysis
    30–31, 35
Clinical audit 263, 264
  meaning of 264
  weaknesses of 264
  see also Cost-effectiveness analysis
Clinical freedom
  accountability and 280–81
  clinical autonomy and 278
  cost-effectiveness and 273–81

363